ELECTRONIC INTEGRATED CIRCUITS AND SYSTEMS

MICROELECTRONICS SERIES

Electronic Integrated Circuits and Systems, *Franklin C. Fitchen*

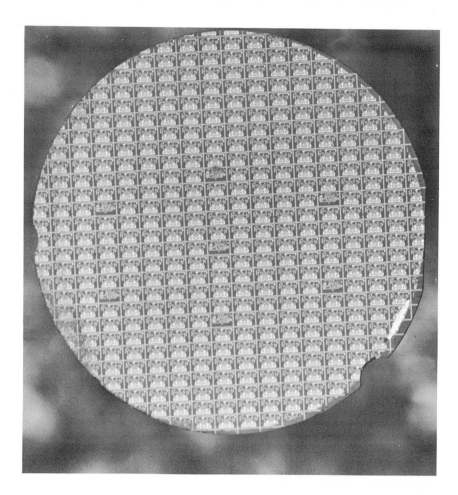

A semiconductor wafer about one inch in diameter that includes about three hundred monolithic integrated circuit chips. Each chip contains about forty circuit elements. The circuit is an emitter-coupled logic gate. Included on the wafer are seven test sites for resistivity and parameter measurements. (*Photo courtesy Texas Instruments, Inc.*)

ELECTRONIC INTEGRATED CIRCUITS AND SYSTEMS

FRANKLIN C. FITCHEN

Electrical Engineering Department
South Dakota State University

CONTRIBUTORS

VIRGIL G. ELLERBRUCH
AELRED J. KURTENBACH
DUANE E. SANDER
Electrical Engineering Department
South Dakota State University

MICROELECTRONICS SERIES

 VAN NOSTRAND REINHOLD COMPANY
New York / Cincinnati / Toronto / London / /Melbourne

To
Roy and Jeanne,
Sally and Bobby

Van Nostrand Reinhold Company Regional Offices:
New York   Cincinnati   Chicago   Millbrae   Dallas

Van Nostrand Reinhold Company Foreign Offices:
London   Toronto   Melbourne

Copyright © 1970 by Litton Educational Publishing, Inc.

Library of Congress Catalog Number Card: 70–122668

ISBN: 0-442-22406-0

Manufactured in the United States of America

Published by Van Nostrand Reinhold Company
450 West 33rd Street, New York, N.Y. 10001

Published simultaneously in Canada by
Van Nostrand Reinhold Ltd.

15   14   13   12   11   10   9   8   7   6   5   4   3   2

PREFACE

This volume has been written for use by practicing engineers and students as an introduction to the new era in electronics fashioned by integrated circuit technology. It can be used at the college level, in industrial courses, or for self-study. A knowledge of electric circuit theory is assumed, and the calculus is used where helpful.

The emphasis in this book is placed upon the circuit and system aspects of integrated circuits, rather than upon their fabrication. In this context, liberal use is made of examples of available IC's in order to provide the reader with a broad background in the capabilities and limitations associated with integrated circuit engineering.

The text material is primarily concerned with silicon monolithic integrated circuits, and nearly all concepts are illustrated using actual IC's. Research laboratory developments are not discussed; only those aspects of IC technology that have reached a form of success through availability are considered.

The book is arranged in three parts. The first four chapters constitute an introduction to IC's. Definitions, representations, and fabrication methods are considered in Chapters 1 and 2.

The heart of any integrated circuit is its transistors. Chapter 3 reviews transistor circuit analysis techniques. The reader familiar with the material may wish to omit that chapter. A more thorough treatment is given in the author's *Transistor Circuit Analysis and Design*.

Linear integrated circuits are characterized by the abundant use of negative feedback. Chapter 4 is a standard treatment of feedback principles for those readers unfamiliar with this important phase of electronics.

Part II, "Linear Integrated Circuits and Systems," Chapters 5 through 8, treats linear IC building blocks, the operational amplifier, and communications circuits useful in analog systems.

The final four chapters form Part III, "Digital Integrated Circuits and Systems." Gates, flip-flops, and registers are discussed in detail for they form the basis for complex digital communication, computation, and control systems. Analog-to-digital conversion and multiplexing operations are also considered.

The problems included with each chapter have been written for several purposes: to review standard material not in the text; to increase comprehension of the text material; to introduce new concepts, circuits, and systems. For class learning situations, design problems have been included, for they serve as an excellent vehicle in the understanding and motivational aspects of engineering education.

Because of the approach taken in reviewing fundamentals in the text and problems, it is possible for those whose exposure time is limited to use this volume as a self-contained introduction to modern electronics. The book can also be used successfully following a semester or a quarter devoted to conventional electronics.

In keeping with the teaching of electronic subjects, the physical basis for electronics is not emphasized in this book. That material is often considered separately. Also it must be mentioned that the field of microwave applications of integrated circuits is not considered for that same reason.

Chapter 10 was written by Virgil Ellerbruch; Chapter 11 by Aelred Kurtenbach; and Chapter 12 by Duane Sander. Thanks are also extended to Dean John E. Lagerstrom of the College of Engineering of South Dakota State University for his continuing cooperation, to Dr. Sanjit K. Mitra of University of California, Davis, for his excellent assistance, and to the contributing industrial concerns.

Brookings, South Dakota   FRANKLIN C. FITCHEN
March, 1970

CONTENTS

PART ONE

INTEGRATED CIRCUIT FUNDAMENTALS

1
INTRODUCTION TO INTEGRATED CIRCUITS

Progress in electronics is generally associated with considerations of enhancement of electrical performance, improvement in reliability, and reduction in cost. The vacuum tube gave way to the transistor because the latter device clearly displayed improved reliability and efficiency and lent itself to miniaturization of electronic equipment. The integrated circuit (IC) for digital signals was readily accepted for applications where arrays of similar circuits are utilized as in digital computer systems. Considerable cost reductions were made possible through the elimination of great numbers of interconnecting wires needed in such systems. Large-scale integration (LSI) achieves further economies by reducing the number of steps required in the manufacture of complex electronic equipment.

The initial consideration in this chapter will be "What is an IC?" After gaining a feeling for the answer to that question, we turn our attention to methods for representing the behavior of IC's. Finally, this chapter discusses some general considerations involved in systems design: the effects of terminal impedance levels upon system performance, and the differential (nongrounded) system.

1-1 IC TYPES

The older or conventional type of electrical circuit, where the elements are separately manufactured and, when assembled, are joined by wires or plated conductors, is now referred to as a *discrete circuit.* The word discrete literally means separately distinct. From a fabrication standpoint, discrete circuits and integrated circuits are opposites.

Integrated circuits are electrical networks formed upon or within a *substrate.* A function of the substrate is support, and substrates commonly are made from semiconductor material or from insulating material.

An *element* of an integrated circuit is a constituent part of the IC that contributes directly to its operation. Examples of elements are diodes, transistors, resistors, and capacitors.

An *integrated circuit* may be defined as a combination of interconnected circuit elements inseparably associated on or within a continuous substrate.[1]* The distinctive features of circuits produced by IC technologies are

    (a) inseparability of circuit elements

    (b) embodiment on a common substrate

    (c) fabrication in situ (in its natural position).

The definition of an integrated circuit given in the preceding paragraph is rather broad and therefore encompasses several different assembly techniques employed in their manufacture.

The *monolithic integrated circuit* is an IC whose elements are formed in situ upon or within a semiconductor substrate with at least one of the elements formed within the substrate. Other modifiers may be used; a dielectric-isolated monolithic IC uses dielectric isolation to accomplish separation between elements in the substrate. A monolith is a single stone. Monolith in this regard represents "single crystal" structure; the semiconductor material is the crystal. The monolithic IC is the type considered most frequently in this book.

The monolithic IC is thus an inseparable single entity containing substrate, elements, and interconnection pattern. This pattern of thin deposited aluminum, usually called *metallization,* serves in place of wires or the printed circuit connections used in an assembly of discrete elements. Because a monolithic IC is a small slice of processed silicon with dimensions often smaller than 0.1 × 0.1 inch and 0.025 inch thick, it is called a *chip.* The frontispiece to the present book clearly shows a wafer that contains many semiconductor chips.

In a *multichip integrated circuit*, the elements are formed on or within two or more semiconductor chips that are separately attached to a substrate.

An IC whose elements are formed upon an insulating substrate, such as glass or ceramic, is called a *film integrated circuit.* An integrated circuit consisting of a combination of two or more integrated circuit types, monolithic or film, or one IC type together with discrete elements is a *hybrid integrated*

---

* Superior numbers refer to References, following Chapter 12.

circuit.   This addition of technologies can extend the performance capability of the circuit beyond that attainable using a single monolithic or film IC. The manufacturing cost, however, may also be increased.

The film IC may use a pattern of passive elements (resistors and capacitors) deposited upon its substrate.   To make a resistance, a film of material of known resistivity is deposited with the proper dimensions to achieve the required resistance.   The lead pattern is also deposited.   Discrete transistors, diodes and inductors are added separately; they are specially manufactured for the use in a thin-film IC.   Figure 1-1 depicts a film hybrid IC.

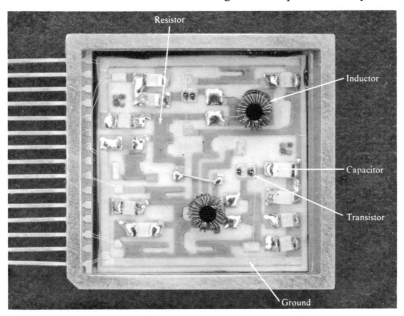

Fig. 1-1   Film hybrid 30 MHz IF amplifier for communications systems. Package is about 1″ × 1″ × 0.15″. (*Courtesy Sylvania Electric Products, Inc.***)**

Practical integrated circuits are, in the main, fabricated almost exclusively from semiconducting materials.   Silicon in its crystalline form has the necessary semiconductor properties for successful IC manufacture.   The technological knowledge necessary for successful IC manufacture was developed during the early 1960s in the production of Planar transistors.

Since all IC's are simply combinations of familiar circuit elements such as transistors and resistors, it can be said that the IC represents an evolutionary step in electronics.   Neither alchemy nor witchcraft is involved.   However, IC manufacture does impose important limitations upon circuit design. These limitations are concerned with heat dissipation, value of components, and parasitic effects. The effect of these limitations is seen in the circuits that have been integrated. Certain successful discrete circuit designs cannot be provided in IC form. Because the integration processes to some extent dictate the circuit design, we must become aware of the constraints imposed by this technology.

Unless otherwise specified, *all IC circuit examples given in this book will employ monolithic construction.* Circuit and system analysis is often independent of construction techniques employed, so that much of the material will apply to film and hybrid IC's.

1-2 IC LIMITATIONS

A monolithic IC is not an "only child." The decision to manufacture a given design implies considerable expenditure for manufacturing equipment, and thus it is anticipated that thousands or perhaps millions of identical circuits will be made. Circuits produced in this manner have certain advantages over circuits that use discrete components:

(a) interconnection errors are nonexistent;
(b) temperature differences between parts of a circuit are small;
(c) close matching of components (such as resistance ratios) and temperature coefficients is likely;
(d) transistors and diodes may be generously used, for they are more economical than passive elements;
(e) economies are achieved in the costs of manufacturing and of interconnecting elements of an electronic system.

The designer and the user of the monolithic IC are restricted in several important ways:

(a) capacitors and resistors are limited in maximum value;
(b) low tolerances on capacitors and resistors and small temperature coefficients are difficult to obtain;
(c) the basic process of making *npn* transistors does not lend itself to manufacture of high performance *pnp* units;
(d) inductors are not successfully integrated;
(e) parasitic, undesired elements are formed when isolating desired elements from each other;
(f) saturation resistance of transistors tends to be large;
(g) circuit power dissipation is limited;
(h) special considerations, such as low noise or high voltage operation, are not easily obtainable.

Developments in techniques of IC manufacture will undoubtedly eliminate the effects of certain of the limitations noted above. Until such time as the limitations are removed completely, it will be the mission of the integrated-circuit designer and the system engineer to circumvent such limitations. Much of this book is devoted to circuit and system design under the constraints imposed by IC technology.

A frontier in IC engineering is obviously concerned with the total number of elements and, therefore, functions that can be included on a single chip. Elements, such as resistors and capacitors, take up significant surface areas. Even after surface cleaning and treatment, physical defects remain that affect the *yield*, the number of good individual circuits on a silicon wafer. For example, if twenty such defects exist on a wafer composed of two-hundred circuits, the yield is potentially 90 percent. The yield drops to 60 percent if the area of each circuit is increased four times, with fifty circuits per wafer.

The term *large-scale integration* (LSI) refers to relatively large-area chips containing many hundreds of circuit elements. The cost reduction associated with maximizing the number of system connections made on the chip by the metallization pattern represents an attractive incentive to produce complex IC's. Multilayer metallization is used to allow increased component density. Layers of interconnection metallization, separated by silicon dioxide insulation, are stacked on the chip.

Medium-scale integration (MSI) is a term sometimes used to represent large IC's containing fewer than one hundred gate circuits.

Because large numbers of identical circuits are employed in digital computers, MSI and LSI are currently the concerns of computer system designers. Time will tell if these capabilities are extended to other areas, such as consumer electronics.

1-3 ANALOG AND DIGITAL SYSTEMS

Integrated circuits are often classified as either *analog* or *digital*, according to the type of signal-processing system for which they are intended. In this section we discuss in a general way the basis for this classification.

An electrical signal-processing system employs *transducers* at the signal source and at the output or display ends of the system. A well-known pair of transducers is the microphone and loudspeaker, as used in a public address system. The function of a transducer, then, is to transform energy from sound (or light or heat) to the electrical form, and vice versa. Between transducers, signal information is in the form of electrical energy.

The information contained in an electrical signal is represented by a *code*. The code may be related to the amplitude of the signal, or it may depend upon the frequency of the signal or on any other time-dependent function. Two general classes of codes are used to carry information: analog and digital.

If instantaneous values of a characteristic of the coded signal and the original signal are directly related, the system uses an analog code. AM and

FM radio broadcasts use analog codes, for signal amplitude and frequency, respectively, conform to characteristics of original signals generated in the studio.

A digital system may use the number, duration, or frequency of pulses to carry information. An example is the pulse duration code important to telegraphy (dots and dashes). Digital computers are fed instructions and data in a digital code.

Analog and digital circuits and systems differ in a large number of ways. Analog information is lost when it is passed through a circuit that cannot respond perfectly to all characteristics of the incoming wave. Digital information is lost when it is badly delayed in time or when its amplitude is insufficient to trigger a regeneration circuit to restore its wave shape.

Analog systems are often referred to as *linear* systems; digital systems, as *nonlinear.* It is interesting to note, however, that signal processing in an analog system may rely upon nonlinearities in the electronic devices used. Processes such as modulation, detection, and mixing are examples; these are considered in Chapter 8, " Linear IC's for Communications."

Figure 1-2 shows five digital IC's along with discrete components that form a subsystem in a digital computer. These are shown mounted on a printed circuit board.

Certain information concerning electrical circuit elements is fundamental to their use in either analog or digital systems. We shall therefore study

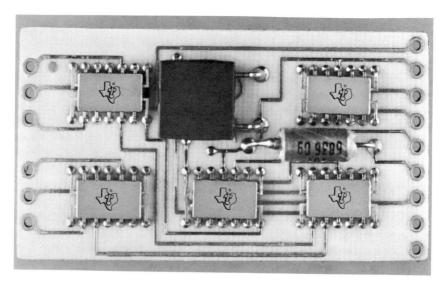

Fig. 1-2 Five digital IC's along with discrete elements mounted on a printed circuit board about four square inches in area. (*Courtesy Texas Instruments, Inc.*)

methods of representation of devices, their physical structure, operation principles, fabrication methods, feedback, and coupling techniques in the first four chapters of this book; the four subsequent chapters are concerned with linear circuits and systems; the final four chapters are devoted primarily to digital IC's.

1-4 CHARACTERISTICS

Semiconductor devices exhibit both linear and nonlinear behavior. In a linear device such as a wirewound resistor, we expect terminal voltage and current to be linearly related; since the resistance is essentially unaffected by voltage or current level, we say that the device has a constant resistance, expressed in ohms. It is redundant to plot V versus I for such a device; the constant slope of the graph would not be interesting.

Nonlinear behavior, on the other hand, means that the ratio of voltage to current may differ at different operating points, or voltage or current levels, and therefore a single description (such as a constant ohmic value) is not possible. Such devices are usually described by *graphical characteristics*.

An IC is a system or subsystem because it is made up of a number of elements performing diverse functions. Therefore, the most useful graphical characteristics are taken at the external terminals, and relate those currents and voltages that the system designer must deal with. Figure 1-3 shows the

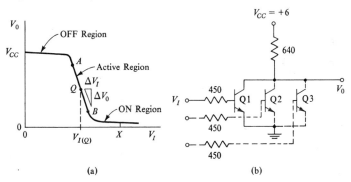

Fig. 1-3 IC gate: (a) transfer characteristic; (b) circuit with typical element values.

transfer characteristic of an IC gate and the gate circuit. The gate is simply a switch to be used in either the ON or the OFF mode. The curve relates an output quantity, in this case, output voltage, V_O, to an input quantity, here input voltage V_I. Thus the transfer characteristic gives an indication of the processing that the circuit will perform upon a signal supplied to its input terminals.

The transfer characteristic is plotted for a single transistor, such as Q1.

Consider that the other transistors are turned OFF and do not affect Q1. We note from the characteristic that this circuit has a linear region and two nonlinear regions. Between points A and B this circuit behaves linearly, because the output V_O and input V_I are simply related ($\Delta V_O/\Delta V_I = K$, a constant). On the other hand, for inputs less than A or greater than B, the system is said to be nonlinear. To be more specific, the circuit is said to be in its OFF state for low values of V_I, and to be ON for large values of input voltage.

Reasons for the behavior shown graphically are easy to understand. When V_I is too low in value, transistor Q1 is OFF, no current flows from V_{CC} and the voltage at V_O is equal to the supply, $+6$ V. An increased level of V_I will turn that transistor ON, with full ON corresponding to values of V_I beyond point B. When fully ON, Q1 is a very low resistance, and the current from V_{CC} to ground through Q1 will be as large as possible, almost equal to 6/640 or about 9.4 mA. The voltage across the transistor is practically zero, so that the output V_O level is zero volts.

The *active* region of operation is between points A and B. Nonlinearities may also exist here, although they are not as abrupt as those associated with the ON or OFF conditions. A large input signal, such as a pulse extending from 0 to X volts, would cause the output voltage to traverse all three regions and, because of the nonlinearities, the output waveform might be distorted when compared to the input.

Suppose that an *operating point* were to be established at point Q by supplying the input terminal of the gate with an appropriate direct current or voltage. For very small input voltage excursions about that point, as is possible by adding a small ac signal to $V_{I(Q)}$, variations in the output would be faithful reproductions of the input variations. To an infinitesimal input, the device or circuit will behave linearly. We reach an important conclusion in the study of electronics: *linear operation always results when signals are sufficiently small.*

As discussed in conjunction with the wirewound resistance, linear operation need not be described graphically. Rather, an *electrical equivalent circuit* or mathematical *model* is often employed. The model can be studied with greater ease and more accurate results than can a graphical representation. Occasionally a model may be available for studying large-signal or nonlinear behavior, and is preferred over graphical descriptions when available. However, the parameters of large-signal models are often functions of circuit currents and voltages and may be very complicated.

In addition to transfer characteristics, other graphical representations are possible. For the circuit of Figure 1-3(b) we could plot input voltage versus input current or output voltage versus output current, for example. Where useful, such representations are employed.

It follows, then, that the *linear IC* is designed to be used in the linear portion of its transfer characteristic to provide the behavior or function for which it was intended. Examples of linear IC's are operational amplifiers, video amplifiers, and radio-frequency (RF) amplifiers. On the other hand, an IC designed to be used in a nonlinear mode of operation may be referred to as a nonlinear or *digital IC*. Examples are gates, flip-flops, and other circuits that rely upon a switching form of operation where the output quantity is not necessarily proportional to the input.

1-5 SYMBOLS

A system of symbols to designate circuit quantities in this book is both desirable and necessary, for a great many abbreviations will be encountered by the reader. Circuit notation follows the several general rules listed here. Symbols have been chosen to conform with current practice.[2]

1. DC values of quantities are indicated by capital letters with capital subscripts (I_B, V_{CE}). Direct supply voltages have repeated subscripts (V_{BB}, V_{EE}).
2. Rms values of quantities are indicated by capital letters with lower-case subscripts (I_c, V_{cb}).
3. The time-varying components of currents and voltages are designated by lower-case letters with lower-case subscripts (i_x, v_{cb}).
4. Instantaneous total values are represented by lower-case letters with upper-case subscripts (i_C, v_{BE}).
5. Maximum or peak values are designated similarly to rms values but bear an additional final subscript m (V_{cm}, V_{bem}).
6. Circuit elements are given capital letters (R_1, C_p); device parameters have lower-case symbols (r_e, h_{11}).

Fig. 1-4 may help in understanding this notation system. In the diagram, i_I is the instantaneous total value of the wave, and i_i is the instantaneous

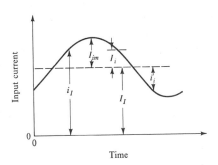

Fig. 1-4 Example of notation.

value of the time-varying component of i_I. I_I designates the average or dc value, I_{im} the peak, and I_i the effective value of the alternating component.

To designate a constant-current source or generator on circuit diagrams, the symbol of Fig. 1-5(a) will be employed. No distinction is made in the

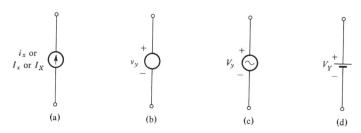

Fig. 1-5   Symbols for sources.

symbol between dc and time-varying currents; should a confusing situation arise, the time function will be specified. The arrow indicates the instantaneous direction of the current supplied by the source.

Constant-voltage sources are symbolized in (b), (c), and (d) of Fig. 1-5. The general symbol is in (b) with instantaneous polarity as shown. When the supply is specifically sinusoidal, a "cycle" may be employed, as in (c). Direct-voltage supplies are characterized by the familiar battery symbol as in (d).

1-6   AMPLIFICATION

A *passive* circuit is one composed exclusively of passive elements: resistances, capacitances, inductances. These passive elements can, if desired, be connected in such a manner as to provide current or voltage amplification, as discussed in Problem 1-2. However, no combination of passive elements can provide an amplification of power level; to accomplish this, we must use *active* elements; transistors, vacuum tubes, relays, magnetic amplifiers. Since power amplification is easily accomplished with active elements, they are universally used for voltage and current-gain applications as well.

Active elements have one thing in common—they must be *biased*. Biasing means that power must be supplied from a source other than the signal to be amplified. Therein lies the secret of amplification: *amplification is control of a high power level derived from the bias source by the low power from the signal source.* The difference between load power and signal power is made up for by the power or bias supplies.

Of course, signal amplification or gain is an important specification when considering linear IC's. In the digital IC, amplification is also required. All

IC types will be used in conjunction with dc power supplies. Supplies of both polarities will often be necessary.

The amplification properties of a device or circuit can be specified by the numerical ratio of the magnitudes of output to input quantities or by the *decibel*, a logarithmic unit of power ratio. By definition, power gain in decibels is

$$gain \text{ } in \text{ dB} = 10 \log \frac{P_o}{P_i} \qquad (1\text{-}1)$$

where the subscript *o* represents an output or load quantity and the subscript *i*, the quantity supplied to the circuit by a signal source. Should we wish to extend this definition to voltages and currents, then Eq. (1-1) becomes

$$gain = 10 \log \frac{V_o^2/R_o}{V_i^2/R_i} \text{ dB}$$

or

$$gain = 10 \log (I_o^2 R_o/I_i^2 R_i) \text{ dB.}$$

If $R_o = R_i$,

$$gain = 10 \log (V_o/V_i)^2 = 20 \log (V_o/V_i) \text{ dB} \qquad (1\text{-}2)$$

and

$$gain = 20 \log (I_o/I_i) \text{ dB.} \qquad (1\text{-}3)$$

Eqs. (1-2) and (1-3) are correctly employed only for identical resistance levels. In many circuits the levels may be vastly different. Nevertheless, it is customary to use the voltage and current equations in practice regardless of resistance levels.

As an example of the use of the decibel, it can be shown that a power gain ratio of 10,000 is 40 dB, 200 is 23 dB, and 40 is equivalent to 16 dB.

1-7 SMALL-SIGNAL REPRESENTATION

To describe the small-signal or linear characteristics of a device or a circuit, it was mentioned in Section 1-4 that an electrical equivalent circuit or model may be employed. There are two basic types of models. The *physical model* is a representation of the physical processes going on in the device. Such a model may be extremely complicated; it is most useful to the device designer who is concerned with predicting the change in behavior of a device that occurs when a physical dimension or material is changed. Physical models for the conventional transistor and the field-effect transistor are presented in Chapter 3.

The *functional* or *mathematical model* is not concerned with internal processes, only with terminal properties. It is often simpler than the physical model. In this section we discuss the most common functional models for two-port networks, the matrix representations.

The newcomer to the field of electronics must be aware that *small-signal models for active devices or for IC's assume that biasing has been accomplished and that the models apply to small-signal or low-level ac operation exclusively.*

Values of the small-signal parameters that describe a circuit or a device in general apply only at a particular dc operating point; if the operating point is changed from its reference coordinates, the small-signal parameters may take on new values. This can be seen if one measures the short-circuit current gain of a single transistor in the common-emitter connection. This parameter, referred to as β or as h_{fe}, will vary by as much as 2 : 1 over the range of operating points in the active region.

The sets of two-port parameters commonly used to describe discrete transistors can be used to represent entire networks. Thus the IC, made up of many circuit elements, can be represented in its entirety by considering terminal quantities, input and output currents and voltages. The complete IC is often designed in such a way that operating points for the elements are not easily changed. The variation in parameters with operating point may not be as important for a complete IC as it is when considering a single transistor.

Matrix Parameters

A two-terminal-pair or two-port network may be treated as a black box, and general equations written relating the terminal quantities V_1, I_1, V_2, and I_2. A general representation is shown in Fig. 1-6. Within the box is a linear,

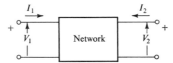

Fig. 1-6 General network showing reference directions for external quantities.

active network. Because the external conditions are measurable, the device within the box can be characterized by a set of four parameters; these parameters are the coefficients in the pair of simultaneous equations that may be written to relate the external quantities. For example, the box of Figure 1-6 can be described by

$$V_1 = z_{11}I_1 + z_{12}I_2$$
$$V_2 = z_{21}I_1 + z_{22}I_2 .$$

(1-4)

Eqs. (1-4) can be written in matrix form:

$$\begin{bmatrix} V_1 \\ V_2 \end{bmatrix} = \begin{bmatrix} z_{11} & z_{12} \\ z_{21} & z_{22} \end{bmatrix} \begin{bmatrix} I_1 \\ I_2 \end{bmatrix}. \tag{1-5}$$

The elementary rules of matrix algebra, as given in Appendix I, provide for an expansion of Eq. (1-5) that yields Eqs. (1-4).   For some purposes, a shorthand form of Eq. (1-5) is useful:

$$[V] = [z][I]. \tag{1-6}$$

Returning to Fig. 1-6, we can note that five other equation pairs may be written to relate the terminal quantities:

$$I_1 = y_{11} V_1 + y_{12} V_2$$
$$I_2 = y_{21} V_1 + y_{22} V_2 \tag{1-7}$$

$$V_1 = h_{11} I_1 + h_{12} V_2$$
$$I_2 = h_{21} I_1 + h_{22} V_2 \tag{1-8}$$

$$I_1 = g_{11} V_1 + g_{12} I_2$$
$$V_2 = g_{21} V_1 + g_{22} I_2 \tag{1-9}$$

$$V_1 = a_{11} V_2 - a_{12} I_2$$
$$I_1 = a_{21} V_2 - a_{22} I_2 \tag{1-10}$$

$$V_2 = b_{11} V_1 - b_{12} I_1$$
$$I_2 = b_{21} V_1 - b_{22} I_1. \tag{1-11}$$

These relationships may also be written in matrix form (see Problem 1-6). *The positive directions of currents and voltages are defined as shown in Fig. 1-6.*

For each of the above equation pairs, it is possible to draw an equivalent electrical circuit, a circuit that could be considered to be the contents of the box of Fig. 1-6.   The equivalent circuit that satisfies Eqs. (1-4) is shown in Fig. 1-7, and has been called the z equivalent.   In a like manner, equivalent circuits for the other equations can be drawn; many variations are possible, particularly when one considers the equivalence of sources.

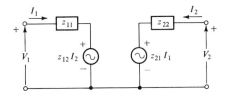

Fig. 1-7   z-parameter equivalent circuit.

Further study of the z equivalent is of value. Each of the z parameters must have the dimensions of impedance, but it is to be noted that $z_{12}I_2$ and $z_{21}I_1$ are *dependent* or *controlled* voltage sources. If I_2 were to be caused to equal zero, by open-circuiting the output-terminal pair, z_{11} could be referred to as the *input impedance with output open-circuited*. Mathematically,

$$z_{11} = \frac{V_1}{I_1}\bigg|_{I_2=0}. \tag{1-12}$$

In a similar manner, with open terminations, the other z parameters are referred to as transfer and output impedances. Because the z parameters are determined from open-circuit measurements, and it is virtually impossible to provide such terminations in semiconductor circuits while retaining the required dc levels necessary for biasing, the z parameters have not been widely employed.

Hybrid Parameters

The *hybrid* or h parameters often are used to describe the characteristics of the transistor. They are the coefficients in Eqs. (1-8), repeated here:

$$V_1 = h_{11}I_1 + h_{12}V_2$$
$$I_2 = h_{21}I_1 + h_{22}V_2. \tag{1-8}$$

The h-parameter equivalent circuit can take on the form of Fig. 1-8 in order to satisfy Eqs. (1-8). Since the defining equations must obey Kirchhoff's Laws, h_{11} must be an impedance, and h_{22} an admittance, while h_{12} and h_{21} are dimensionless. *For low-frequency analysis of transistor circuits, h_{11} and h_{22} will be resistive and h_{12} and h_{21} will be real.*

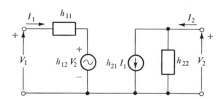

Fig. 1-8   *h*-parameter equivalent circuit.

The ease or measurement of the h parameters has contributed to their widespread adoption. If the output terminals are ac short-circuited, then $V_2 = 0$ and

$$h_{11} \equiv \frac{V_1}{I_1}\bigg|_{V_2=0}. \tag{1-13}$$

Also

$$h_{21} \equiv \frac{I_2}{I_1}\bigg|_{V_2=0}. \tag{1-14}$$

Opening the ac input circuit reduces I_1 to zero and gives

$$h_{12} \equiv \frac{V_1}{V_2}\bigg|_{I_1=0} \tag{1-15}$$

and

$$h_{22} \equiv \frac{I_2}{V_2}\bigg|_{I_1=0}. \tag{1-16}$$

Therefore h_{11} is called the "input impedance with output short-circuited," and h_{22} is the "output admittance with input open-circuited." Likewise, h_{12} is the "voltage feedback ratio with input open-circuited," and h_{21} is the "current amplification with output short-circuited."

The short and open circuits referred to in the preceding paragraph for the measurement of transistor parameters may be accomplished in the laboratory by the insertion of suitable capacitors and inductors. A large-valued capacitor across an output terminal pair will short-circuit an ac signal but will not disturb quiescent conditions. Likewise, a large-valued choke in an input bias current supply will essentially open that circuit to ac.

In order to standardize transistor circuit nomenclature, the Institute of Electrical and Electronic Engineers recommends the following parameter symbols:

$h_i = h_{11}$ (*input impedance*)
$h_r = h_{12}$ (*reverse voltage feedback ratio*)
$h_f = h_{21}$ (*forward current transfer ratio*)
$h_o = h_{22}$ (*output admittance*)

y Parameters

The defining equations were given in Eqs. (1-7). In matrix form,

$$\begin{bmatrix} I_1 \\ I_2 \end{bmatrix} = \begin{bmatrix} y_{11} & y_{12} \\ y_{21} & y_{22} \end{bmatrix} \begin{bmatrix} V_1 \\ V_2 \end{bmatrix}. \tag{1-17}$$

All of the y parameters are obtained from short-circuit measurements. They are primarily applicable at high frequencies where obtaining a valid open circuit is especially difficult because of the effects of stray capacitances upon measured parameter values.

The y equivalent circuit is shown in Fig. 1-9. Two dependent current generators are apparent in the diagram. Should a two-voltage-generator model be desired, source interchanges could be employed, and the result

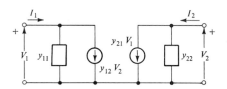

Fig. 1-9   y-parameter equivalent circuit.

would be analogous to Fig. 1-6 and the z equivalent. The y parameters are often employed to describe a transistor or an IC at a particular frequency. For example, at 10 MHz, supply voltage $= +9$ V, the following values apply to a sample type CA3028A cascode IC amplifier:

$$y_{11} = (0.5 + j1.5)10^{-3} \text{ mho}, \qquad y_{21} = (100 - j15)10^{-3} \text{ mho},$$
$$y_{12} = (0.5 - j0.3)10^{-6} \text{ mho}, \qquad y_{22} = (-j0.2)10^{-3} \text{ mho}.$$

This cascode amplifier is discussed in Section 8-1.

Several sets of parameters have been introduced for describing the small-signal operation of a device or a network. The sets must be interrelated, for a given network can be described by any or all sets. A listing of interrelationships is given in Appendix II. Equations for calculation of network gains and impedance levels using these parameters are discussed in Section 3-8.

1-8   LARGE-SIGNAL REPRESENTATION

It has been mentioned that electrical behavior can be described graphically; graphical characteristics contain information about behavior under both small and large signal conditions. Unfortunately, simple graphical data are generally taken by direct current and voltage measurements and do not show reactive effects from capacitances and inductances. It is usually more convenient to work with a set of numerical parameters, if available, than with graphs.

When signal sizes cannot be considered to be small, it is necessary to provide parameters that are not only functions of frequency but are also amplitude dependent. For parameters to contain all of this information is clumsy although some success has been reported.

It is not always convenient to find a simple equation to represent experimental behavior. One example of successful agreement between theoretical and experimental operation is the "diode equation." The V-I curve for a semiconductor diode is shown in Fig. 1-10.   This curve is very nonlinear.   A good mathematical representation is

$$I = I_S(\epsilon^{qV/kT} - 1),\qquad(1\text{-}18)$$

where

I = current in amperes,
I_S = saturated value of reverse current in amperes,
q = charge on an electron, 1.602×10^{-19} coulomb,
V = potential difference in volts,
T = absolute temperature in degrees Kelvin,
k = Boltzmann's constant, 1.380×10^{-23} joule per °K.

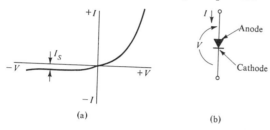

(a)   (b)

Fig. 1-10   Semiconductor diode: (a) static characteristic; (b) circuit symbol with quantities defined.

Equation (1-18) applies for both polarities of voltage, and agrees well with experimental data.   The constants q/kT have the value of about 40 at room temperature.   This factor is given the symbol Λ in the pages that follow. To use the equation, one must know I_S, and the equation will give values of I when values of V are assumed.

Because transistors are made from semiconductor diodes, Eq. (1-18) can be extended to represent transistor operation.   See Fig. 1-11.   The emitter-base diode voltage and base current are related by

$$I_B \cong I_{BS}(\epsilon^{\Lambda V_{BE}} - 1),\qquad(1\text{-}19)$$

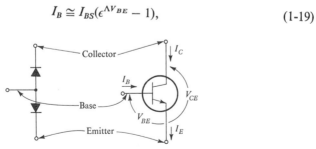

Fig. 1-11   npn transistor showing diode analogy and circuit symbol with terminals defined.

where I_{BS} is the reverse diode current.   Eq. (1-19) assumes that the collector voltage and current have negligible effect upon the emitter-base diode.   This is usually the case.

A diode analogy for the transistor has merit when considering biasing the device.   However, the transistor does not operate as two back-to-back diodes because of its thin base width.   This construction allows relatively large current to flow through the normally reverse-biased collector-base junction.

A useful parameter for junction transistors is the "dc beta," defined as follows:

$$\beta_{dc} \equiv \left. \frac{I_C}{I_B} \right|_{V_{CE}=const.} . \tag{1-20}$$

From this definition and Eq. (1-19), we can relate collector current I_C to V_{BE}:

$$I_C \cong \beta_{dc} I_{BS}(\epsilon^{\Delta V_{BE}} - 1). \tag{1-21}$$

This is a form of transfer characteristic, for an output quantity I_C is given in terms of an input quantity, V_{BE}.

Equations (1-18) through (1-21) are dc relations; they are also applicable to time-varying quantities.   If the frequencies considered are low so that reactive effects are unimportant, then it is possible to substitute total values into the equations noted.   For example, suppose that the total emitter-base voltage is composed of a direct voltage $V_{BE(Q)}$ plus a time-varying component $v(t)$.   In place of V_{BE} in Eq. (1-21) we substitute $V_{BE(Q)} + v(t)$.   The resulting collector current will be composed of a dc part $I_{C(Q)}$ and a time-varying component, $i_c(t)$.

1-9   TERMINAL IMPEDANCES

An IC cannot function alone.   It must process a signal supplied to its input terminals and feed a load connected to its output terminals.   The importance of the input and output impedance levels of the IC cannot be overestimated, for they seriously affect system steady-state and transient performance.   For an example, we consider these influences upon a linear IC network.   Terminal impedances are defined in Fig. 1-12(a).

Input Impedance

In Fig. 1-12(b), a signal source with open-circuit voltage V_g and internal impedance Z_G is feeding the IC with input impedance Z_i.   By voltage division, it can easily be appreciated that the signal voltage actually available at input terminals *a-b* is

$$V_{ab} = V_g \left[\frac{Z_i}{Z_G + Z_i} \right]. \qquad (1\text{-}22)$$

To maximize the fraction in brackets, it is necessary for Z_i to be large-valued or for Z_G to be very small (resonance not being considered). Since the impedance of the signal source is often not under the control of the circuit designer, in general the most desirable characteristic is for Z_i to be large.

Another consideration is that of the actual current-delivering capabilities of transducers. Certain microphones, for example, cannot function properly when feeding a low-resistance load. The linearity of some transducers may be adversely affected by a low-impedance load on the device.

On the other hand, high-impedance circuits can also cause difficulties. The adverse effects of shunt capacitance are to be considered. Capacitance between wires a and b in Fig. 1-12(b) will reduce the fraction of V_g reaching the IC at high frequencies compared to the low-frequency behavior of the circuit. This conclusion is apparent from observation of Eq. (1-22). Circuits designed for high-frequency use usually operate at low impedance levels. Fifty ohms is a standard.

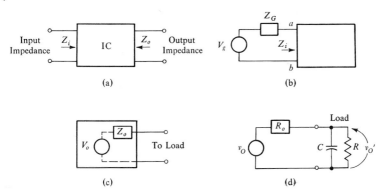

Fig. 1-12   (a) definition of Z_i and Z_O; (b) (c) (d) circuits for discussion.

Output Impedance

The output impedance of the IC shown in Fig. 1-12(c) is of concern because its value affects the signal power that can be supplied to a load. It is generally desired, therefore, that Z_o be low-valued.

When considering transient response, the v_O waveform will be reproduced faithfully at the load in Fig. 1-12(d) only when R_o is zero. In response to a step of voltage available at v_O of amplitude ΔV, the voltage across the load becomes

$$v_O' = \frac{\Delta V R}{R + R_o} (1 - \epsilon^{-(R + R_o)t/RR_oC}). \qquad (1\text{-}23)$$

Eq. (1-23), when plotted as v_O' versus t, yields a rising exponential curve starting with $v_O' = 0$ at $t = 0$ and ending with $v_O' = \Delta VR(R + R_o)$ as $t \rightarrow \infty$. The rate of rise of this waveform is determined by the *time constant*, the reciprocal of the coefficient of t.   A large value of the time constant means that the curve rises slowly toward its steady state value.   Thus a large value of R_o will slow down the rate of rise of a pulse being delivered to the load.

1-10   IC SYSTEM EXAMPLE: TV SOUND CHANNEL

As an example of a system that can be integrated, we consider the sound channel of a television receiver.   We assume that the video (picture) information has previously been separated from the audio and has been fed to the video channel for processing.

Sound information is a frequency-modulated (FM) analog waveform in commercial TV.   The FM wave varies in its frequency with time, and those variations are representative of the audio intelligence.   The major functions of the system are to amplify the rather weak signal and to transform the frequency variations into amplitude variations suitable for the output transducer, a loudspeaker.   The system also provides volume control and *deemphasis.*

High-frequency audio signals are accentuated by a *preemphasis* network in the FM transmitter to improve the signal strength relative to noise.   The FM receiver must include a response-shaping or deemphasis network to reduce the strength of high frequency components.

Eight functions of a typical sound channel are shown in Fig. 1-13(a). The input signal is referred to as an IF (intermediate frequency) wave.   In TV, this is a 4.5 MHz carrier, frequency modulated.   The *IF amplifier* stage raises the signal level.   The *limiter* removes any unwanted amplitude variations, since they may adversely affect operation of certain *discriminator* types.   The discriminator block removes the IF, and turns frequency variations into amplitude variations.   The discriminator output is a weak audio voltage.   At this point, the manual *volume control* may be added along with deemphasis to reduce the strength of high frequency components.   *Audio amplifier* stages raise the signal level.   The *power amplifier* is capable of handling the load power, and performs the required impedance match between the electronics and the loudspeaker.

The functions of blocks A, B, and C can be integrated, with the exception of coils that may be required for discrimination.   The conventional volume control cannot be made in integrated form because it is mechanical, and the deemphasis network usually requires a rather large capacitor that must be discrete.   Also, any capacitors required for blocking dc that must pass audio frequencies must be discrete.   These functions are shown in Fig. 1-13(b).

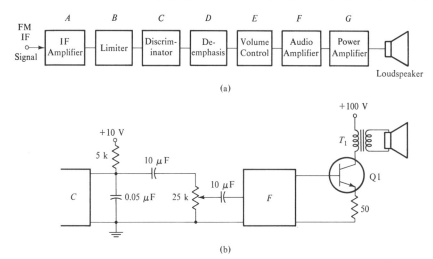

Fig. 1-13   (a) TV sound channel; (b) details of blocks *D*, *E*, and *G*.

The 10 μF units are for blocking, and the 25 kΩ potentiometer is the volume control.

The audio amplifier function may be integrated, but the power amplifier stage would usually be discrete or hybrid because of the relatively high power dissipation at transistor Q1.   Transformer T_1 provides an impedance match between the low-impedance loudspeaker and the higher output impedance of Q1.

The contents of blocks, A, B, C, and F, when monolithically integrated, represent 70 percent of the components in the channel.   By using hybrid methods, a higher percentage is possible.   An IC for this type of service is Fairchild Semiconductor Type μA719 Multi-Purpose Amplifier.   Further discussion of the gain blocks is given in Sections 8-4 and 8-7.

1-11   MEASUREMENT CONSIDERATIONS

The characteristics of analog signal sources and transducers vary greatly. By using Thevenin's Theorem, such sources can be represented by a voltage generator in series with an equivalent impedance, V_g and Z_G in the system shown in Fig. 1-14.   The amplifier shown is considered to have a very high

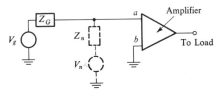

Fig. 1-14   Effect of pickup upon operation of simple amplifier.

input impedance, and in this circuit one terminal of the signal source and one terminal of the amplifier are grounded. The amplifier feeds a load that may be a recorder, a loudspeaker, or some other type of end device.

Also shown in Fig. 1-14 in dashed lines is a representation of electrostatic and electromagnetic "pickup," symbolized by generator V_n and impedance Z_n. This pickup is the result of inductive coupling between the circuit being studied and power lines, fluorescent lights, and so on, and of electrostatic coupling existing because of capacitance from this circuit to other energized circuits in the vicinity. While usually a distributed phenomenon, this form of noise is shown lumped in the figure. It is referred to as noise, but this voltage bears no similarity to the internal noise generated inside conductors and semiconductors caused by the physics of the devices and the processes involved. As with internal noise, this external noise or pickup is unwanted.

It is evident from the Fig. 1-14 representation that if the signal source resistance can be made to approach zero, all of V_n will drop across Z_n and no noise will therefore enter terminals a-b and thus become amplified and trans-mitted to the load. The first rule for successful measurement of small signals is for the signal source impedance to be small. Likewise, we would wish for the resistance of the wires connecting the signal source to the ampli-fier to be low-valued.

Unfortunately, it is not always possible to control the transducer charac-teristics in order to achieve a low resistance and therefore a favorable voltage division of the noise voltages. This is especially true in the measurement of biological voltages.

Let us consider the possibility of not grounding both source and amplifier input. Instead, we use a pair of wires between V_g and a *differential amplifier* with floating input as shown in Fig. 1-15. Symbols Z_1 and Z_2 represent source and wire impedance, and Z_3 and Z_4 are noise source impedances. Again assuming that the amplifier has infinite input impedance, we conclude from the figure that the voltage appearing at the amplifier terminals, V_{ab}, can be expressed by

$$V_{ab} = V_g \left(\frac{Z_4}{Z_1 + Z_4} \right) + V_n \left[\frac{Z_1}{Z_1 + Z_4} - \frac{Z_2}{Z_2 + Z_3} \right] \qquad (1\text{-}24)$$

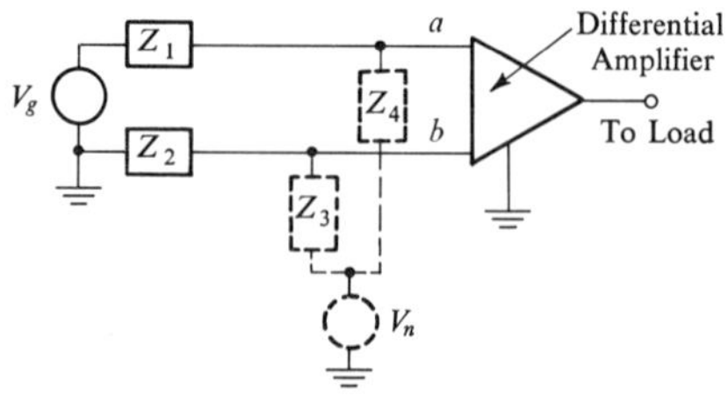

Fig. 1-15   Effect of pickup upon operation of differential amplifier.

V_{ab} is now referred to as the *normal mode voltage*.   It is clear from Eq. (1-24) that the noise voltage will not affect circuit operation if $Z_1 = Z_2$, and $Z_3 = Z_4$, for then the second term in the equation becomes zero.   Note that it is not necessary for Z_1 or Z_2 to be zero or even low-valued to eliminate V_n. *Because V_n is fed essentially in phase to both input lines, it is referred to as the common-mode voltage.*

For systems using a differential amplifier, a large *common-mode rejection ratio* (CMRR) is desired.   This factor is the ratio of the gain of the amplifier to a differential or normal signal to the gain of the amplifier to a common-mode signal.   Usually expressed in dB, CMRR for IC op amps is typically 80 to 100 dB.   Further discussion of differential amplifiers and CMRR is contained in Chapters 5, 6, and 7.

We have discussed in a general way IC characteristics, limitations, and representations.   Next, in Chapter 2, we consider the circuit elements of the IC in detail; their principles of operation, methods of fabrication, and the reasons behind their limitations.   Later, elements will be combined to form simple circuits that are the building blocks for complicated integrated circuits. The IC's, then, are utilized to perform required functions in communications, in computation, in control, and in measuring systems.

Problems

1-1   A schematic diagram of the hybrid IC of Fig. 1-1 is shown.   Make a sketch of Fig. 1-1 and identify circuit elements on the sketch.   Pin 1 is at the bottom of the figure.

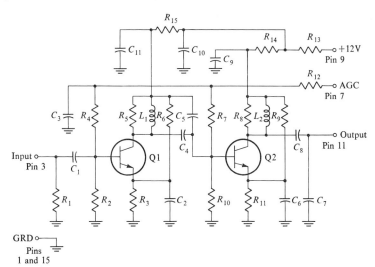

Problem 1-1

1-2   The L-C circuit of the figure is supplied from an adjustable-frequency signal generator. Show that the circuit can provide voltage amplification ($V_o/V_i > 1$). Over what frequency band does the gain exceed unity?

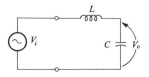

Problem 1-2

1-3   Express in dB:
(a) power ratios of 1000, 500, 5;
(b) voltage ratios of 10,000, 50, 1.

1-4   An ideal constant-current source of value $i_o = 5 \cos 10^6 t$ is paralleled by internal source resistance of $10^5 \Omega$ and feeds a 2000-ohm load connected across the terminals of the source.

(a) What percentage of the source current reaches the load: (1) under steady-state conditions; (2) under transient conditions?

(b) Perform a source interchange. What is the value of the equivalent voltage source, and the new series value of source resistance? (Do not transform the load resistance.)

1-5   Write Eqs. (1-7), (1-8), (1-9), (1-10), and (1-11) in matrix form.

1-6   Draw an equivalent circuit for: (a) Eq. (1-9); (b) Eq. (1-10); (c) Eq. (1-11).

1-7   Convert the numerical set of cascode amplifier y parameters given in Section 1-7 into a corresponding set of h parameters.   Use Appendix II.

1-8   In applying Thevenin's Theorem at a terminal pair in an electrical network, one normally "shorts" all voltage generators and "opens" all current generators to obtain the equivalent impedance.   However, when generators are "dependent" or "controlled," they must be retained.   Show that the Thevenin

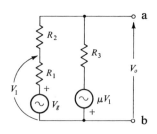

Problem 1-8

equivalent for terminals a-b of the circuit shown is

$$R_{TH} = \frac{(R_1 + R_2)R_3}{R_1 + R_2 + R_3 + \mu R_1}$$

when V_g is shorted but generator V_1 is retained. This can be accomplished by connecting a generator at V_o and noting that $R_{TH} = V_o/I_o$.

1-9 Prove the interrelationships between h and y parameters given in Appendix II. Do this by equating the input, output, and transfer admittances of the two equivalent circuits.

1-10 Use the y parameters given in Section 1-7 for a cascode amplifier to predict the magnitude and phase angle of voltage gain A_r of that amplifier feeding a 1000-Ω load. Use equations given in Appendix II.

1-11 The ratio of collector current to base current of a transistor has been defined as β_{dc}. For $\beta_{dc} = 100$, calculate I_B and I_E for $I_C = 2$ mA.

1-12 A transfer characteristic for a transistor relating output to input currents is shown in the figure.

(a) Determine β_{dc} at $I_B = 5, 10, 15,$ and $20\ \mu A$.

(b) The small-signal or ac-current-amplification factor β is the slope of this characteristic. Estimate β at $I_C = 2$ mA.

(c) How would you convince a colleague that β and h_{21} are the same "animal"?

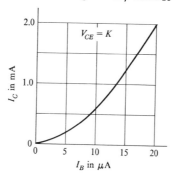

Problem 1-12

1-13 The constant Λ symbolizes q/kT. Evaluate at $T = 300°K$.

1-14 For the circuit shown in Fig. 1-12(b), consider that $Z_G = Z_i = 10^4$ ohms resistive. Show that wiring and stray capacitance of 5pF between lines a and b will affect the V_{ab}/V_g ratio by

(a) deriving a relation for that ratio;

(b) determining the frequency where the ratio has declined to 0.707 of its low-frequency value from observation that the denominator of the ratio has equal real and imaginary parts.

1-15 Derive Eq. (1-23).

1-16 Derive Eq. (1-24).

2
PHYSICAL BASIS FOR INTEGRATED CIRCUITS

The transistor is the most important element in a monolithic integrated circuit; the methods used in the fabrication of diffused transistors are directly applicable to the formation of the other circuit elements of the IC, resistances, capacitances, and diodes.

This chapter begins with a brief discussion of the physical principles of transistor operation. The reader who is unfamiliar with the transistor is referred to Reference 3 for a more complete discussion. The techniques of impurity diffusion and epitaxy are discussed in order to understand the methods used in manufacturing a single transistor and an entire integrated circuit. Latter sections of this chapter are devoted to integrated resistances, capacitances, diodes, and field-effect transistors.

2-1 TRANSISTOR OPERATION

The common junction transistor operates under the principle that a stream or current of charge carriers between terminals called the *emitter* and *collector* can be controlled by relatively minor variations in the voltage applied to an intermediate terminal called the *base*. The principle of *control* is also the basis for operation of other active devices, such as the vacuum tube and the field-effect transistor.

The current being controlled is labeled i_C in Fig. 2-1; it flows through the circuit load resistance, R_L. Amplification of the signal $v(t)$ is

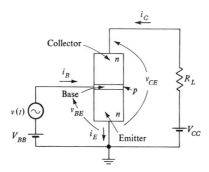

Fig. 2-1 Simple junction transistor circuit.

achieved because the signal power dissipated in the load is greater than the control power supplied to the input or base terminal.   Power supplies V_{CC} and V_{BB} provide the difference between load power and source power.

The transistor is a structure composed of n- and p-type semiconductor materials.   Silicon is the semiconductor most often used. n-type material is pure silicon to which has been added a small amount of an impurity, such as phosphorus.   Phosphorus, called a *donor* impurity, adds free electrons to the silicon crystal, reducing its resistivity.   The process of adding impurities to silicon at a high temperature is called *doping*. p-type material is prepared by adding an *acceptor* impurity, such as boron.   The result is a crystal having voids or *holes* in its electronic bonding structure.   Both holes and electrons can act as charge carriers at normal room temperature.

When n-type and p-type materials are joined in a continuous crystal, diode properties result.   The *pn* junction permits current to flow in one direction across its boundary, but blocks current in the reverse direction when an applied potential is connected with its positive on the n material.

Transistor operation is achieved by forward biasing the emitter-base *pn* junction and reverse biasing the collector-base *pn* junction.   The number of carriers crossing the emitter-base junction from the emitter is determined by the potential applied to that junction, $V_{BB} + v(t)$.   When the charge carriers reach the narrow base, *they then can continue to the collector, because the collector-base junction voltage appears as a forward bias to these carriers.*

Carrier travel through the base region may be the result of *diffusion* or of *drift*.   Diffusion current results because of the movement of charge carriers from regions of high concentration to regions of lower concentration.   Drift is carrier movement resulting from a potential difference, that is, an electric field.   In modern transistors, because of the graded impurity profile in the base region that results in an accelerating field, drift, as well as diffusion, accounts for carrier movement through the base.   The impurity profile will be explained in Section 2-4.

Majority carriers (electrons in *n* material) are injected into the base of an *npn* transistor from the emitter region because of the forward bias on that junction. While the *p*-type base is not heavily doped, it does have a preponderance of holes, and some recombination of holes with electrons will occur in that region. This loss of carriers through recombination must be kept low to maximize the efficiency of the transistor.*

The carrier stream or collector current flows through the load. The waveshape of this current is directly related to the waveform of the base-emitter voltage that controls conduction across the input junction.

Discrete and integrated transistors can be made in the form of an *npn* "sandwich," as shown in Fig. 2-1, or a *pnp* structure in which the base is *n*-type material. Circuits using both *npn* and *pnp* devices are called *complementary-symmetry* circuits. In IC's the *npn* type is most prevalent; *pnp* structures are used where advantageous.

2-2 SILICON

Integrated circuits are fabricated by means of technology developed in the manufacture of discrete silicon transistors, and use silicon as the basic building material. The element silicon is classified in group or column IV of the periodic table of elements; it has four electrons in its outermost orbit or shell, and in its pure crystalline state, is an electrical insulator. Silicon will melt at a temperature of about 1420°C; most of the processing is done below the melting point, at 1300°C or less. The electrical properties of silicon are determined primarily by the concentration of impurity atoms within it.

For the purification of silicon, silica, or common sand, is chemically purified and then zone-refined to eliminate remaining unwanted impurities. To obtain transistor characteristics, the silicon must be in crystal form. Crystals of silicon are usually *grown*. Crystal growth is the process of adding material to a small seed crystal in such a way that atomic planes are in a regular order. To grow a bar of silicon, a seed is touched to the surface of a bath or *melt* of molten silicon. The seed is slowly withdrawn from the melt, with cooling and hence solidification occurring at the seed-melt interface. The finished bar, the result of this growing operation, is typically six inches in length and about $1\frac{1}{4}$ inches in diameter.

By means of diamond-faced saws, the grown crystal is sliced into wafers 10 to 15 mils thick. (1 mil = 0.001 inch). These slices are put in lapping machines to remove any damaged regions of the surface. Finally, the wafers are polished to a mirror finish. It is from these wafers that transistors and IC's are fabricated.

---

* Recombination is simply the joining of an electron with a hole. Since a hole is a void, joining annihilates both as carriers.

A concept often used in IC design and manufacture to describe the electrical conduction characteristic of either intrinsic (pure) or extrinsic (impure) material is *sheet resistance* or *resistance per square*. A familiar relation for the resistance of a uniform layer of material of length *l* and cross-section *t* by *w* as shown in Fig. 2-2 is

$$R = \rho \frac{l}{tw},$$
(2-1)

where ρ signifies the resistivity of the material in ohm-meters, and the other dimensions are in meters. Equation (2-1) can be modified to

$$R = R_s \left(\frac{l}{w}\right).$$
(2-2)

The sheet resistance R_s is equivalent to ρ/t and has the dimensions of ohms. This is also referred to as *ohms per square*. The term in parenthesis, l/w is the dimensionless length-to-width ratio, and is often called the *number of squares*.

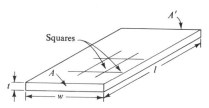

Fig. 2-2   Sheet resistance dimensions; A and A′ are symbols for the front and rear faces.

The number of squares refers to the top or bottom face of the geometric solid shown in Fig. 2-2. We know that resistance between front (*A*) and rear (*A′*) surfaces of the figure would increase with length *l* because of series resistance, and decrease when more parallel paths, as represented by larger *w*, are present. For the case when $l = w = 1$, we say that one square, exists. For $l = 2$, $w = 3$, the resistance is less than for one square, because the number of parallel paths has increased more than the effect of increased length. Hence the number of squares is 2/3. A square, then, has no fixed dimensions; its size is solely determined by the l/w ratio; and that ratio directly affects the resistance between *A* and *A′*.

The primary source of the charge carriers in a silicon sample is the impurity used in doping the melt. As has been mentioned, the type of impurity determines whether holes or electrons will predominate. Another source of carriers, usually undesired, is related to the temperature of the specimen. Heat energy can break some of the covalent bonds in the crystal lattice causing both a free electron and a hole to be liberated for every broken bond. Since, under the influence of an applied voltage, these carriers represent opposite charges that drift in opposite directions, the resistivity of the semiconductor may decline as temperature increases.   Even near room temperature, thermally derived carriers cause measurable variations in the characteristics of IC elements.

The thermal generation of hole-electron pairs is a reversible process, provided the crystal is not damaged by heat.   Upon cooling, holes and electrons will recombine.

2-3 DIFFUSION

It is necessary to *dope* the silicon, to add impurities in order to alter the basic electrical characteristics of the pure material.   While several methods are available for this purpose, the solid-state *diffusion* process has emerged as standard.

The basic physical laws of diffusion are referred to as Fick's Laws.   When an impurity is dissolved in silicon and the impurity has a nonuniform concentration (it is bunched in a certain region), the first law of diffusion tells us that the impurity will tend to spread out until the concentration is uniform throughout the sample.   Thus, a flow or diffusion of impurity atoms from regions of high concentration to regions of lower concentration will take place.   Mathematically, in one dimension, x, this process can be expressed by

$$J = -D \frac{dN}{dx}, \qquad (2\text{-}3)$$

where

J = rate of impurity flow through a plane normal to the concentration gradient;

D = diffusion constant, with units of area/time;

$\frac{dN}{dx}$ = impurity concentration gradient at the plane.

Since the direction of the flow is opposite to the direction in which N is increasing, a negative sign is included in Eq. (2-3).   A pictorial representation of the quantities of Eq. (2-3) is given in Fig. 2-3.

Solid-state diffusion is practical at elevated temperatures.   Usually impurities from a gaseous atmosphere diffuse into the surface of the silicon wafer.   Because of the high temperature, atoms of silicon will be highly excited and many will leave their lattice sites and take up new positions in the crystal.   At the same time, atoms of the impurity diffuse into the silicon and can occupy vacancies in the lattice caused by the wandering silicon atoms.

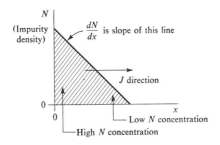

Fig. 2-3   Illustration for Eq. (2-3); special case where dN/dx is constant.

Equation (2-3) is concerned with the direction of the flow of the impurity, but does not explicitly consider time.   The second law of diffusion brings this time dependence into play:

$$\frac{\partial N}{\partial t} = D\,\frac{\partial^2 N}{\partial x^2}. \tag{2-4}$$

We use Eq. (2-4) in order to predict impurity density versus distance and time, that is, $N(t, x)$.   In order to solve that equation, certain boundary conditions must be imposed.   Therefore, let us consider a practical example.

A p-type wafer of silicon is placed in a diffusion oven in which a gas containing an n-type impurity, such as phosphorus, can be introduced.   The wafer is heated to about 1200°C and exposed to the gaseous atmosphere. Phosphorus atoms will proceed to diffuse into the silicon at its surface, and the concentration gradient will support this flow according to Eq. (2-3).

If we assume that the impurity concentration of the supply gas will remain constant during the process and that no phosphorus existed in the wafer at $t = 0$, a solution to Eq. (2-4) is:

$$N_x = N_o\, \text{erfc}\left(\frac{x}{2(Dt)^{1/2}}\right), \tag{2-5}$$

where

N_x is the concentration of impurity in the wafer at a distance x from the surface;

N_o is the equilibrium surface concentration (depending upon the gas concentration) at the wafer surface;

erfc is the *error function complement*, which may be determined from books of tables once $x/2\sqrt{Dt}$ is evaluated.   This quantity represents the evaluation of an integral that occurs sufficiently often to list in tables.   Fig. 2-4 shows N_x/N_o versus $x/2\sqrt{Dt}$ information.

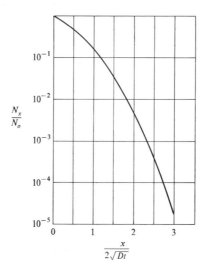

Fig. 2-4   Error function compliment (erfc).

The quantity x in Eq. (2-5) represents depth below the surface of the silicon specimen.   The natural result of impurity diffusion into silicon is to change the resistivity of the wafer because of the additional charge carriers introduced.   At a particular depth the diffused impurity density will override the basic impurity density of the original silicon sample.   *A* pn *junction is formed* at the depth where the impurity concentrations are equal.   Thus for the example given here, n diffusion into p material, the material would continue to be p type for values of x beyond the junction, and is n type for values of x less than the distance from the surface to the junction.

Temperature is an important consideration in the diffusion process.   However, it is not explicitly given in any of the preceding equations.   Temperature primarily affects the diffusion constant, D.   Elevated temperatures impart energy to diffusant atoms that increases their speed of travel through the semiconductor crystal.   Because D is so greatly affected by temperature in the 1200°C vicinity, the temperature of a diffusion furnace is carefully controlled.

The diffusion process is ended when the wafer is removed from the furnace.   No further diffusion takes place after cooling, and the impurity concentration within the solid remains.

A characteristic of the dopant called *solid solubility* places an upper limit on the amount of the impurity that can be dissolved in a unit volume of solid silicon and hence may affect the value of N_o.   At 1200°C, phosphorus, for example, cannot contribute more than 1.5×10^{21} atoms/cm^3.   At that temperature aluminum as a dopant is limited to about 2×10^{19} atoms/cm^3.

Gold is sometimes used as an additional dopant in IC manufacture.   The reason for its use is not to create new conductivity regions.   Rather, it has been found that gold enhances the recombination of carriers.   This in effect can result in faster switching speeds in digital circuits.

2-4   EPITAXY

While impurity diffusion forms the backbone of IC manufacture, it does have limitations.   Diffusion always affects the area close to the surface more than it does internal areas of a specimen.   Diffusion-derived junctions are always of the gradient type, and the mobilities of holes and electrons are adversely affected by the multiple diffusions needed in the fabrication of transistors. *Epitaxy* is an allied process used in IC manufacture.

The main application of epitaxy is the growing of low-resistance films on silicon crystals.   The word essentially means "arranged upon."   When an epitaxial layer is grown upon a silicon substrate, the grown layer possesses the same crystal orientation as the substrate.

To grow an epitaxial film, silicon wafers are heated to about 1200°C in an evacuated reaction chamber.   After surface cleaning by etching, hydrogen, silicon chloride ($SiCl_4$), and phosphine (PH_3) gases are admitted.   These react with the surface of the wafer, and an *n*-type layer of silicon grows at the rate of approximately one-half micron per minute (1 micron $= 10^{-6}$ meter).

While the epitaxial process has not been perfected to date exclusively to fabricate complete transistors or IC's, it is used to supplement diffusion in the manufacture of these devices.

2-5   SILICON DIOXIDE

Silicon dioxide (SiO_2), a transparent glass, grows spontaneously on pure silicon when exposed to air.   At room temperature, only a very thin oxide layer grows. SiO_2 is an effective electrical insulator, and is also used to great advantage to shield portions of a silicon wafer from chemical reactions. Therefore, relatively thick oxide layers are purposely grown on silicon at elevated temperatures, between 1000°C and 1200°C, by using an atmosphere of steam or pure oxygen.

Silicon dioxide is purposely formed during certain steps in the manufacture of IC's, and must be selectively removed during later steps.   The removal process will now be described.   A photosensitive compound called *photoresist* is applied on top of the oxide.   Through a mask, ultraviolet light polymerizes the photoresist.   Unpolymerized photoresist can be washed away, selectively exposing part of the oxide.   Dilute hydrofluoric acid is then used to dissolve the exposed oxide and expose the silicon beneath.

2-6   IC FABRICATION

Although heat dissipation and material properties are to be considered, the minimum size of an integrated circuit is actually limited by control of physical dimensions. The diffusion or epitaxy process determines below-surface depth, but other dimensions are determined by *photolithography*. In the photolithographic process, master drawings are made for masks, and then the drawings are photographically reduced. These small masks or stencils are used to determine areas on the IC surface that will be processed. Limitations encountered are changes of dimensions with temperature and humidity, alignment of masks on wafers, and lens distortions in photographic reduction.

We now consider the fabrication of an IC transistor. As previously mentioned, a single crystal wafer of doped silicon about one inch in diameter and about 7 mils thick is the start of IC manufacture. For this discussion, consider that the wafer is *p* material, doped to a resistivity of 12 Ω-cm. It has been lapped and highly polished.

To assure that all surface damage has been removed, an acid etch using hydrochloride gas (HCl) at 1100°C, is made. Then, in a diffusion furnace at perhaps 1100°C, the surfaces are exposed to steam in order to grow an oxide layer of approximately 6000 Å.*

Photoresist is applied to the entire surface. A glass mask is now brought in contact with the wafer surface, and the assembly is exposed to ultraviolet light to polymerize the photoresist. The unpolymerized material is dissolved exposing selected SiO_2 areas. Dilute hydrofluoric acid is then used to remove oxide and expose portions of the silicon surface.

The steps noted in the above paragraph are in preparation for *buried-layer diffusion*, the purpose of which is to place an n^+ layer under the transistor collector region. The plus sign on n^+ denoted a very high impurity content. This layer will result in lowered saturation resistance for the device. The wafer is placed in a diffusion furnace, and, by using an *n*-type dopant, such as arsenic, the n^+ region is diffused. The resulting sheet resistance may be about 8 $\Omega/\square$. The wafer now has a cross section as shown in Fig. 2-5(a).

All oxide may now be etched away and an epitaxial *n*-layer 15 microns thick grown on the entire surface, as shown in Fig. 2-5(b). This layer, which has a resistivity of 0.5 Ω-cm, will form the collector of the completed transistor.

Again an oxide layer is grown on the entire wafer, 10,000 Å thick. This layer is selectively removed for the isolation diffusion. This p^+ diffusion

---

* Measurements of minute lengths and distances in diffused structures usually use the mil (one-thousandth of an inch), the *micron* (μ) (one-millionth of a meter), and the *Angstrom* (Å) $((1/10)^{10}$ of a meter). Thus 1 μ = 0.0000394 in., and 1Å = 3.94×10^{-9} in.

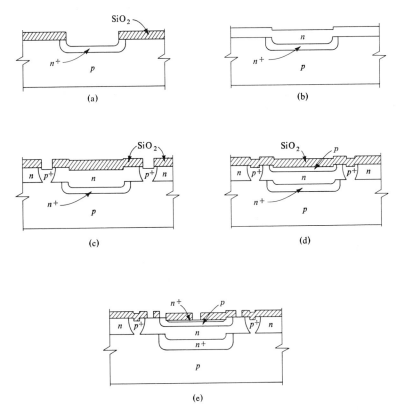

Fig. 2-5 Steps in the manufacture of an IC transistor: (a) diffused n^+ layer in p substrate; (b) growth of epitaxial n layer; (c) isolation diffusion results in p^+ channels; (d) p-type base diffusion; (e) n^+ layer forms emitter.

will result in isolated islands of n-type material that are surrounded by p material.   See Fig. 2-5(c).

The oxide may now be regrown and, again, selectively etched in preparation for the base diffusion.   The base will naturally be p-material diffused into the large n region; however, mechanics of this operation are usually somewhat different from the isolation diffusion step previously discussed. Base diffusion is an extremely important step, because it determines the current gain or beta of the transistor and it determines the resistance of integrated resistance elements being diffused at the same time at other locations on the wafer.   Two steps are often involved.   In the *predeposition* step, a heavily doped boron glass is grown on the silicon at a relatively low temperature.   This fixes the total number of impurities that are available for diffusion in the *drive-in* cycle.   This second step is controlled, so that the pn junction is formed about 3 microns below the surface. The result is shown in Fig. 2-5(d).*

Emitter diffusion is next. The n-type emitter region is diffused into the base, and the resulting base width may be about 0.7 micron. The emitter region is the center n^+ in Fig. 2-5(e). Typical impurity concentration information is given in Fig. 2-6. Note that the *junctions are locations where net impurity level shifts from one type to the other.*

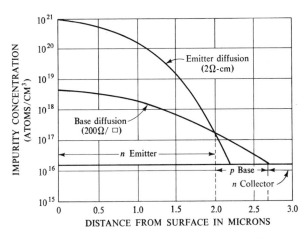

Fig. 2-6 Impurity profiles for an IC transistor.

It is necessary to etch holes in the oxide in order to prepare the wafer to accept electrical connections. After etching and cleaning, the wafer is placed in a vacuum and heated sufficiently so that aluminum vapor will alloy with silicon and with its oxide. This process, referred to as vapor deposition, is terminated when an aluminum layer about one micron thick covers the entire surface.

The wafer is again coated with photoresist, and the aluminum is selectively removed in order that the interconnecting leads and bonding pads remain. A bonding island or pad is a relatively large area of metal located near the perifery of the IC.

* Diffusion taking place with the total number of impurity atoms fixed yields a solution to the diffusion equation different from Eq, (2-5). Thus

$$N_x = \frac{Q}{\sqrt{\pi Dt}} \, \epsilon^{-(x^2/4Dt)},$$

with Q the initial surface impurity concentration in atoms/cm³. This equation is referred to as a Gaussian distribution. It does not differ greatly from the erfc function.

A diamond stylus is scribed across the wafer in the x and y directions, and the wafer cracked and broken along the scribe lines. The result is the separation of the individual circuits or chips, the dimensions of which may be 40 mils square.

The chip may next be bonded to the final enclosure, and fine wires bonded to the aluminum pads. These wires connect the chip to pins of the can or enclosure. The necessary tests are performed before and after encapsulation.

The basic method for the formation of diffused components relies upon diffusion in a vertical direction; lateral diffusion concurrently takes place. This accounts for at least one important benefit in transistor fabrication; pn junctions form below a surface of undisturbed SiO_2, and are therefore immune to contaminants. It has been shown that the lateral-to-vertical ratio of diffusion influence has a value of 0.6 to 0.8.[4]

2-7 ISOLATION

The components of an integrated circuit—resistances, capacitances, transistors, and diodes—are made simultaneously. The masks determine the kind of element at each location. Since the circuit elements adjoin one another on the wafer, a means for isolating the elements is necessary. Otherwise, if they were not isolated, part of one device—for example, the collector of a transistor—would be physically joined to the collectors of all adjacent transistors, and circuits would be very limited in their performance.

Junction isolation is shown in Fig. 2-7. At the right of the center p^+ region we have a transistor, and to the left of that region is a diffused resistance. The resistance value is proportional to the length of the p region between points marked A and B. The isolation diffusion accomplishes the following: it provides for a reverse-biased diode between the transistor collector C and the substrate p^+. The resistance is isolated because its active portion (the p region) is not in any way connected to the transistor. Two series pn junctions exist between the resistance and the substrate. Either of these

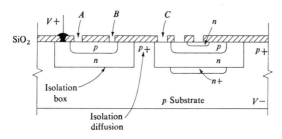

Fig. 2-7 Isolation between adjacent IC elements; transistor at right, resistor at left.

can be reverse-biased to provide the junction isolation.   In the figure, the *isolation box*, or *n* region, is connected to the most positive point in the circuit, V^+.   In this type of structure it is also possible to leave the box unconnected or floating, because the resistance is connected to positive potentials and the substrate to the most negative potential, so that one of the two series *pn* junctions is effectively reverse-biased.

Dielectric isolation methods have been proposed in order to eliminate parasitic junctions within the system, thereby reducing undesired capacitances and eliminating leakage paths, and providing improved design freedom.   In the design of IC's that are resistant to ionizing radiation, which is present in severe environments, dielectric isolation has proven advantageous.

Consider Fig. 2-8(a).   An n^+ epitaxial layer has been grown on an *n* substrate.   Moats (*A*) are etched into the structure, and SiO_2 grown on one

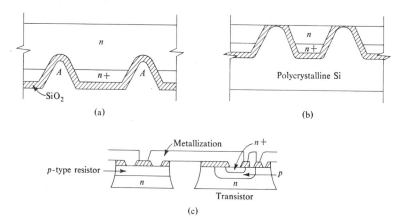

Fig. 2-8 (a) (b) Steps in IC manufacture using dielectric isolation; (c) beam lead structure.

surface.   In (b) of the figure, a thick layer of polycrystalline silicon has been deposited on the oxide for mechanical strength.   The *n* region has been lapped so that it does not extend above the grooves.   Thus it is evident that the *n* and n^+ regions in the center are isolated from other portions of the circuit.   The n^+ region can form the low-resistance collector path for a transistor, as previously discussed in Section 2-6.   Further diffusions can create devices in the *n* and n^+ regions.

Air gap isolation is also a solution to the problem of parasitic effects. This isolation method is usually referred to as the *beam-lead* technique. Elements are formed in the same manner as previously discussed in Section 2-6.   An extra heavy metallization pattern is deposited.   All superfluous silicon is then removed, and the result is an array of devices interconnected and supported by the semirigid metallization. The final form is sketched in Fig. 2-8(c).

2-8 DIFFUSED TRANSISTORS

Integrated-circuit transistors may be fabricated in one of several ways. We shall not attempt to discuss all of the design ramifications possible; a representative procedure that yields high-quality devices was discussed in Section 2-6. This transistor has been referred to as the *four-layer epitaxial structure.*

The discrete planar transistor is also a four-layer structure. Built on an n^+ substrate, its collector terminal is made at the bottom surface of that substrate. This geometry is not possible in present-day IC transistors, for no contacts are made on the bottom surface of the wafer. Also, it is to be remembered that IC transistor manufacture must include provisions for isolation between devices. Thus the IC unit is built on a *p*-substrate, and the *n*-type collector is brought out to contacts at the *top* of the wafer. The *pn* junction between collector and substrate extends both laterally and vertically and provides the required isolation.

Figure 2-9(a) gives typical dimensions, impurity data, and resistivities pertaining to the four-layer transistor. The epitaxially grown collector region

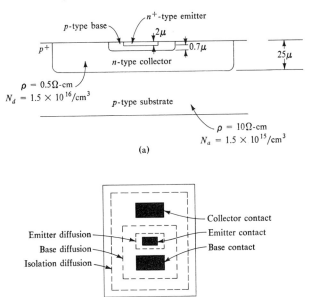

(a)

(b)

Fig. 2-9 IC transistor: (a) vertical profile; (b) geometry.

is broken into separate pads on the surface of the substrate by the p^+ isolation diffusion.   Impurity profiles are given in Fig. 2-6.   Because the base and emitter regions are diffused, their impurity concentrations are not constant through the material but, instead, vary with depth from the surface.

From the top surface of the wafer, a typical IC transistor may appear as shown in Fig. 2-9(b).   This is just one of many possible geometric configurations.   The entire transistor will easily fit on a surface 10 mils by 10 mils.

Physical dimensions and impurity levels are responsible for the performance characteristics of transistors: the saturation resistance, voltage breakdown, magnitudes of leakage currents, current gain, and gain-bandwidth product or transition frequency f_T (the frequency at which current gain is unity).   The designer of IC transistors must be cognizant of the many interrelationships and compromises necessary to achieve transistor characteristics according to the particular set of requirements.   The mathematical relations that apply to transistor device design will not be given here; they are available in the literature.[5,6]

Saturation Resistance

The collector of the IC transistor is buried in the structure.   However, contact to the collector region is made at the top face of the wafer.   Collector current must flow downward, to the side, and then upward to reach the terminal.   The resistance of this path is important because it is in series with the collector terminal.   When the device is fully ON, and thus the collector current is at its maximum value, this "saturation resistance" represents a major voltage and power loss in the device.   As mentioned previously, an n^+ buried layer may be used to reduce this resistance.

In the transistor shown in Fig. 2-10, the collector current path is designated by R_a, R_b, and R_c.   There is additional resistance in the n^+ collector contact, if used, but it will be neglected.   Using typical dimensions, we shall calculate the sum of these resistances.

Let the collector resistivity be 0.5 Ω-cm, and the buried layer be 20 Ω/□.

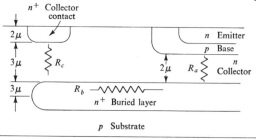

Fig. 2-10   Model for calculation of saturation resistance.

Consider the R_a and R_c paths to have effective areas of 0.5 mil^2 and the length/ width ratio (number of squares) in the buried layer to be unity.   We conclude that $R_b = 20\ \Omega$; and

$$R_a = (0.5\ \Omega\text{-cm}) \frac{(2\mu)}{(0.5\ \text{mil}^2)} = 31\ \Omega;$$

and

$$R_c = (0.5\ \Omega\text{-cm}) \frac{(3\mu)}{(0.5\ \text{mil}^2)} = 47\ \Omega.$$

The total saturation resistance of the path is the sum, $20 + 31 + 47 = 98\ \Omega$.

pnp Transistors

Up to this point, the *npn* transistor structure has been discussed exclusively. The *npn* device has the inherent advantage of better high-frequency performance over its complement, the *pnp* transistor.   The mobility or drift velocity of electrons as charge carriers is decidedly superior to that of holes, and electrons are the important carriers in the *npn* device.   However, there are benefits to be derived from being able to fabricate and utilize both transistor types.

No severe problem exists in making an all-*pnp* IC.   The fabrication procedure is similar to that discussed in Section 2-6, except that the starting point is a wafer of *n*-type silicon for use as the substrate, instead of the *p*-type previously considered.

Occasionally an application is found for the *pnp* structure to be found in every common IC, the one shown in Fig. 2-9(a) *with the substrate its collector*. This transistor is referred to as a *substrate pnp*.   The emitter of the substrate transistor is the base region of the original *npn*, and the base of this transistor is the collector of the original *npn*.   The *pnp* thus formed would usually be singly employed, for it would have a common collector region with any other substrate *pnp*.   Its beta is usually extremely low-valued because of the wide base of the device, and the large collector area restricts its high-frequency performance.

The *npn* and *pnp* transistors described in the two preceding paragraphs are referred to as *vertical structures* because the collector and the base and emitter regions are parallel horizontal layers.   If additional processing steps are economically feasible, a vertical *pnp* can be fabricated on a *p* substrate, as shown in Fig. 2-11(a).   This structure has been referred to as a five-layer transistor.   Because of the additional steps required in its manufacture, the production yield of this type of transistor would be low and the corresponding costs could possibly be prohibitive.

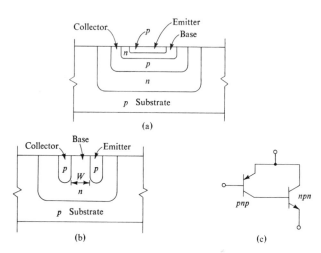

Fig. 2-11   (a) Five-layer *pnp* transistor; (b) lateral *pnp* transistor; (c) composite *pnp* structure.

When the dielectric isolation method is used, *pnp* transistors may be fabricated without isolation problems.   The result is a four-layer structure.

All transistor types discussed up to this point have been vertical structures wherein the primary transistor action is perpendicular to the surface of the wafer.   A *lateral pnp* transistor can be successfully fabricated without adding processing steps to the normal all-*npn* process, and has become widely accepted.   The base of the lateral transistor is formed from the *n* epitaxial region, as shown in Fig. 2-11(b).   The emitter and collector are made simultaneously with a *p* diffusion (the same diffusion used for bases of *npn* devices).   The base width is therefore the lateral distance *W* in the figure; it is determined by mask dimensions and by the lateral diffusion of the emitter and collector regions.

Electrical characteristics of lateral transistors are far from ideal.   The base width tends to be wide, thereby limiting the high-frequency performance and the current gain.   However, by combining the lateral *pnp* with a standard *npn*, as shown in Fig. 2-11(c), the result is a high-gain structure that effectively acts like a *pnp* as far as the three external terminals are concerned.   This composite *pnp* is found useful where the complement of an *npn* is called for in circuit design.

Multiple-Emitter Transistors

The form of logic gate circuit known as T^2L, and discussed in Section 9-9, uses a multiple-emitter transistor.   This transistor has one collector, one base, and several separate emitter regions. The physical structure is illustrated in Fig. 2-12.

Multiple-emitter transistors have been successfully fabricated for a number of years. Their use is limited to gate circuits where several identical parallel signal paths are required.

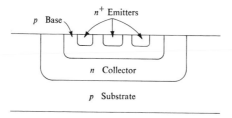

Fig. 2-12   Multiple-emitter transistor with three emitters.

2-9 DIODES

The fabrication of diodes in integrated circuits is usually based upon the philosophy that the *pn* junctions made for transistors can be used to attain diode behavior. According to this philosophy, it follows that five diode configurations are possible, as seen in Fig. 2-13:

1. Emitter-base junction–collector floating ($I_C = 0$).
2. Emitter-base junction–collector shorted to base ($V_{CB} = 0$).
3. Collector-base junction–emitter floating ($I_E = 0$).
4. Collector-base junction–emitter shorted to base ($V_{EB} = 0$).
5. Collector-base and emitter-base junctions connected in parallel ($V_{CE} = 0$).

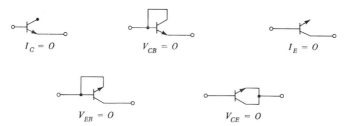

Fig. 2-13 IC diode connections.

In order to conveniently discuss these five cases, we use the equations in parentheses as names for the connections. The five methods of diode fabrication actually result in different characteristics, and we cannot immediately say that one is obviously better than the others.

If reverse breakdown voltage is important, $V_{EB} = 0$ and $I_E = 0$ provide the higher voltage, for these connections have the characteristic of a reverse-biased collector junction. That junction can normally withstand reverse potentials of 30 volts or more, while the breakdown of an emitter junction is more likely to be from 5 to 10 volts. Breakdown voltages are primarily determined by impurity concentrations. Breakdown of the collector-substrate diode presents no problem, because it will usually be in the vicinity of 70 volts.

Because of the small area of the emitter-base junction, $V_{CB} = 0$ and $I_C = 0$ will have the lowest reverse leakage currents, and $V_{CE} = 0$ will have the largest. (In Eq. (1-18) I_S is the reverse current).

In some applications, the capacitances associated with the diode are important to circuit behavior. $I_C = 0$ yields about one-half of the largest capacitance. The worst case occurs for $V_{CE} = 0$.

In digital applications, diode recovery time is indicative of the amount of stored charge that must be dissipated before a change in state from ON to OFF can be accomplished. The $V_{CB} = 0$ connection has the shortest recovery time, while $V_{CE} = 0$ has the longest recovery time because both junctions are simultaneously forward-biased and this results in the largest amount of stored charge.

The lowest forward voltage drop at a given level of forward current is observed in diodes with $V_{CB} = 0$. All other connections have about the same drop, 10 percent higher than the best performance.

Parasitics

A *parasitic pnp* transistor is formed in many of the diode connections noted here. In Fig. 2-14, the substrate acts as the collector of the unwanted transistor, and the regular collector and base become base and emitter of the parasitic. The case shown is for $V_{EB} = 0$. For forward operation, the diode is

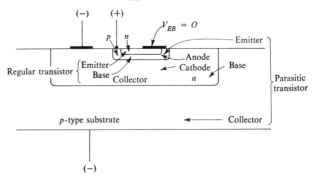

Fig. 2-14   A parasitic *pnp* transistor is formed when diodes are fabricated from transistor structures.

biased with positive on the regular p-type base and negative on the regular n-type collector.   As usual, the substrate is connected to the most negative voltage available.   Now, if we assume that the diode action is primarily holes moving from the regular p-type base to the regular n-type collector, we observe that some of those holes may continue to the substrate and become the collector current of the parasitic transistor.

The worst connections for parasitic transistor action are $V_{EB} = 0$ and $I_E = 0$.   From 50 percent to 70 percent of the diode current may be diverted to the substrate under certain conditions. $V_{CB} = 0$ is not affected by this parasitic.

In general, $V_{CB} = 0$ and $I_C = 0$ are most often used in digital IC's.

Voltage Breakdown

The collector-base junction of a transistor, operating as it normally does under a large reverse bias, must not exhibit electrical breakdown.   On the other hand, in certain diode applications, nondestructive reverse breakdown is used as circuit protection.

Avalanche breakdown, the most common form of breakdown, occurs when the junction voltage is so high that the electric field produces ionization when free high-energy carriers collide with atoms of the lattice.   This phenomenon is responsible for large reverse currents in the device, but the device is not destroyed if the allowable power dissipation level is not exceeded.   The breakdown region of a diode characteristic is seen in Fig. 2-15(a).

Because breakdown is a reversible and repeatable process, it can be used to protect certain other devices and circuits from destruction.   The *gate* portion of a field-effect transistor (*FET*) is easily voltage damaged; the SiO_2 insulation layer below that electrode may be permanently punctured.   A breakdown diode that parallels the voltage-sensitive path can provide the protection necessary to eliminate damage by breaking down at a voltage level below the damage threshold of the gate.

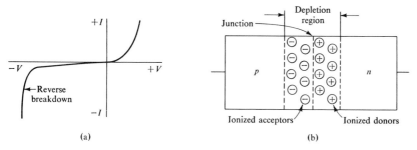

Fig. 2-15   (a) Diode characteristic showing reverse breakdown; (b) depletion or space-charge region.

Breakdown, avalanche, or *Zener* diodes, as they are called, may also be used to provide a known and constant direct voltage difference between two points in a circuit, independent of the current in that branch.   The resistance of a diode in its breakdown mode is very low; therefore its terminal voltage is not seriously changed by the amount of current it passes.

The breakdown voltage of a *pn* junction depends upon the width of the *depletion layer* at the junction.   Under reverse bias, the depletion layer "holds off" the applied voltage.   Thus, with a wide layer, the electric field intensity measured in volts/cm is smaller than for a narrow layer.   The depletion layer is that space where mobile carriers (holes and electrons) have been removed or depleted leaving behind impurity ions.   This is symbolized in Fig. 2-15(b). Removing electrons yields a positively charged space; hole removal results in a remaining accumulation of negative space charge.   The depleting of this region at an unbiased junction is the result of the migration of the higher energy electrons in the vicinity of the junction from the *n* to the *p* region. When in the *p*-region, they tend to combine with holes.   This process continues until the resulting potential difference or accumulation of charge discourages further migration across the junction.   Equilibrium results in two layers of space charge in the depletion region.

If a voltage is now applied to the diode, the effective width of the depletion region will change; and if this voltage forms a reverse bias, additional immobile charge will be uncovered and the width will increase.   It has been shown that the total width of the depletion region is a function of the dielectric constant $\varepsilon_r \varepsilon_o$, the voltage applied V_A, and the densities of acceptors N_a and donors N_d in the *p* and *n* regions, respectively.[7]   The width may be expressed by

$$x_T = [2\varepsilon_r \varepsilon_o V_A (N_a + N_d)/q N_a N_d]^{1/2}. \tag{2-6}$$

The electric field intensity existing in the junction region is simply V_A/x_T. The avalanche breakdown value of this field, E_{max}, can be considered to be 200,000 V/cm for silicon.   It is possible to express the breakdown voltage in terms of impurity density using Eq. (2-6):

$$V_A = E_{max}^2 [2\varepsilon_r \varepsilon_o (N_a + N_d)/q N_a N_d]. \tag{2-7}$$

According to this equation, the more lightly doped side of the junction determines V_A.   For

   ε_r = relative dielectric-constant of silicon, 12;
   ε_o = permittivity of free space, (8.85×10^{-14}) F/cm;
   $q = 1.6 \times 10^{-19}$ coulomb;
   $N_a = 10^{17}$ acceptors/cm^3;
   $N_d = 10^{21}$ donors/cm^3,

Eq. (2-7) predicts breakdown at 5.3 V.

2-10 RESISTORS

Resistors can be fabricated simultaneously with emitter diffusion, using low-resistivity material, or with base diffusion, using higher resistivity material. The latter is most common.

Figure 2-16(a) shows a fully integrated resistance. The upper p-type layer has been formed during a base diffusion cycle. The n-type layer, formed on the p-type substrate, is for isolation. If 200 $\Omega/\square$ material is used, a 2000 Ω resistance requires 10 squares, or a length-width ratio of 10. The resistor can be 10 mils long and one mil wide.

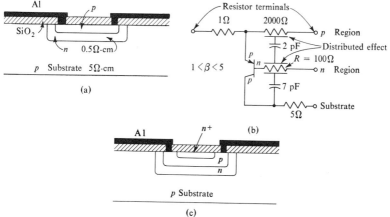

(a)

(b)

(c)

Fig. 2-16 Diffused resistance: (a) vertical profile; (b) 2000-ohm resistance showing parasitic effects; (c) pinch resistor.

We also have, as can be observed from the figure, a *pnp* sandwich with the isolation n region as its base. This is an undesired parasitic element. The substrate is normally connected to the lowest potential in the circuit, so to minimize current in this transistor, the n-region should be kept at a high positive voltage.

Other parasitic elements are shown in Fig. 2-16(b). The one-and five-ohm values represent contact and substrate bulk effects, respectively. The 2-pF and 7-pF values refer to distributed capacitances between layers. These parasitics adversely affect the high-frequency value of diffused resistances.

Because of imprecision in the diffusion, etching and mask-manufacturing techniques, integrated-circuit resistors are not made with tight tolerances on their absolute resistance values. A tolerance of ±10 percent is typical. However, if two resistors are made simultaneously, the ratio of the two values can be held to a much tighter tolerance—±3 percent is realistic. Large resistance values cannot be practically obtained. These facts dictate circuit design philosophies that will be noted in succeeding chapters when specific circuits are discussed.

Integrated resistors are temperature-dependent. They usually exhibit a positive temperature coefficient (rise with temperature rise) of 500 to 3000 parts per million/°C. A one-thousand-ohm element would therefore become 1050 to 1300 ohms if the temperature increased by 100°C. The amount of increase depends upon sheet resistivity.

A variation to the basic structure of Fig. 2-16(a) is the "squeezed" resistor shown in Fig. 2-16(c). This device is often referred to as a *pinch resistor*. An n^+ diffusion has reduced the cross-sectional area of the *p*-type resistance region, increasing the resistance between contacts. Sheet resistance of up to 10,000 $\Omega/\square$ may be possible, but because accurate control is difficult, large tolerance on the value of the resistor is expected.

To achieve particular operational specifications in an IC design, it is sometimes necessary or convenient to utilize resistance derived from methods other than solid state diffusion. Diffusion is limited to about 50 kΩ. The low-current source discussed in Section 5-1 approximates a high resistance by using a semiconductor diode in an interesting manner. The gate circuit of Section 9-11 uses a MOSFET with two of its terminals shorted to give resistance values of several hundred thousand ohms. Thin films of silicon-chromium have been used for resistances of several megohms. The material is deposited upon the silicon wafer near the end of the IC manufacturing process. It does not degrade the performance of the material beneath, and has good temperature properties.

2-11   CAPACITANCES

Integrated capacitances are of two types: (1) *pn* junction; (2) SiO_2 dielectric. With either method of manufacture, the total capacitance value of a single element is limited.

Junction Capacitance

We make use of the depletion layer that exists at a *pn* junction. It was noted in Section 2-9 that the width of that layer is dependent upon applied voltage. Since the charges in the depletion region are bound or immobile, these charge layers are equivalent to a capacitor. When the junction is made with an abrupt change from N_d donors per cm^3 in the *n* region to N_a acceptors per cm^3 in the *p*-region, it is referred to as a step junction. The junction capacitance for a step junction is most dependent upon the more lightly doped side. For $N_d \ll N_a$, we obtain

$$C = \left(\frac{q\varepsilon_r\varepsilon_o N_d}{2V_A}\right)^{1/2} \quad \text{farad/cm}^2, \tag{2-8}$$

where V_A = applied potential in volts; the other quantities have been defined previously. As an example, consider that $N_d = 6 \times 10^{16}$ donors/cm³ and $V_A = 2$ volts. Then Eq. (2-8) predicts 5×10^{-8} F/cm². This number is equivalent to 0.32 pF/mil². A junction area of 10 mil² yields a capacitance at 2 volts of 3.2 pF.

The junction capacitance shown in Fig. 2-17(a) has several disadvantages. As can be seen from the example given, the capacitance is low-valued, it varies with applied voltage, and the higher the reverse voltage, the lower the capacitance. In addition, a junction capacitance is somewhat difficult to isolate by the conventional method of diode isolation because such isolation also exhibits junction capacitance.

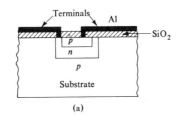

(a)

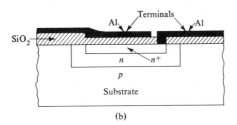

(b)

Fig. 2-17 (a) Vertical profile of junction capacitor; (b) oxide capacitor has aluminum and n^+ as plates and SiO_2 as dielectric.

Oxide Capacitance

An oxide capacitance is simply two conductors separated by SiO_2. The bottom conductor is usually an n^+-layer, as shown in Fig. 2-17(b); the top conductor is aluminum metalization. With an oxide thickness of 800 Å to 1000 Å, the capacitance of this structure is about 0.30 pF/mil².

The fundamental formula for the capacitance of two parallel plates separated by a suitable dielectric is

$$C = \frac{\varepsilon_r \varepsilon_o A}{d} \qquad \text{farad,} \qquad (2\text{-}9)$$

ε_r = relative dielectric constant of the insulation

ε_o = permittivity of free space, 8.85×10^{-14} farad/cm

A = plate area in cm^2

d = separation between plates in cm.

Typical values for ε_r for SiO_2 lie between 2.7 and 4.2.

The oxide capacitance is said to be nonpolar because it does not require a specific polarity of applied voltage as does a reverse-biased diode. Further, the capacitance value is not voltage-sensitive. Both the diode and the oxide types of capacitance are subject to parasitic effects, such as additional capacitance to the substrate caused by an undesired *pn* junction. From the information given in this section, it is possible to predict the wafer area required for an integrated capacitance. Area is certainly a limitation on the value of C to be included in an IC.

2-12 FIELD-EFFECT TRANSISTORS

Operation of the *field-effect transistor* (FET) is based upon the principle that the effective resistance of a conducting channel of silicon can be controlled by an electric field that is established by a potential applied partway along the channel. Two types exist; the *junction* FET, and the *insulated-gate* FET.

Terminals of FET's are referred to as *gate*, *source*, and *drain*. These loosely correspond to base, emitter, and collector of a junction transistor. Source and drain are often interchangeable.

The junction variety or JFET is shown in Fig. 2-18(a). The control *pn* junction is located directly under the gate terminal. The conducting channel is the *n* region between drain and source. A positive potential is applied to the drain terminal, and $I_D \cong I_S$ in the circuit of Fig. 2-18(b). The potential applied between gate and source forms a reverse bias for the junction. The magnitude of that reverse bias affects the current-carrying ability of the channel because the depletion layer in the channel will widen with increased reverse bias. Since a depletion layer is a high-resistance path, channel current will be impeded in inverse relation to the width of the layer. Should the reverse gate bias become sufficiently great, source and drain currents will be reduced to essentially zero.

Because the input is a reverse-biased diode, the input impedance of the JFET is high, perhaps 10^{10} ohms.

The insulated-gate device, or IGFET, is most commonly called a MOSFET. This name refers to its layered structure: *m*etal, *o*xide, *s*emiconductor. The gate is often a plate of conducting metal, SiO_2 forms the dielectric between gate and channel, and the channel may be of *n*-type silicon, as shown in Fig. 2-18(c). If the structure has *p*-type areas below source and drain, as in the figure, it is considered an " enhancement " type of MOSFET. Let us consider

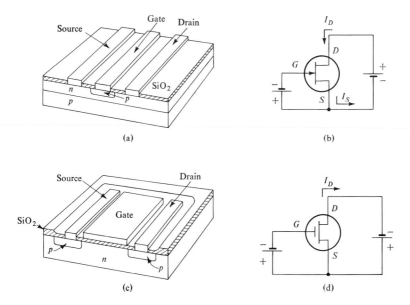

Fig. 2-18   FET types: (a) *n*-channel JFET; (b) biasing for (a); (c) enhancement MOSFET; (d) biasing for (c).

the mechanics of operation of this device.   The drain-source voltage is negative.   Little or no channel current will flow, and depletion layers exist in the *n* substrate around the *p* regions.   By applying a negative potential to the gate, the source-substrate *pn* junction becomes forward-biased, supplying holes to form a conducting channel below the gate.   This channel is holes, now minority carriers in an *n* region, that are free to continue to the drain terminal because the drain-substrate junction is reverse-biased, but that condition impedes the flow of majority carriers exclusively.

Another form of MOSFET, the "depletion" type, is fabricated without the *p*-type diffusions shown in the figure.   The depletion type behaves analogously to the JFET.   It is normally in an ON state, for a significant drain current exists when the gate voltage is zero.   Gate bias tends to reduce the level of drain current.

Problems

2-1   In crystal growth, a seed is required to start the process.   Where do you think the first silicon seed may have come from?

2-2   When a semiconductor is doped in order to provide a suitable density of charge carriers, explain why the specimen remains electrically neutral.

2-3   It has been mentioned that some carriers are lost through recombination while passing through the base of a transistor.   Explain briefly how recombination affects the β_{dc} parameter of a transistor.   What must be true regarding the magnitudes of I_C, I_B, and I_E?   Try to relate these currents in an equation.

2-4   If the base of a transistor is made very wide, the device is said to degenerate to simply two diodes.   With $V_{CC} = 10$ V, $V_{BB} = 0.7$ V, and a wide base, comment upon the relative current magnitudes in Fig. 2-1.

2-5   Determine the number of microns in an inch, and also the number of angstroms in an inch.

2-6   A diffusion furnace operates at 1100°C.   Boron atoms from an infinite supply are maintained at the silicon surface at a concentration of 5×10^{19} atoms/cm^3.   At this temperature, $D = 4.3 \times 10^{-13}$ cm^2/sec.   After nine hours of diffusion, find the penetration distance x where the boron concentration is 10^{17} atoms/cm^3.

2-7   If the silicon in the preceding problem is initially doped with an n-type concentration of 10^{16} atoms/cm^3, find the junction depth after nine hours of diffusion of boron to form a p-type layer near the surface.

2-8   Instead of boron, aluminum is used as the dopant in the preceding problem.   The diffusion constant D at 1100°C is 2×10^{-12} cm^2/sec.   Determine the junction depth under the conditions given.

2-9 through 2-11   Sketch a top view of an IC chip showing component layout for manufacture of each of the circuits given in the accompanying figures.   Show connections between elements, but avoid overlapping lines.   Transistor geometry can be as shown.   Use p-type substrate for all circuit designs.   Connections can be made at any place on a surface.   Bring connecting lines to pads around the perifery of the chip, and number the pads to correspond to the numbers in this diagram.

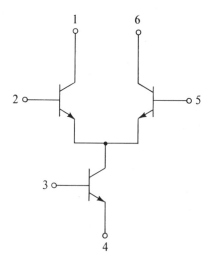

Problem 2-9

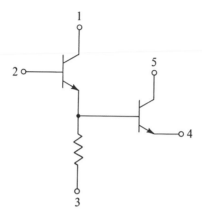

Problem 2-10

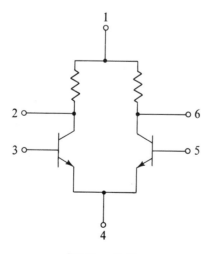

Problem 2-11

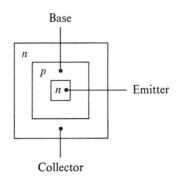

Transistor geometry for Problem 2-9 through 2-11.

2-12   Design an integrated circuit consisting of three elements as shown in the figure by showing cross-sectional view.   Use isolation diffusion between elements.   Also show connections between elements.

2-13   Calculate the breakdown voltage of a *pn* diode in which $\varepsilon_r = 12$, $N_d = 10^{21}$ donors/cm^3, $E_{max} = 200{,}000$ V/cm and $N_a = 5 \times 10^{10}$ acceptors/cm^3.

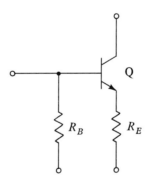

Problem 2-12

2-14   The device shown is a *Zener* diode.   It is operated in its reverse breakdown mode.   Consider that its terminal voltage remains constant at 20 V, regardless of the current passing through it.   The load is 1000 Ω and invariant, but V will vary from 23 to 28 V.   Select R so that the diode current will not exceed 60 mA when V is maximum.   What will be the diode current when $V = 23$ V?

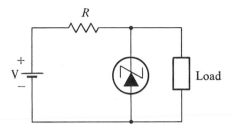

Problem 2-14

2-15   The capacitance of a parallel plate structure is given by Eq. (2-9).   If the oxide thickness is 1000 Å, and the relative dielectric constant is taken as 3.0, determine the plate surface area in square mils of a 30 pF capacitor.

3
TRANSISTOR CIRCUIT ANALYSIS

Integrated circuits are transistor circuits. This chapter is a compilation and condensation of fundamental information concerning the analysis of linear transistor circuits. For some readers, it represents a review of already-familiar topics. Concepts not applicable to IC's have been omitted. A more thorough treatment of the subject is available in many texts, including one by the present author.[3]

The chapter briefly considers static characteristics and biasing. Small-signal parameters and the derivation of gain and impedance relations warrant considerable attention. A modified hybrid-π representation for an IC transistor is presented. Discussions of coupling, large signals, high-frequency operation, and noise are included.

3-1 STATIC CHARACTERISTICS

Consider a junction transistor circuit instrumented as shown in Fig. 3-1(a). Intentional variations in supplies V_{BB} and V_{CC} will cause variations in the metered quantities. For common-emitter (CE) operation, those quantities of interest are plotted in Fig. 3-1(b). This graph is referred to as the *static collector characteristics.*

A great deal of information is portrayed in the figure, even though only static or dc measurements are involved. For example, the dc amplification factor can be obtained. This factor, referred to as β_{dc} or h_{FE}, was defined in Chapter 1 as simply the ratio of I_C to the corresponding I_B

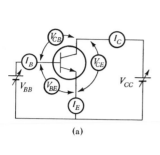
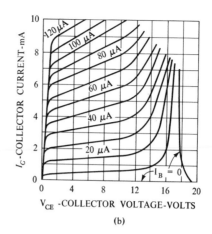

Fig. 3-1 Circuit for measurement of static characteristics, and common-emitter collector characteristics.

at a particular value of V_{CE}. β_{dc} takes on many values because of its dependence upon current and voltage levels. For example, at $V_{CE} = 10$ V, $I_B = 70$ μA and $I_C = 7$ mA, so that $\beta_{dc} = 100$. At $V_{CE} = 4$ V and $I_B = 50$ μA, the corresponding $I_C = 4$ mA, so that $\beta_{dc} = 80$ at that point.

The collector current corresponding to $I_B = 0$ is referred to as I_{CEO}, and is a measure of I_C when the base is unconnected. Since I_{CEO} is a small current, it is not possible to read its value from the graph at normal operating voltages below 12 V.

The region of collector current lower than I_{CEO} is referred to as the *cutoff* state of the transistor. Actually one can never reduce I_C to zero—there will always be some leakage current from collector to base. That junction is normally reverse-biased, and its minimum leakage is referred to as I_{CBO}.

At low values of V_{CE} in Fig. 3-1(b), the lines of constant base current fuse into what appears to be a single nearly vertical line. This is referred to as the *saturation* region, and the slope of that line is indicative of a parameter called the saturation resistance, R_{CS}. This resistance represents the resistance of the collector-emitter path when the transistor is fully ON.

For small ac signal excursions about a suitable operating point, the parameter of importance is

$$\beta \equiv \lim_{\Delta i_B \to 0} \frac{\Delta i_C}{\Delta i_B}\bigg|_{v_{CE} = \text{const}} = \frac{\partial i_C}{\partial i_B}\bigg|_{v_{CE} = \text{const.}} \tag{3-1a}$$

The practical measurement of β employs very small alternating currents, with the collector-to-emitter path ac short-circuited. Then

$$\beta = \frac{I_c}{I_b}\bigg|_{V_{ce} = 0}. \tag{3-1b}$$

This β is identical with h_{21} for common-emitter studies and is often referred to as h_{fe}. In practice, β takes on values up to 500.

A simple mathematical expression may be used to approximate the collector characteristics. If we assume that the lines of constant base current are horizontal (no V_{CE} dependence) and that their vertical separation is constant (β is constant), the relation is

$$I_C \cong \beta I_B + I_{CEO}. \tag{3-2}$$

It is possible from knowledge of β and I_{CEO} to sketch the characteristics using Eq. (3-2). Although Eq. (3-2) mixes an ac parameter β with dc quantities, it will cause no great amount of difficulty if one remembers that β is being used here in the sense of $\Delta I_C/\Delta I_B$.

Because the emitter-base junction is forward-biased in normal operation, the static input characteristics curve for those terminals is a diode-like curve relating I_B to V_{BE}. *When the transistor is ON and thus biased in its active or its saturation region, the value of the voltage drop between base and emitter is in the neighborhood of 0.7 for a silicon device.* This level is verified by the diode Eq. (1-21) and by Fig. 3-2. Only a minor change in V_{BE} is experienced when the collector current or voltage experiences a large change.

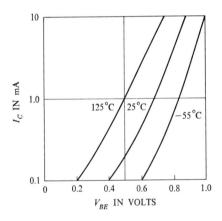

Fig. 3-2 Input characteristics for typical silicon transistor showing variation with temperature.

In the analysis of transistor circuits, we shall often assume that $V_{BE} \cong 0.7$ V. If a large voltage is applied—say, 2 V—Eq. (1-21) predicts such a high level of I_C that the transistor will fail.

Figure 3-2 also indicates that the V_{BE} drop is highly sensitive to temperature. This must be borne in mind when considering the transistor for dc amplifying service, and it is a cause of difficulty in biasing network design where stable operating point coordinates are desired.

3-2 BIASING

To establish a quiescent operating point in the active region, away from saturation and cutoff, it is necessary *to forward-bias the emitter-base junction and
reverse-bias the collector-base junction.* Under these conditions, emitter
current will flow in the direction of the arrow in the transistor symbol.
Since collector and base currents are smaller in magnitude than I_E, they will
flow *into* an *npn* transistor. The magnitudes are related according to

$$I_E = I_C + I_B.\qquad(3\text{-}3)$$

Because β_{dc} can take on values up to perhaps 500, I_B is very small and often
can be neglected in comparison with the other two terminal currents. Many
practical biasing circuits, such as the one shown in Fig. 3-3(a), essentially set a
constant base current. The base current for that circuit is exactly given by

$$I_B = \frac{V_{CC} - V_{BE}}{R_B}.\qquad(3\text{-}4)$$

For silicon transistors, $V_{BE} \cong 0.7$ V and therefore may be much smaller than
V_{CC}. Consequently

$$I_B \cong V_{CC}/R_B.\qquad(3\text{-}5)$$

It can be seen that I_B is set to be as constant as V_{CC} or R_B.

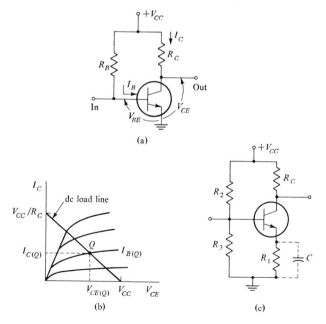

(a)

(b)

(c)

**Fig. 3-3 Biasing for the common-emitter stage: (a) fixed base current;
(b) construction of dc load line; (c) circuit for stabilized operating point.**

The collector current in Fig. 3-3(a) is

$$I_C = \frac{V_{CC} - V_{CE}}{R_C}. \qquad (3\text{-}6)$$

If V_{CC} and R_C are fixed, this equation specifies neither I_C nor V_{CE}, merely a relation between those two quantities.   On the static collector characteristics, Eq. (3-6) could be plotted as a straight line between $I_C = 0$ when $V_{CE} = V_{CC}$ and $V_{CE} = 0$ when $I_C = V_{CC}/R_C$.   This line, referred to as the *dc load line*, represents the locus of possible operating or quiescent points.   Now, if the base current is set, as noted by Eq. (3-5), then the operating point is the intersection of the dc load line and the line of constant base current, shown as Q in Fig. 3-3(b), and the coordinates of this point are $I_{C(Q)}$ and $V_{CE(Q)}$.

The relation between $I_{C(Q)}$ and $I_{B(Q)}$ is β_{dc}.   Since this parameter is temperature sensitive and varies greatly from unit-to unit because of manufacturing tolerances, fixed base current biasing is not a good way to achieve a stable operating point for a common-emitter stage.   It has been shown that the circuit of Fig. 3-3(c) provides a more stable operating point.[3]   Resistance R_1 provides negative dc feedback to cancel a portion of the effect that causes the operating point to shift.   This can be appreciated from the following example.   If β_{dc} increases because of temperature or replacement so that I_E takes on a new, larger quiescent value, the $I_E R_1$ drop is increased and is of such a polarity as to *reduce the base current*, thereby tending to lower the emitter current and returning Q toward the original coordinates.

Shown in dashes in the figure is a bypass capacitor.   The purpose of this element is to ac-ground the emitter so that the stage performs as a regular *CE* amplifier.   This bypassing function is seldom employed in IC technology, for C is necessarily large-valued to assure good low-frequency operation, and large-valued capacitors are not suitable for integration.   Biasing for linear IC common-emitter and common-collector (*CC*) stages usually employs constant-current sources, as discussed in Section 5-2.

3-3 SMALL-SIGNAL MODELS

The hybrid and y-parameter representations for two-port networks were discussed in Section 1-7.   These models are of course valid for a single transistor as well as for an entire IC.

When a set of four hybrid parameters is available to describe a common-emitter connected transistor, and it is desired to use that same device in a common-collector or common-base configuration, parameter interrelations are required.   Small-signal matrix parameters have a second subscript to signify the connection; thus h_{ie}, h_{re}, h_{fe}, and h_{oe} pertain only to the common-emitter connection. To change this set to h_{ib}, and so forth, or to h_{ic}, and so forth, for common-base and common-collector studies, one can use Table 3-1. Methods for derivation of this information are discussed in the literature[3] (see also Problems 3-9 and 3-10).

TABLE 3-1   APPROXIMATE RELATIONS AMONG h PARAMETERS

Relations between common-base and common-emitter parameters	Relations between common-base and common-collector parameters
$h_{ie} = h_{ib}/(1 + h_{fb})$	$h_{ic} = h_{ib}/(1 + h_{fb})$
$h_{re} = \dfrac{h_{ib}h_{ob} - h_{rb}h_{fb} - h_{rb}}{1 + h_{fb}}$	$h_{rc} = 1$
$h_{fe} = -h_{fb}/(1 + h_{fb})$	$h_{fc} = -1/(1 + h_{fb})$
$h_{oe} = h_{ob}/(1 + h_{fb})$	$h_{oc} = h_{ob}/(1 + h_{fb})$

Current Gain

Current gain parameters β and β_{dc} have been discussed. These parameters, also referred to as h_{fe} and h_{FE}, pertain to the common-emitter connected transistor. Several of the IC's discussed in later chapters include common-base stages. The applicable current gain parameters are α and α_{dc} :

$$\alpha \equiv \left.\frac{\partial i_C}{\partial i_E}\right|_{v_{CB}=\text{const}} \qquad \alpha_{dc} \equiv \left.\frac{I_C}{I_E}\right|_{V_{CB}=\text{const.}} \tag{3-7}$$

These are also referred to as h_{fb} and h_{FB}. Values for α and α_{dc} range from about 0.95 to 1.0.

It follows from Table 3-1 relations that

$$\beta = \frac{\alpha}{1 - \alpha} \qquad \text{and} \qquad \alpha = \frac{\beta}{\beta + 1}. \tag{3-8}$$

Both α and β are usually considered to be positive numbers by circuit designers. Numerical values for h_{fb} and h_{fc} are negative.

Parameter Variations

Transistor parameters are dependent upon the physical characteristics of the materials used in manufacture, such as the conductivities of the various portions of the structure, and upon the important dimensions of the device, such as base width. It is known that junction temperature affects material properties, that collector voltage influences the effective base width, and that quiescent current determines the density of carriers in the base region; it is natural to expect that transistor parameters will be dependent upon T, V_{CE}, and I_C.

Because parameters are sensitive to biasing conditions, parameter values are meaningful only when an operating point is specified. Nominal parameter values for low-power transistors are often given for operation at a point such as at $I_C = 1$ mA and V_{CE} of 5 or 6 V; should the application require operation at a different point (as it usually does), the nominal parameter values should be multiplied by the correction factors supplied by the manufacturer or determined by test. Typical correction information is shown in Fig. 3-4.

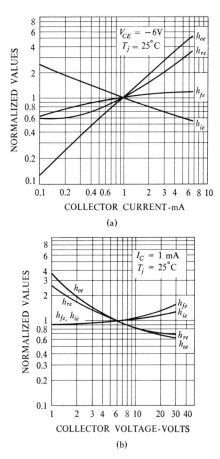

Fig. 3-4 Parameter variations for a typical transistor: (a) variations with $I_{C(Q)}$; (b) variations with $V_{CE(Q)}$.

EXAMPLE.   It is desired to operate a particular transistor at $I_C = 5$ mA, $V_{CE} = 2$ V.   The parameter-correction information of Fig. 3-4 applies. Consider the nominal parameters to be as follows: $h_{ie} = 3000$ Ω, $h_{fe} = 100$, $h_{re} = 10^{-3}$, $h_{oe} = 10^{-5}$ mho.

The corrected values of the parameters are:

$$h_{ie} = (3000)(0.6)(0.92) = 1655 \ \Omega;$$
$$h_{fe} = (100)(1.2)(0.92) = 110;$$
$$h_{re} = (10^{-3})(2.8)(1.7) = 4.75 \times 10^{-3};$$
$$h_{oe} = (10^{-5})(4.0)(2.0) = 8 \times 10^{-5} \ \text{mho}.$$

A most troublesome variation in parameters occurs because of production tolerances.   The spread in possible values for the parameters of a single transistor type when it leaves the manufacturers is often quite large; 3 : 1 variations in values are encountered.   To successfully design circuits to overcome such unit-to-unit variations negative feedback is usually employed. This topic is considered in Chapter 4.

3-4   HIGH-FREQUENCY PARAMETERS

The parameters of some transistor types begin to take on complex form at frequencies beyond the upper end of the audio-frequency spectrum.   Low-frequency parameters are simply real numbers; for high-frequency analysis, the parameters also have reactive portions that represent physical capacitance effects and the distributed nature of some of the internal functions.

Because of the diffusion capacitance, the magnitude of the common-base current amplification factor h_{fb} is found to vary with frequency according to the approximate relation

$$h_{fb} = h_{fbo}/[1 + j(f/f_{hfb})]. \tag{3-9}$$

f_{hfb} is referred to as the alpha-cutoff frequency; it is also symbolized by f_{ab} and by f_a.   At the cutoff frequency, h_{fb} has decreased to 0.707 of its reference or low-frequency value, h_{fbo}.

For CE operation we are interested in h_{fe} ; h_{fe} is a function of h_{fb} according to the relations given in Table 3-1,

$$h_{fe} = -h_{fb}/(1 + h_{fb}). \tag{3-10}$$

Therefore by substitution of Eq. (3-9) into Eq. (3-10) we obtain

$$h_{fe} = \frac{-h_{fbo}}{1 + h_{fbo} + j(f/f_{hfb})}. \tag{3-11}$$

h_{fe} in Eq. (3-11) will be down by 3 dB from its low-frequency value when real and imaginary parts of its denominator are equal; then

$$1 + h_{fbo} = f/f_{hfb}$$

this occurs when

$$f = (1 + h_{fbo})f_{hfb}.$$

We now have an expression for the beta cutoff frequency:

$$f_{hfe} = (1 + h_{fbo})f_{hfb}. \tag{3-12a}$$

The $(1 + h_{fbo})$ term has a nominal value of less than 0.1.   It may be concluded that the common-emitter configuration is inferior to common-base circuitry when high-frequency current gain is considered, for the former configuration experiences a decline in its current-amplification parameter at a much lower frequency.   Nevertheless, because it is basically a higher-gain configuration, common-emitter stages are used in most high-frequency applications.

Test data taken on transistors manufactured by various methods are not in complete agreement with Eq. (3-12a), and suggest the equation be modified to the form

$$f_{hfe} = K_{\Theta}(1 + h_{fbo})f_{hfb}. \tag{3-12b}$$

The value of K_{Θ} is never greater than unity and ranges as low as 0.6 with certain transistor types.

A parameter especially applicable to the common-emitter stage is f_T, the *current-gain bandwidth product* or the *transition frequency*. This parameter is the frequency at which $|h_{fe}|$ equals unity.   Obviously f_T is very much higher in the spectrum than f_{hfe}.

If Eq. (3-10) is employed with $|h_{fe}| = 1$ at $f = f_T$, the following relation may be obtained by using the fact that $f_T \gg f_{hfe}$:

$$f_T = h_{feo}f_{hfe}. \tag{3-13}$$

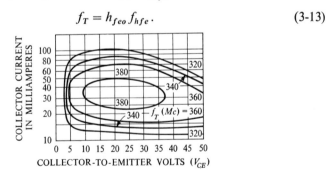

Fig. 3-5   Contours of constant f_T for silicon planar transistor.

Upon insertion of Eq. (3-13) for $f_{h_{fe}}$ into Eq. (3-12a), we conclude that the value of f_T is slightly smaller than $f_{h_{fb}}$.

The measured variation of f_T with operating-point coordinates for a typical silicon planar transistor is shown in Fig. 3-5.   From the figure it is evident that an optimum locus of operating points exists, at least in regard to attaining a maximum value of f_T.

3-5   THE HYBRID-π MODEL

When studying transistor-circuit operation at higher frequencies it is some-times inconvenient to work with the set of h or y parameters that are complex functions of frequency.   It would be desirable to represent the transistor by a model that separates the frequency-variant parameters from those that show no changes with frequency.   In addition it is sometimes of value to examine the device by representing it in terms of a set of parameters that are directly related to the physical processes present in normal operation, related to operating-point coordinates in a known manner, and related to one another if possible.   The *hybrid-π* equivalent circuit developed by Giacoletto is a valuable representation of many transistor types, for its parameters may be considered to be frequency invariant up to vicinity of alpha cutoff.

To develop the hybrid-π circuit for common-emitter applications from physical considerations, we start by making the assumption that I_c and I_b are linearly related.   If signal size is sufficiently limited, I_c and V_{be} will also be linearly related.   The proportionality constant is g_m, the *transcon-ductance*.   Therefore

$$I_c = g_m V_{be}. \tag{3-14}$$

An expression for the diode-like static transfer characteristic is

$$i_C \cong K(\epsilon^{qv_{BE}/kT} - 1). \tag{3-15}$$

Because g_m is the slope of the transfer characteristics, it may be obtained from Eq. (3-15) by differentiation:

$$g_m \equiv \partial i_C / \partial v_{BE}. \tag{3-16}$$

It follows that

$$g_m \cong (q/kT)I_C = \Lambda I_C. \tag{3-17}$$

The magnitude of the direct collector current is apparent in Eq. (3-17). Note that g_m *is independent of the transistor type being considered.*

Between base and emitter, the transistor can be simulated by the low-frequency input resistance $r_{b'e}$, and the diffusion and space-charge capaci-tances, taken together to be $C_{b'e}$.

The collector and base may be joined by space-charge capacitance and by resistance and capacitance representing the mechanism of *base-width modulation,* wherein the effective width of the transistor base is dependent upon collector voltage.

In addition to the aforementioned parameters, the collector is joined to the emitter by the parameter r_{ce}, also representing the base-width modulation effect. All of these elements are depicted in Fig. 3-6(a). The bulk

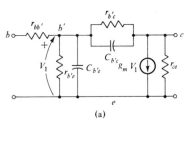

(a)

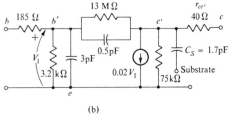

(b)

Fig. 3-6 Hybrid-π models: (a) general circuit; (b) modification useful for IC transistors with typical values.

resistance of the lightly doped base material is called the *base-spreading resistance $r_{bb'}$,* and joins the internal or intrinsic base terminal b' to the actual available base terminal b.

The original hybrid-π developed for discrete transistors has been slightly modified to be a useful representation for integrated transistors. As shown in Fig. 3-6(b) C_S has been added to represent collector-to-substrate capacitance. Also, $r_{cc'}$ represents collector series resistance, as discussed in Section 2-8. A set of typical parameter values is given in the figure.

For-low frequency analysis, the capacitances have high reactances and may be replaced by open circuits.

Because the regular hybrid-π needs five low-frequency parameters, while only four are used in the matrix equivalent representations, it is necessary to know an additional parameter for the latter networks in order to convert to the hybrid-π. This extra parameter is the base-spreading resistance. Using the normal tools of circuit theory, the relations may be derived; Table 3-2 lists some of the possible equalities.

TABLE 3-2 APPROXIMATE RELATIONS AMONG PARAMETERS. *

$$r_{b'e} = h_{ie} - r_{bb'}$$

$$r_{b'c} = \frac{h_{ie} - r_{bb'}}{h_{re}}$$

$$g_m = \frac{h_{fe}}{h_{ie} - r_{bb'}}$$

$$\frac{1}{r_{ce}} = h_{oe} - \frac{h_{fe} h_{re}}{h_{ie} - r_{bb'}}$$

* Low-frequency relations.

3-6 HYBRID-π PARAMETERS

Let us briefly investigate the behavior of the hybrid-π parameters with operating-point coordinates and with temperature. Derivations of some of the equations given in this section are not reproduced here; they are available in the literature.[3]

Analytical expressions for the base-spreading resistance $r_{bb'}$ for simple transistor geometries show inverse linear dependence upon base conductivity σ_b and W the base width:

$$r_{bb'} = K/\sigma_b W. \tag{3-18}$$

From this information we may predict the behavior of $r_{bb'}$ with changes in operating-point coordinates and temperature. One would expect that with increased collector voltage, which causes a widening of the depletion layer within the base region, $r_{bb'}$ should also increase. An increase in the level of emitter current should result in reduced $r_{bb'}$ because recombination will occur after a shorter distance of travel for majority carriers in the base region. It is also to be expected that $r_{bb'}$ will decrease somewhat with frequency, although not predicted by Eq. (3-18), because a portion of this parameter is of a distributed nature in certain transistor types.

The value of the $r_{b'c}$ element of the complete hybrid-π is a very large and often may be considered to be an open circuit. Under this assumption, at low frequencies, the input-resistance parameter h_{ie} of the circuit of Fig. 3-6(a) is

$$h_{ie} \cong r_{bb'} + r_{b'e}. \tag{3-19}$$

Because the model must exhibit a short-circuit current gain of h_{fe}, it follows that

$$h_{fe} = g_m r_{b'e}. \tag{3-20}$$

Equation (3-20) is a very useful relation; while the current-amplification factor h_{fe} is not a basic parameter of the hybrid-π, it is implied in $r_{b'e}$ as noted in that equation.   Since h_{fe} does not show a substantial variation with operating point, it follows from the inverse relation between $r_{b'e}$ and g_m that $r_{b'e}$ varies as I_C^{-1}.

The intrinsic base-to-emitter capacitance $C_{b'e}$ is found to vary directly with I_C, at least for relatively low current levels.   This capacitance can be expressed in terms of the f_T parameter.   Refer to Fig. 3-7.   With the output

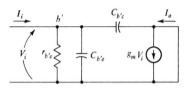

Fig. 3-7   Simplified hybrid-π with load short-circuited.

terminals shorted and $r_{b'c}$ omitted, the input admittance of the transistor at terminal b' is

$$I_i/V_i = g_{b'e} + j\omega(C_{b'e} + C_{b'c}). \tag{3-21}$$

At any frequency

$$h_{fe} = I_o/I_i \cong g_m(V_i/I_i) = g_m/[g_{b'e} + j\omega(C_{b'e} + C_{b'c})]. \tag{3-22}$$

The short-circuit current amplification factor will be down by 3 dB from its low-frequency value when real and imaginary parts of the denominator are equal:

$$f = f_{hfe} = 1/[2\pi r_{b'e}(C_{b'e} + C_{b'c})]. \tag{3-23}$$

If we consider that $C_{b'e} \gg C_{b'c}$ and solve Eq. (3-23) for $C_{b'e}$, we obtain

$$C_{b'e} \cong 1/\omega_{hfe} r_{b'e}. \tag{3-24}$$

Recognizing that $h_{feo}\omega_{hfe} \cong \omega_T$, Eq. (3-24) can be written

$$C_{b'e} \cong g_m/\omega_T. \tag{3-25}$$

3-7   FET EQUIVALENT CIRCUITS

To develop a small-signal circuit for the junction FET, we make use of the information available from the static characteristics given in Fig. 3-8 and from the discussion of physical principles in Chapter 2.   An equivalent circuit for common-source operation is shown in Fig. 3-9.   The input is a reverse-biased junction, and so leakage-resistance and depletion-layer capa-

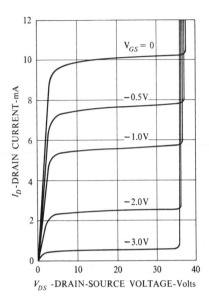

Fig. 3-8   JFET common-source drain characteristics for a typical *n*-channel device.

citance elements linking the gate to each of the other terminals are appropriate.   A constant-current generator of transconductance g_m is dependent upon V_1, and is paralleled by the incremental resistance of the bar, r_d.   The bulk resistance of the channel is of much smaller value than values of the leakage elements r_a and r_b and has been omitted from consideration; r_d, essentially the output resistance of the device, is of intermediate value.   It is immediately obvious that this equivalent bears a high degree of similarity to the complete hybrid-π discussed earlier.   The element $r_{bb'}$ is not present in Fig. 3-9, for no such parameter is important to JFET operation.

Important static JFET parameters are: (a) V_P, *the gate pinch-off voltage, the potential required to reduce the drain current to zero*; and (b) I_{DSS}, *the drain current flowing when gate is shorted to source.*

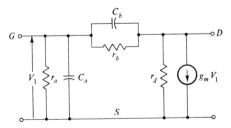

Fig. 3-9   Simple JFET equivalent circuit.

Richer and Middlebrook have reported that the JFET transfer characteristic in the pinched-off region can be described faithfully by the power-law relation

$$I_D = I_{DSS}(1 - |V_{GS}/V_P|)^n. \tag{3-26}$$

Sevin has shown that when the exponent $n = 2$, the resulting parabola fits well with experimental evidence obtained on diffused devices. The transconductance may be mathematically determined by taking the derivative of Eq. (3-26); with $n = 2$, one obtains

$$g_m = \partial I_D / \partial(-V_{GS}) = (2I_{DSS}/V_P)[1 - |V_{GS}/V_P|] \tag{3-27}$$

For operation in the active region, it usually suffices to represent the JFET by the circuit of Fig. 3-9 with r_a and r_b omitted.

The model shown in Fig. 3-10(a) is an accurate lumped representation of a MOSFET connected as a common-source amplifier that is useful over a

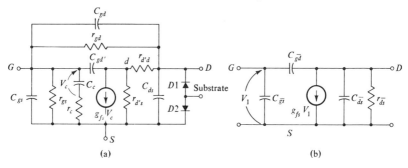

Fig. 3-10 MOS equivalent circuits: (a) complete; (b) approximate for general low-frequency use.

wide frequency range when the device is operated in the active region.   A brief discussion of the parameters of this model follows.

$D1$ and $D2$ are diodes representing the pn junctions between the substrate and drain and source.   The drain-substrate diode is normally reverse-biased; consequently this element will add little parallel conductance to the output terminal pair.   The capacitances associated with $D1$ and $D2$ can be lumped into C_{ds}.

C_{gd}, C_{gs}, and C_{ds} represent interlead capacitances.   Typical values for these elements are 0.1, 0.9, and 2.0 pF, respectively.

r_{gs} and r_{gd} symbolize the resistance of leakage paths through the insulating oxide and around its edges.   Generally these resistances are greater than 10^{15} Ω.

C_c and r_c represent the distributed network associated with the metal gate and the active channel.   The voltage across C_c performs the charge control, and r_c represents the resistance of the channel between C_c and source, and may be several hundred ohms in value. C_c may be 5 pF, for example.

g_{fs} or g_m, the transconductance, relates drain current to charge control voltage.   Because V_c is the important control voltage, the generator is $g_m V_c$. For low-frequency analysis, $g_m V_{gs} \cong g_m V_c$.

$r_{d's}$ or r_d, the active channel resistance, is the output resistance of the device, and may be determined from the slope of the output characteristics. The value of this parameter is highly dependent upon the operating point.

$r_{d'd}$ is the resistance of that portion of the channel which is not modulated by the voltage V_c.   In a unit designed with a partial of offset gate so that the gate electrode does not fully cover the entire channel, $r_{d'd}$ may be several hundred ohms.   In a full-gate unit, this parameter is negligible.

$C_{gd'}$ in partial gate units may have a value such as 0.1 pF, an order of magnitude smaller than $C_{gd'}$ of full-gate types.

For low-frequency analysis, the equivalent of Fig. 3-10(b) is valuable. $C_{\overline{gs}} = C_{gs} + C_c$, $C_{\overline{gd}} = C_{gd} + C_{gd'}$, and $r_{\overline{ds}}$ is the parallel equivalent of $r_{d's}$ and diode resistance.   Capacitances are used between terminals in this low-frequency model because resistances associated with these paths are very large-valued.   The capacitances are most important even at low audio frequencies.

FET Biasing

JFET's and depletion type MOSFET's of the p-channel variety require that the gate be supplied with a positive voltage when the drain is fed negatively. The opposite polarities apply to n-channel devices.   This combination of voltages can be easily obtained from one power supply if a resistance is used between source terminal and ground.   A sample circuit for biasing a p-channel MOSFET is shown in Fig. 3-11.   The substrate terminal has been connected to the source terminal.

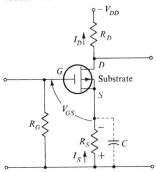

Fig. 3-11   Self-biasing for p-channel MOSFET.

To show the effect of self-biasing, consider I_S to exist and equal I_D. The drop across R_S, $I_S R_S$, acts as a voltage supply for the loop that also includes R_G and V_{GS}. Since direct current cannot flow in this loop because of the insulated gate, there must be no current through R_G, and hence the entire $I_S R_S$ voltage forms V_{GS}. The polarity of $I_S R_S$ as shown is consistent with that required to turn the transistor ON, and if R_S is chosen properly the FET will be biased in the active region.

Element R_S will have a degenerative effect upon the gain of this stage (see Problem 3-24). Therefore bypass capacitor C may be used to ground the source terminal for ac operation. As noted with the junction transistor, C must be large-valued if good low-frequency operation is necessary.

3-8 GAIN AND IMPEDANCE RELATIONS

We now consider that the active device is fed from a signal source and that it feeds an electrical load. Formulas will be derived for the prediction of the voltage, current, and power amplification provided by the network, and for the terminal impedance levels presented to the signal source and to the circuit load by the transistor's input and output ports or terminal pairs. In the analysis of simple transistor and FET circuits, it is necessary to represent the terminations in terms of equivalent electrical parameters. Thus a 500 Ω resistive load may physically represent a small motor, or a light bulb, or the input resistance of another stage; as far as the transistor being studied is concerned, these terminations are equivalent.

h Parameters

The circuit shown in Fig. 3-12 is the general h-parameter network terminated in a resistive load R_L and fed from a signal source V_g of internal resistance R_G. This is a special case of the more general circuit with impedance terminations Z_L and Z_G. It can be seen from the figure that network parameters h_i and h_o are also being considered as resistive elements. Additionally, we may assume elements h_f and h_r to be described by real, rather than complex,

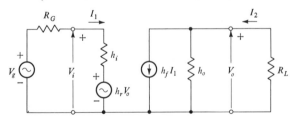

Fig. 3-12 Equivalent circuit for the general configuration with source and load terminations.

numbers. *The equations to be derived using Fig. 3-12 pertain to the conventional transistor in any configuration* over a rather wide frequency spectrum. For the more general complex case, refer to Appendix II.

To describe this circuit, we use the general *h*-parameter equations given in Chapter 1:

$$V_i = h_i I_1 + h_r V_o$$
$$I_2 = h_f I_1 + h_o V_o. \tag{3-28}$$

Note that V_o can be eliminated from Eqs. (3-28) by making the substitution

$$V_o = -I_2 R_L.$$

Thus

$$V_i = h_i I_1 - h_r R_L I_2$$
$$0 = h_f I_1 - (1 + h_o R_L) I_2.$$

The currents are considered to be unknown. The determinant of coefficients is

$$D = -h_i(1 + h_o R_L) + h_r h_f R_L.$$

Then

$$I_1 = \frac{\begin{vmatrix} V_i & -h_r R_L \\ 0 & -(1 + h_o R_L) \end{vmatrix}}{D} = \frac{-V_i(1 + h_o R_L)}{D}$$

$$I_2 = \frac{\begin{vmatrix} h_i & V_i \\ h_f & 0 \end{vmatrix}}{D} = \frac{-V_i h_f}{D}.$$

The *voltage gain* of this circuit is

$$A_v = \frac{V_o}{V_i} = \frac{-I_2 R_L}{V_i} = \left(\frac{V_i h_f R_L/D}{V_i}\right) = \frac{-h_f R_L}{h_i + R_L \Delta^h}, \tag{3-29}$$

where

$$\Delta^h = h_i h_o - h_r h_f$$

Note that the definition of voltage gain used here is V_o/V_i. *If* V_o/V_g *is of interest, add* R_G *to* h_i. The phase reversal present in common-emitter stages appears algebraically in Eq. (3-29). When h_f is positive, as is h_{fe}, A_v retains its negative sign. This indicates that a positive-going V_o results from a negative-going V_i. The absence of phase reversal in common-base and common-collector stages is apparent, for h_{fb} and h_{fc} are negative quantities. It is valuable for the reader to remember that *only the common-emitter connection provides phase reversal.*

Current gain:

$$A_i = \frac{I_2}{I_1} = \frac{V_i h_f/D}{V_i(1 + h_o R_L)/D} = \frac{h_f}{1 + h_o R_L}.$$ (3-30)

Power gain:*

$$G = A_i \cdot A_v = h_f{}^2 R_L/[(1 + h_o R_L)(h_i + R_L \Delta^h)]$$ (3-31)

Input resistance:

$$R_i = \frac{V_i}{I_1} = \frac{V_i}{-V_i(1 + h_o R_L)/D} = \frac{h_i + R_L \Delta^h}{1 + h_o R_L}$$ (3-32)

To find the output resistance of the network, V_g is short-circuited and a signal source V_o connected at the load terminals in place of R_L. The equations describing this situation are

$$0 = (h_i + R_G)I_1 + h_r V_o.$$
$$I_2 = h_f I_1 + h_o V_o.$$ (3-33)

Current directions and voltage polarities are as shown in Fig. 3-12. The ratio V_o/I_2 is the output resistance of the network. From Eqs. (3-33) we obtain

$$I_2 = h_f[-h_r V_o/(h_i + R_G)] + h_o V_o$$

and thus

$$R_o = V_o/I_2 = (h_i + R_G)/(R_G h_o + \Delta^h)$$ (3-34)

Although V_g is short-circuited in obtaining the expression for R_o, the "dependent" generators $h_r V_o$ and $h_f I_1$ must be retained.

With the preceding equations, it is possible to calculate the performance of a common-base stage by simply inserting h_{ib}, h_{rb}, h_{fb}, and h_{ob}, whereas to predict the circuit operation of a common-emitter-connected transistor, we should use h_{ie}, h_{re}, h_{fe}, and h_{oe}. The common-collector configuration would be treated accordingly. Table 3-3 is a convenient summary of the formulas derived.

The relations of this section could be derived by using other methods. Source interchanges could be made, and loop equations could be written, and the resulting two equations solved for the unknown currents.

* The product of A_i and A_v is the power gain of the network when all parameters are real.

**TABLE 3-3   FORMULAS FOR THE
HYBRID EQUIVALENT**

Voltage gain, A_v	$\dfrac{-h_f R_L}{h_i + R_L \Delta^h}$	(3-29)
Current gain, A_i	$\dfrac{h_f}{1 + h_o R_L}$	(3-30)
Power gain, G	$\dfrac{h_f{}^2 R_L}{(1 + h_o R_L)[h_i + R_L \Delta^h]}$	(3-31)
Input resistance, R_i	$\dfrac{h_i + R_L \Delta^h}{1 + h_o R_L}$	(3-32)
Output resistance, R_o	$\dfrac{h_i + R_G}{\Delta^h + R_G h_o}$	(3-34)

$\Delta^h = h_i h_o - h_r h_f$

The values used for R_L and R_G in the equations given in this chapter must include any biasing elements that may be connected across the output and input ports.   An example is given in Section 3-12.

y Parameters

The y-parameter network is connected to a load Y_L and signal source I_g paralleled by Y_G in Fig. 3-13.   Admittances are used at the terminations be-

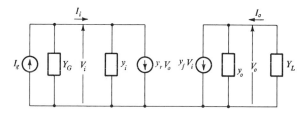

Fig. 3-13   The terminated y-parameter network.

cause the resulting network then contains only two nodes and therefore may be simply analyzed.   The nodal equations for the transistor are:

$$I_i = y_i V_i + y_r V_o$$
$$I_o = y_f V_i + y_o V_o.$$ (3-35)

If the substitution $V_o = -I_o Z_L$ is used, the equations for gains and impedances presented in Table 3-4 may be derived.

TABLE 3-4   FORMULAS FOR THE y EQUIVALENT

Voltage gain, A_v*	$\dfrac{-y_f Z_L}{1 + y_o Z_L}$	(3-36)
Current gain, A_i†	$\dfrac{y_f}{y_i + \Delta^y Z_L}$	(3-37)
Input impedance, Z_i	$\dfrac{1 + y_o Z_L}{y_i + \Delta^y Z_L}$	(3-38)
Output impedance, Z_o	$\dfrac{1 + y_i Z_G}{y_o + \Delta^y Z_G}$	(3-39)

$\Delta^y = y_i y_o - y_r y_f$
* Ratio of V_o to V_i in Fig. 3-13.
† Ratio of I_o to I_i in Fig. 3-13.

3-9   HIGH-FREQUENCY CONSIDERATIONS

If the appropriate parameters are available in complex form, either the h- o
y-equivalent circuit may be used to analyze the performance of a transisto
stage at a particular frequency by using the equations presented in Appendi
II.

   With the help of the hybrid-π model, additional study will be made of the
common-emitter connection in this section.

   The beta-cutoff frequency f_{hfe} has been defined as the frequency at which
the short-circuit current amplification factor has declined to 0.707 of its
low-frequency reference value.   With most practical values of load resistance
A_i and β are synonymous.   Therefore, with many transistor types one finds
current gain falloff at a relatively low frequency.   However, *the upper-cutoff
frequency for voltage gain may extend far beyond f_{hfe}.*   We may appreciate
this statement by writing the general expression for voltage gain:

$$A_v = A_i R_L / Z_i$$

If Z_i and A_i behave similarly with frequency, we expect little change in A_v
This is precisely what happens with the junction transistor; Z_i and A_i decline
with increased frequency because of input capacitance.

   To assist in the understanding of the frequency behavior of A_v, consider
the circuit of Fig. 3-14 to represent the quantities of interest in the input circuit of a junction transistor stage.   The element $R_{G'}$ can represent the sum
of source resistance R_G and base-spreading resistance $r_{bb'}$.   The elements C
and R_i are the remaining input quantities.   An expression for the voltage
transfer function of this simple network is

$$V_b / V_g = R_i / (R_i + R_{G'} + j\omega C_i R_{G'} R_i). \qquad (3\text{-}40)$$

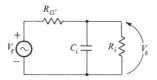

Fig. 3-14 Input circuit of junction transistor.

The voltage available at the base V_b has decreased by 3 dB when the real and imaginary portions of the denominator of this expression are equal. The frequency at which this reduction occurs is the upper cut-off frequency, f_U. For this circuit,

$$f_U = 1/2\pi C_i R_T, \qquad (3\text{-}41)$$

where R_T is the parallel equivalent of all resistive elements. Here $R_T = R_i \| R_{G'}$. Equation (3-41) also describes the frequency at which the current through R_i will be down by 3 dB from its low-frequency reference value.

An important conclusion obtainable from Eq. (3-41) *is that the voltage-gain upper cutoff frequency is determined to a large extent by source resistance and base-spreading resistance.*

Miller Effect

The element R_i in Fig. 3-14 is approximately $r_{b'e}$ for a junction transistor, but C_i is not simply the $C_{b'e}$ element of the hybrid-π representation. In circuits using conventional transistors, FET's or vacuum tubes, additional input capacitance is apparent and is referred to as the *Miller effect*. The Miller effect is significantly increased input capacitance caused by a mutual capacitance element linking the output to input portions of an amplifying circuit. To show this effect, let us find the input admittance of the "boxed" portion of Fig. 3-15. We write

$$I_i = I_m + I_s.$$

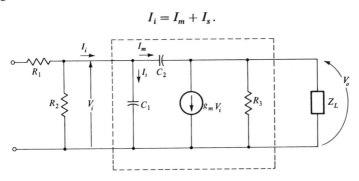

Fig. 3-15 Circuit analyzed for input capacitance.

Now

$$I_s = j\omega C_1 V_i,$$

and

$$I_m = -j\omega C_2(V_o - V_i).$$

Z_L is generally a small impedance compared with the other parallel paths. Consequently,

$$V_o \cong -g_m V_i Z_L.$$

A combination of the foregoing equations yields

$$I_i/V_i = Y_i = j\omega[C_1 + C_2(1 + g_m Z_L)]. \tag{3-42}$$

Notice that C_2 has effectively been multiplied by a term $(1 + g_m Z_L)$ considerably greater than unity. This, then, is the Miller effect; the total input capacitance is much greater than C_1.

Should the load impedance Z_L be inductive, it is easy to prove that Y_i will have a negative real part; a negative resistance physically means that power is being fed back from the output circuit through C_2, and under certain conditions oscillation may occur. Neutralization, discussed later, can be used to minimize this internal feedback.

EXAMPLE. A numerical example is in order. It is desired to find the voltage gain upper cutoff frequency for a stage coupling a 500 Ω source to a 1000 Ω load. Assume a transistor with parameters:

$$f_T = 60 \text{ Mz} \qquad g_m = 0.04 \text{ mho}$$
$$h_{fe} = 50 \qquad r_{bb'} = 100 \text{ ohms}$$
$$C_{b'c} = 5 \text{ pF}$$

Calculations yield

$$r_{b'e} = h_{fe}/g_m = 1250 \ \Omega$$
$$C_{b'e} = g_m/\omega_T = 106 \text{ pF}$$
$$C_i = C_{b'e} + C_{b'c}(1 + g_m Z_L) = 311 \text{ pF}$$

From Eq. (3-41), the upper-cutoff frequency is

$$f_U = 1.26 \text{ MHz}$$

3-10   COUPLING

To couple or join transistor stages in a cascade amplifier, it is common practice to use coupling or blocking capacitances. In the circuit of Fig. 3-16(a) C_1 and C_2 are "blocking" dc from the signal source and the load. The major purpose of these elements is to isolate portions of a circuit from the direct

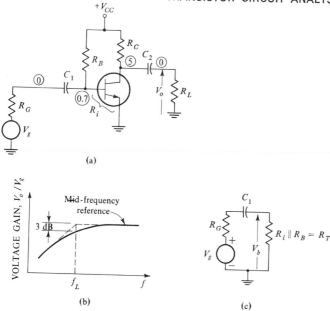

(a)

(b)   (c)

Fig. 3-16 Figures for discussion of interstage coupling.

voltages that may exist in adjacent portions while passing ac signals with a minimum of attenuation.

A capacitance-coupled amplifier is not able to amplify low-frequency signals. Not only is dc blocked by the capacitors, but low frequencies are seriously attenuated because of the large value of series reactance, $X = 1/2\pi fC$. Thus coupling capacitance will result in low-frequency gain rolloff; the gain at low frequencies will be lower than that available at a mid-frequency or reference frequency. This behavior is shown in Fig. 3-16(b).

To calculate the lower cutoff frequency caused by C_1, we make use of the circuit shown in Fig. 3-16(c). The transistor input resistance R_i is effectively in parallel with the base-biasing element R_B. By voltage division the amount of V_g that reaches the base is given by

$$V_b = V_g \frac{R_T}{R_T + R_G - jX_{C1}}. \tag{3-43}$$

The lower cutoff frequency f_L is defined as the frequency where voltage transmission has been reduced by 3 dB or equals 0.707 of the mid-frequency reference value. At this frequency, the real and imaginary parts of the denominator of Eq. (3-43) are equal. Using that denominator, we conclude that

$$f_L = \frac{1}{2\pi C_1 (R_T + R_G)}. \tag{3-44}$$

For $R_T = 5\,\text{k}\Omega$, $R_G = 5\,\text{k}\Omega$, $C_1 = 1\,\mu\text{F}$, this equation predicts that $f_L = 16\,\text{Hz}$.

The transistor acts as a constant-current source.   Therefore, to determine f_L for the *load* circuit of Fig. 3-16(a), the transistor becomes I_o paralleled by output resistance R_o and by R_C.   Voltage V_o will be down by 3 dB at

$$f_L = \frac{1}{2\pi C_2(R_L + R_{C'})},$$
(3-45)

with $R_{C'} = R_C \parallel R_o$.   For $R_{C'} = 5\ k\Omega$, $R_L = 5\ k\Omega$, $C_2 = 1\ \mu F$, we get $f_L = 16$ Hz.

In the numerical examples given above, both values of f_L happened to be equal because of the numbers cited.   For this case, the lower cutoff frequency of the entire amplifier form V_o to V_g would *not* be at 16 Hz because *each* capacitance is causing 3 dB loss at that frequency.   Equations (3-44) and (3-45) give the total or overall f_L only when the two frequencies are adequately separated.

Because of the sizes of C_1 and C_2 necessary for good low frequency operation, linear IC's *do not* employ coupling capacitors between stages.   However, they are often used externally to join the IC to its signal source and its load.   Thus the limit on frequency response is apparent, and may be calculated by the method of this section.

3-11   LARGE-SIGNAL OPERATION

When signal amplitudes are sufficiently large, analysis using small-signal parameters is no longer valid.   Graphical methods are useful to assist in visualizing behavior.   For convenience, names are ascribed to the various modes of operation.   Waveforms are shown in Fig. 3-17.

Operating Classes

Class A. The output electrode current (i_C of a conventional transistor) flows throughout the cycle of input signal.   Small-signal operation is exclusively class A.

Class AB. Collector current flows less than 100 percent but more than 50 percent of the time.

Class B. Collector current flows 50 percent of the time.

Class C. Collector current flows less than 50 percent of the time.

Class A operation results in the largest standby power dissipation (product of $I_{C(Q)}$ and $V_{CE(Q)}$).   Therefore it will exhibit the lowest power efficiency. However, when using Class B push-pull output stages to handle higher power levels with greater efficiency, a pair of transistors is required to faithfully reproduce the signal at the load.

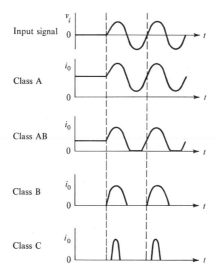

Fig. 3-17 Definitions of classes of operation.

Dynamic Load Line

Section 3-2 discussed the drawing of a static or dc load line on the output characteristics of the active device. Summation of all resistance (ΣR) in the emitter-collector path dictates the slope of the line; the line originates at $V_{CC} + V_{EE}$ and terminates at $(V_{CC} + V_{EE})/\Sigma R$. The quiescent or operating point is established by circuitry providing the desired base or collector current.

The dynamic path of operation or dynamic load line is the locus of all corresponding values of instantaneous collector current and voltage during a cycle of the signal. The ac load on a stage can be established by summing the impedance of all circuit elements in the emitter-to-collector circuit. This task is usually simplified by considering that all bypass and coupling capacitors are short circuits to an ac signal.

To illustrate the handling of loads and the drawing of load lines, consider the circuit of Fig. 3-18(a). The dc loading is comprised of the series combination of R_e and R_C. The ac load is R_e plus the parallel combination of R_C, R_2, and R_i of the second stage.

If the parameters of the circuit of Fig. 3-18(a) are

$$R_e = 220 \ \Omega \qquad R_i = 10 \ \text{k}\Omega$$
$$R_C = 10 \ \text{k}\Omega \qquad R_2 = 120 \ \text{k}\Omega,$$

the dc load line has a slope that is the negative reciprocal of $220 + 10$ k, or $10,220 \ \Omega$, and the slope of the ac line is the negative reciprocal of

$$R_{ac} = R_e + R_C \, \| \, R_2 \, \| \, R_i \, ,$$

or 4800 Ω.   *Ac load lines must pass through the Q point and are not establishe*
by axes intersections.

In Fig. 3-18(b), dc and ac load lines are drawn on output characteristics
After the dc line linking V_{CC} and V_{CC}/R_{dc} is positioned and the Q point located
the ac line may be drawn through Q with the appropriate slope.   Witl
capacitance coupling, the dynamic line will always exhibit the greater slope
because it is dependent upon the paralleling of two or more elements.

For some purposes a *dynamic transfer characteristic* is valuable.   Projec
tion of the points of intersection of the ac load line and lines of constant bas
current to an I_C-I_B curve is easily accomplished as noted in Fig. 3-18(c)
Linearity of operation is easily studied from the transfer characteristic, an(
operating-point suitability therefore clearly established.   The characteristi
is said to be dynamic, for it is derived from the load line and does not assum
a constant collector potential.   A static characteristic at potential V_1 is als(
shown in the figure.

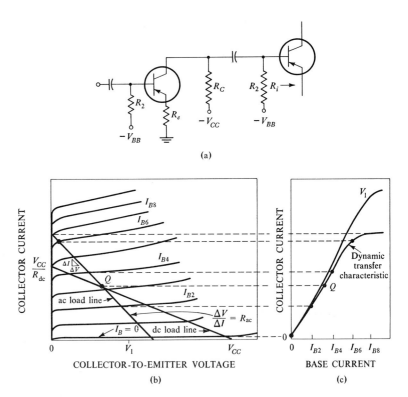

**Fig. 3-18   (a) Circuit for example; (b) dc and ac load lines on outpu
characteristics; (c) construction of dynamic transfer characteristic.**

3-12 TRANSISTOR CIRCUIT ANALYSIS EXAMPLE

A numerical example illustrating many of the topics discussed in this chapter is now presented. The goal is to analyze a given circuit to determine its performance. See Fig. 3-19.

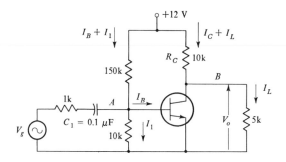

Fig. 3-19 Figure for circuit analysis example.

Given Information

Nominal parameters: $h_{ie} = 4000$, $h_{fe} = 50$, $h_{re} = 10^{-4}$, $h_{oe} = 10^{-4}$, $\beta_{dc} = 50$. Correction information is given in Section 3-3.

Dc Analysis

To find the dc load-line equation:

$$V_{CE} = 12 - (I_C + I_L)10 \text{ k},$$

but

$$I_L = V_{CE}/5 \text{ k},$$

so

$$V_{CE} = 4 - 3.33 \times 10^3 I_C.$$

Neither I_C nor V_{CE} is known at this point. We may find I_C by first finding I_B. For $V_{BE} = 0.7$ V, $I_1 = 70 \ \mu\text{A}$.

$$(I_B + I_1)(150 \times 10^3) = 12 - 0.7.$$

Then $I_B = 5.3 \ \mu\text{A}$, and $I_C = \beta_{dc} I_B = 0.265$ mA. The other coordinate of the operating point is, therefore, $V_{CE} = 3.12$ V.

Ac Analysis

Consider small-signal operation exclusively. The parameters are corrected
for the calculated Q point:

$$h_{ie} = 4000(1.7)(0.9) = 6100 \ \Omega$$
$$h_{fe} = 50(0.8)(0.9) = 36$$
$$h_{re} = 10^{-4}(0.6)(1.4) = 0.84 \times 10^{-4}$$
$$h_{oe} = 10^{-4}(0.3)(1.5) = 0.45 \times 10^{-4} \ \text{mho.}$$

The ac load is 5 k $\|$ 10 k $= 3.33$ kΩ. The source resistance is 1 k $\|$10 k$\|$ 150 k.
From Table 3-3, we obtain

$$A_v = -17 \qquad Z_i = 6.1 \ \text{k}\Omega$$
$$A_i = 31 \qquad Z_o = 22 \ \text{k}\Omega.$$

These calculations do not give the overall performance because the base
biasing network parallels the calculated Z_i, and R_C parallels the calculated
Z_o. The true input impedance looking to the right of point A is $Z_i \|$10 k$\|$ 150
k:

$$Z_{i'} = 3.8 \ \text{k}\Omega.$$

Output impedance looking to the left of point B is $Z_o \|$ 10 k:

$$Z_{o'} = 6.9 \ \text{k}\Omega.$$

The value calculated for A_v refers to the ratio of V_o to the voltage at the
transistor base. For the V_o/V_g ratio, we must include the voltage division
taking place, $Z_{i'}/(1000 + Z_{i'})$. Therefore

$$A_{v'} = V_o/V_g = \left(\frac{3,800}{4,800}\right)(-17) = -13.5.$$

Signal current is lost in the base-biasing resistances. Considering only the
10 kΩ element, current division takes place according to 10 k/(10 k + Z_i).
Because of R_C, one-third of the collector signal current is not delivered to the
load. Then

$$A_{i'} = \left(\frac{10,000}{16,100}\right)31\left(\frac{10,000}{15,000}\right) = 12.8.$$

Frequency Response

Element C_1 is the only limit on low-frequency behavior:

$$f_L = \frac{1}{2\pi(0.1 \times 10^{-6})(1000 + 3800)} = 330 \ \text{Hz.}$$

The upper cutoff frequency is determined by $C_{b'e}$, $C_{b'c}$, and the Miller effect.

3-13 TUNED CIRCUITS

A *tuned circuit* is basically an inductance and a capacitance connected in parallel.   Sometimes the term refers to a transformer with parallel capacitance at its input port, output port, or at both locations.   For simplicity, the present discussion is concerned with the single L and C arrangement shown in Fig. 3-20(a).

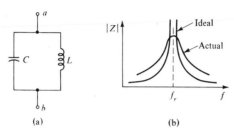

(a)   (b)

Fig. 3-20   (a) Elementary tuned circuit; (b) |Z| vs. frequency of applied voltage.

Tuned circuits have several desirable characteristics.   At the resonant frequency, when $X_L = X_C$, the impedance between points a and b is theoretically infinite.   Thus if the circuit were in the load section of a transistor network, very high voltage gain would be possible since A_v is proportional to Z_L.

At frequencies below and above resonance, the tuned circuit has a finite impedance as shown in Fig. 3-20(b).   An actual circuit has some resistance, usually in the coil winding and behaves as shown.

The tuned circuit acts as a bandpass filter, for when it forms the load in a transistor amplifier, gain of the network will be lower at frequencies above and below resonance.

Direct collector voltage for a transistor can be fed through the coil with nearly zero direct voltage drop, and the associated quiescent power loss is nil. For operation of high frequencies, the inductor can be physically quite small.

These desirable characteristics result in widespread use of tuned circuits in circuits made of discrete elements.   When IC's are used, it is necessary for the tuned circuit to be discrete or, if possible, the tuned circuit is eliminated. A means for simulation of an inductor is discussed in Section 6-4.

Since tuned circuits are fairly common in modern electronic systems, a summary of associated analysis and design tools is given in this section.

Impedance Transformations

For purposes of circuit analysis, it is convenient to be able to mathematically transform series networks into parallel networks and vice versa.

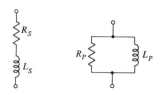

Fig. 3-21   Series and parallel R-L circuits.

Consider the series R-L and the parallel R-L circuits of Fig. 3-21. The impedance of the series circuit is

$$Z_S = R_S + jX_{LS}, \tag{3-46}$$

and that of the parallel circuit is

$$Z_P = jX_{LP} R_P/(R_P + jX_{LP}). \tag{3-47}$$

The S and P subscripts pertain to series and parallel circuits, respectively. If we equate the two impedances, then

$$R_S + jX_{LS} = jX_{LP} R_P/(R_P + jX_{LP}). \tag{3-48}$$

Multiplication of both numerator and denominator of the right-hand side by the conjugate of the denominator, and separation of real and imaginary terms, gives

$$R_S + jX_{LS} = \frac{X_{LP}^2 R_P}{R_P^2 + X_{LP}^2} + \frac{jX_{LP} R_P^2}{R_P^2 + X_{LP}^2}. \tag{3-49}$$

To find the series resistance equivalent to the resistance of the parallel circuit, equate reals; thus

$$R_S = X_{LP}^2 R_P/(R_P^2 + X_{LP}^2), \tag{3-50}$$

and the series inductive reactance that can take the place of the parallel circuit reactance is

$$X_{LS} = X_{LP} R_P^2/(R_P^2 + X_{LP}^2). \tag{3-51}$$

If the admittances of each type of circuit are equated, we obtain

$$R_P = (R_S^2 + X_{LS}^2)/R_S \tag{3-52}$$

and

$$X_{LP} = (R_S^2 + X_{LS}^2)/X_{LS}. \tag{3-53}$$

Equations (3-50), (3-51), (3-52), and (3-53) provide the tools for conversion from one circuit form to the other. These relationships can be employed when the circuits contain R and C rather than R and L. For this use it is necessary only to replace each X_L in Eqs. (3-50) through (3-53) by X_C.

Let us now define the Q or figure of merit of the series circuit previously considered as the ratio of inductive reactance to resistance

$$Q = X_{LS}/R_S \qquad (3\text{-}54)$$

at any designated frequency.   Equation (3-54) is a practical form of the more general relation,

$$Q = 2\pi \left[\frac{\text{energy stored in electric and magnetic fields per cycle}}{\text{energy dissipated per cycle}} \right]$$

The series definition of Q may be used for a parallel circuit by making use of the proved interrelations, Eqs. (3-50) and (3-51).   The division indicated in Eq. (3-54) yields

$$Q = R_P/X_{LP}. \qquad (3\text{-}55)$$

By equating admittances of the two circuits, the following relationships can be proved:

$$R_P = R_S(1 + Q^2) \qquad (3\text{-}56)$$

and

$$X_{LP} = X_{LS}(1 + 1/Q^2). \qquad (3\text{-}57)$$

A single inductance coil is, in reality, a series R-L circuit.   Therefore such a coil is often described by its own Q, the ratio of its series reactance to resistance at a specific frequency.   The discussion of the preceding paragraphs pertains to complete circuits or to single coils; in the material that follows, Q_C will be used to designate the quality factor of a single coil while the symbol Q will be reserved for use as a measure of overall circuit selectivity.

The circuit consisting of a parallel combination of capacitance and a practical inductance coil shown in Fig. 3-22(a) can now be replaced mathematically by the parallel R-L-C circuit of Fig. 3-22(b), the value of R being given by Eq. (3-56).   As an example consider a coil with a Q_C of 100 and R_S of 10 Ω.   Eqs. (3-56) and (3-57) permit us to substitute inductive reactance of 1000 Ω (essentially the same as the series reactance because of the high quality of the coil), and parallel resistance of 100 kΩ.

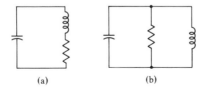

(a)   (b)

Fig. 3-22   Tuned circuit equivalence

We have, therefore, the undamped or coil factor Q_C, and the network factor Q, the latter being defined by Eq. (3-55) as the ratio of parallel resistance to inductive reactance at the specified or resonant frequency. Factors Q and Q_C are equal when no additional R or L is present. Often, in practice, tuned circuits will be "damped" by additional parallel resistance and Q_C may differ significantly from the network Q.

Bandwidth

Bandwidth is defined as the frequency span between half-power points, and half-power points are the frequencies where power gain or power transmission has decreased to one-half the mid-frequency or reference value. The half-power points correspond to a reduction in voltage or current ratio to 0.707 or $1/\sqrt{2}$ of the reference value.

At this stage we seek to find a relation between bandwidth, resonant frequency, and overall quality factor. The circuit of Fig. 3-23 uses single R,

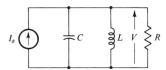

Fig. 3-23   Composite interstage circuit.

L, and C elements to represent the composite of output, coupling, and load quantities. It follows that

$$\frac{V}{I_o} = \frac{j\omega LR}{R(1 - \omega^2 LC) + j\omega L}. \tag{3-58}$$

Load current is proportional to V, although the load resistance is not segregated in Fig. 3-23, and at resonance $V/I_o = R$. Therefore, to examine Eq. (3-58) at the *half-power points where ω is designated as ω_3*, we may write

$$\left|\frac{V}{I_o}\right| = \frac{R}{\sqrt{2}} = \frac{\omega_3 LR}{[R^2(1 - \omega_3^2 LC)^2 + \omega_3^2 L^2]^{1/2}}. \tag{3-59}$$

After making the substitutions that

$$Q = \frac{R}{\omega_r L} \text{ and } \omega_r = \frac{1}{\sqrt{LC}}$$

and rearrangement, Eq. (3-59) may be written as

$$\omega_3^4(Q^2/\omega_r^2) - \omega_3^2(1 + 2Q^2) + Q^2\omega_r^2 = 0. \tag{3-60}$$

The solution for ω_3, if negative frequencies are omitted, is

$$\omega_3 = \left[\frac{1 + 2Q^2 \pm (1 + 4Q^2)^{1/2}}{2Q^2/\omega_r^2}\right]^{1/2}. \tag{3-61}$$

Now bandwidth B is defined by

$$B = f_h - f_l = (\omega_{3h} - \omega_{3l})/2\pi \tag{3-62}$$

Substitute Eq. (3-61) in Eq. (3-62) and square to eliminate the radicals. Finally, take the square root of each side of the resulting equation to obtain

$$B = f_r/Q, \tag{3-63}$$

the required relationship.

EXAMPLE.   A tuned circuit is to be designed with the following specifications:

(a) $f_r = 10^7$ Hz;
(b) bandwidth $= 10^6$ Hz;
(c) $L = 20~\mu H$, $Q_C = 100$;
(d) R_o of driving transistor $= 10$ kΩ.

The required capacitance is found from the equation

$$f_r = \frac{1}{2\pi\sqrt{LC}}$$

to be 12.5 pF.   At the resonant frequency the coil resistance $R_S = 12.6~\Omega$. The parallel equivalent of this is 126 kΩ.   To achieve the network $Q = f_r/B = 10$, additional parallel resistance R_A is needed.   The total equivalent resistance is found to be 12,600 Ω from Eq. (3-55).   Thus $R_A \parallel R_o \parallel R_P = 12,600$. It follows that R_A must be 46.7k Ω.

3-14 NOISE

Noise in electronic devices is any spurious signal, and is almost always unwanted.   In radio and television receivers, noise is apparent to the ear and the eye as " static " and " snow," respectively.   Actually, noise has two general classifications: " external " noise caused by atmospheric disturbances, motor commutation, aircraft and auto ignition, and any sparking device; and " internal " noise generated in the receiver as a result of the physics of the materials and components used.   External noise will not be discussed here because its existence bears no relation to the IC; it can be eliminated or minimized by shielding, antenna location or design, and prayer.

The quality measure of an electronic device or circuit is the *noise figure*; it is customarily given for transistors that are to be employed in low-level circuits. Noise figure F is defined in several ways, probably most widely accepted is the following definition:

$$F = \frac{\text{total available noise power at load}}{\text{available noise power at load due to thermal noise from } R_G}. \quad (3\text{-}64)$$

This definition obviously takes into account the ever present thermal noise attributable to the source resistance R_G. If the amplifying device contributes no noise, F is unity. Naturally, a low value of F is desirable. To express noise figure in decibels, it is simply necessary to multiply $\log_{10} F$ by 10. (Available power assumes impedance matching.)

It has been shown[3] that the overall noise figure of a cascaded network composed of functional blocks or stages signified by A, B, and C is

$$F_{ABC} = F_A + \frac{F_B}{G_A} + \frac{F_C}{G_B}. \quad (3\text{-}65)$$

where F_A and so on, are the noise figures of the blocks and G_A and so on, represent the available power gains of those blocks. *A*, *B*, and *C* need not amplify; they may be simply passive networks. The significance of Eq. (3-65) is simply this: *if the first stage exhibits significant gain, the overall noise figure of a cascaded network is determined entirely by that stage.*

The noise figure of semiconductor networks is usually affected by base-biasing elements, signal source resistance, $r_{bb'}$ and h_{fe}.

Problems

3-1   Use Eq. (3-2) to sketch common-emitter collector characteristics for a transistor with a constant $\beta = 100$ and $I_{CEO} = 0.1$ mA. Let I_B take on values from 0 to 50 μA in 10 μA steps.

3-2   From your solution to Problem 3-1, determine β_{dc} at four different values of collector current.

3-3   Design biasing for the circuit of the figure. It is desired that $I_{C(Q)} = 2$ mA, $V_{CE(Q)} = 3$ V. Given: $V_{CC} = 6$ V, $V_{BE} = 0.7$ V, $\beta_{dc} = 75$. Determine R_B and R_C. If another transistor is substituted, and $\beta_{dc} = 50$, what are the resulting Q-point coordinates?

3-4   In order to stabilize the circuit designed in Problem 3-3 against the effects of changes in β_{dc}, a resistance is to be added between emitter and ground equal to 500 Ω. The Q-point must remain at the desired point; determine new values for R_B and R_C. Now, if β_{dc} again decreases from 75 to 50, how much of a change has occured in the Q-point coordinates?

3-5   For the circuit shown, derive an expression for I_C in terms of V_{CC}, R_B, R_C, V_{BE}, and β_{dc}.   A measure of the sensitivity of $I_{C(Q)}$ to temperature is $S_v \equiv \partial I_C / \partial V_{BE}$.   Derive an expression for S_v.

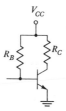

Problems 3-3, 3-4, 3-5

3-6   If V_{BE} changes by -0.002 V/°C, determine the change in $I_{C(Q)}$ caused by a $+50$°C temperature change.   Use the equations derived in Problem 3-5, $\beta_{dc} = 100$, and $R_B = 100$ kΩ.

3-7   Determine β for $\alpha = 0.90, 0.95, 0.98, 0.99, 0.995$.

3-8   A common collector stage has a short-circuit current gain equal to β/α. Express this ratio solely as a function of β and solely as a function of α.

3-9   Prove the relationships shown in Table 3-1 between h_{ie} and h_{fe} and the common-base h parameters.

3-10   Prove the relationships shown in Table 3-1 between h_{re} and h_{oe} and the common-base h parameters.

3-11   Evaluate the *small-signal* hybrid parameters for the device whose characteristics are shown in the accompanying figures.

3-12   Determine the static or dc parameters h_{IE} and h_{FE} for the device of the figure with $V_2 = 10$ V and $I_1 = 10$ μA.   Determine h_{OE} and h_{RE} under the same conditions.

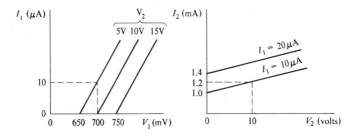

Problems 3-11, 3-12

3-13   Using the information shown in Fig. 3-4, determine Δ^{he} at $I_C = 5$ mA, $V_{CE} = 10$ V, for a transistor with reference parameters of $h_{ie} = 5000$, $h_{fe} = 110$, $h_{re} = 10^{-4}$, $h_{oe} = 10^{-4}$.   What is Δ^{he} at the reference point?

3-14   Consider that the reference parameters given in the preceding problem are valid low-frequency values.   Further, it is known that the transistor has $\omega_T =$ 500×10^6 rad/sec, and a measurement of the input capacitance with collector shorted to emitter yields $C_{in} = 85$ pF. From this information determine all seven parameters of the regular hybrid-π representation. $I_C = 1$ mA.

3-15   A sample silicon transistor exhibits the following parameters at a particular Q point:

$$h_{fe} = 100 \qquad g_m = 0.04 \text{ mho}$$
$$C_{b'c} = 5 \text{ pF} \qquad r_{bb'} = 50 \ \Omega$$
$$f_T = 200 \text{ MHz.}$$

It is used as an amplifier between a 50-Ω load and a 50-Ω source.

(a) Determine $r_{b'e}$ and $C_{b'e}$.
(b) Determine the input capacitance, including the Miller Effect.
(c) Find the voltage-gain upper cutoff frequency.
(d) What is the low-frequency voltage gain (V_o/V_g) for this circuit?
(e) Will the presence of collector-to-substrate capacitance of 5 pF affect your answer to part (c)?   Explain.

3-16   Using the reference parameters given in Problem 3-13, calculate A_v, A_i, R_i, and R_o for an amplifier feeding a 10,000-Ω load from a source of negligible impedance.   The transistor is connected:

(a) common-emitter
(b) common-base
(c) common-collector (emitter-follower).

3-17   Derive Eq. (3-34)

3-18   Derive Eqs. (3-36), (3-37), (3-38), (3-39).

3-19   Use Eq. (3-26) to sketch the static drain characteristics of a JFET with $I_{DSS} = 10$ mA, $n = 2$, $V_p = 5$ V.   Use 1-volt increments for V_{GS}.

3-20   Determine I_{DSS} and estimate V_p from Fig. 3-8.   At $V_{DS} = 20$ V, estimate g_m from $\Delta V_{GS}/\Delta I_D$, where ΔV_{GS} and ΔI_D are distances on the graph.   Do this above and below $V_{GS} = -0.5$ V, and compare this value of g_m with the value obtained by using Eq. (3-27).

3-21   For low-frequency analyses, a simple FET can be approximated by an infinite input impedance and a constant-current generator $g_m V_i$ with resistance r_{ds} as shown in the figure.   When a load R_L is connected between drain and source terminals, and source is common, the gain $V_o/V_g = -g_m R_{L'}$ where $R_{L'} = r_{ds} \| R_L$. Show that the addition of negative feedback in the form of a resistance R_e between source and ground will give

$$A_v = V_o/V_g = -g_m R_{L'}/(1 + g_m R_e).$$

For simplicity, consider r_{ds} to parallel R_L.

3-22   Show that the gain of a *common-drain* stage using the model of the figure is

$$A_v \cong g_m R_L/(1 + g_m R_L).$$

What assumption has been made to arrive at this equation?

3-23   The FET of the preceding problem is to be operated *common-gate* from a signal generator of resistance R_G. Show that

$$A_v = \frac{R_L(1 + g_m r_{ds})}{R_L + r_{ds} + R_G(1 + g_m r_{ds})}.$$

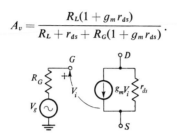

Problems 3-21, 3-22, 3-23

3-24   The FET shown in Fig. 3-11 has the following parameters: $g_m = 0.001$ mho, $r_{ds} = 10$ kΩ.   Consider that $R_L = 10$ kΩ, $R_S = 1$ kΩ.   In parallel with R_S is a capacitance equal to 0.001 μF.   Use Problem 3-21 to find the voltage-gain lower cutoff frequency caused by this bypass capacitor.

3-25   When a transistor feeds a load consisting of $C_{L'}$ and $G_{L'}$ in parallel, the input circuit of the device can be represented by the elements shown in the figure. Show that $C_1 = C_{b'e} + C_{b'c}$, $C = C_{b'c} g_m/G_{L'}$, $R = C_{L'}/C_{b'c} g_m$.

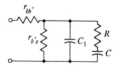

Problem 3-25

3-26   Compare the current-gain frequency response of a particular transistor operating common-emitter with the response of that same transistor operating common-base.   Consider alpha reduction as the only source of gain falloff, and assume that the transistor is feeding a low-resistance load.   Which orientation provides the greater gain at $f_{\alpha e}$ and $f_{\alpha b}$?   (These are often-used symbols for f_{hfe} and f_{hfb}.)

3-27   A particular system uses a direct-coupled video amplifier, such as the SA20 discussed in Chapter 8, to provide the gain needed between a 600-Ω source and a 5000-Ω load.   Coupling capacitors will be used to isolate the amplifier from the terminations (transducers).   The lower cutoff frequency for voltage gain (V_o/V_g) is to be at 50 Hz.   For the amplifier, $R_i = 1.6$ kΩ and $R_o = 5$ Ω.   Determine the size of each capacitance if it alone causes gain falloff.

3-28   The circuit of the figure employs R_1 for bias stabilization, and the purpose of C_1 is to ground the emitter terminal to ac so that R_1 will not lower the gain of the stage.   Consider that $Z_i \cong R_i + (\beta + 1)Z_1$, where $Z_1 = R_1 \| C_1$.   Determine the voltage gain V_{be}/V_g, and show that the lower voltage-gain cutoff frequency is given by

$$f_L = \frac{(\beta + 1)R_1 + R_i + R_G}{2\pi C_1 R_1 (R_i + R_G)} .$$

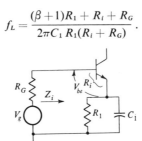

Problem 3-28

3-29   Analysis of the complete voltage transfer function V_{be}/V_g determined in Problem 3-28 indicates that the numerator of that expression is also frequency-dependent.   In order for f_L to be as given in that problem, what must also be true regarding the elements in this circuit?

3-30   Consider that n identical stages are cascaded.   Each may be represented by

$$A = \frac{A_{mid}}{1 + jf/f_U} .$$

Show that the upper cutoff frequency of the n-stage amplifier is given by

$$f_{Un} = f_U(2^{1/n} - 1)^{1/2}.$$

3-31   Examine the change in lower-cutoff frequency caused by cascading of stages.   If each stage of an n-stage amplifier can be represented by

$$A = \frac{A_{mid}}{1 + jf_L/f}$$

show that the overall amplifier has a lower cutoff frequency given by

$$f_{Ln} = f_L/(2^{1/n} - 1)^{1/2}.$$

3-32   Three identical stages are cascaded.   Each has $f_L = 50$ Hz, $f_U = 10^6$ Hz. Use formulas given in the preceding problems to find f_{Ln} and f_{Un}.

3-33   Class B push-pull pairs are responsible for *crossover distortion* in their output waveform.   Explain crossover distortion using a sketch of the waveform. Recall that the input emitter diode has a very nonlinear relation at low voltage levels.

3-34 Show that the resonant frequency of a capacitor paralleling a series R-L branch is given by

$$f_r = \frac{1}{2\pi} \sqrt{\frac{1}{LC} - \frac{R^2}{L^2}} .$$

3-35 Consider an ideal transformer (no losses) being used to transform impedance levels.   When a resistance R is connected at the transformer output port its effect looking into the input port is $a^2 R$.   Show that this reflected resistance level is related to the turns ration N_1/N_2, and comment on the voltage gain of the transformer.

4
FEEDBACK

The parameter variations common to all semiconductor devices are a source of difficulty to the designers as well as to the users of these devices. The variations are caused by ambient and operating temperature, by operating-point excursions, and by signal frequency. Manufacturing tolerances result in significant unit-to-unit differences.

To overcome the expected variations, feedback—the addition of a portion of the output signal to the input signal—is employed. Reduction of the magnitude of the input signal by addition of a feedback signal is called *inverse, degenerative,* or *negative* feedback, while an increase in total input due to this summation is termed *direct, regenerative,* or *positive* feedback. The use of negative feedback is widespread and is the type discussed in detail in this chapter.

Those feedback principles most applicable to IC's are reviewed. Almost every linear IC employs some form of negative feedback, and we shall see that some use several feedback loops to attain the required operating characteristics.

4-1 GENERAL FEEDBACK THEORY

The block diagram of Fig. 4-1 can be used to illustrate feedback in an elementary way. The output voltage V_o of an amplifier with a voltage gain of A supplies a load; V_o is also available to the feedback network, and BV_o, a fraction of V_o, is added

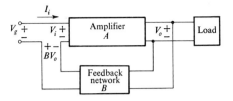

Fig. 4-1   General feedback diagram.

to the input circuit.   This *reference* diagram is the basis for the derivation of many equations.   *In practice, polarities may not be as shown in the figure.*
The nominal amplifier gain is

$$A \equiv V_o/V_i.\qquad(4\text{-}1)$$

If we take polarities into account, A will normally be negative, and may be a complex function of frequency. B will normally be positive, and may also be complex.   In the presence of a feedback signal, a summation at the input yields

$$V_i = V_g + B V_o.\qquad(4\text{-}2)$$

The voltage amplification A_f of the composite network is

$$A_f \equiv V_o/V_g = V_o/(V_i - BV_o) = A/(1 - BA).\qquad(4\text{-}3)$$

Equation (4-3), the fundamental feedback equation, will now be interpreted. In the absence of feedback ($B = 0$), the amplifier exhibits a gain of A.   If the magnitude of the denominator is greater than unity, the overall amplification will be less than A and the feedback is degenerative; the converse of this statement also applies.   If the entire output voltage is supplied to the input ($B = 1$), the composite gain is less than unity.   When $BA = 1$, the resultant gain is infinite, and the circuit has an output independent of any external input voltage.
If $|BA| \gg 1$, then the amplification approaches

$$A_f \to 1/B.\qquad(4\text{-}4)$$

Equation (4-4) signifies that gain is independent of elements in the forward path, usually a *highly desirable condition.*   Because B is often a passive network, A must be very large to satisfy the condition specified by Eq. (4-4). Since B is usually a fraction, A_f can be a large number.   Aging, tolerances, and temperature would certainly have negligible effect upon an amplifier that behaves according to Eq. (4-4).
The price to be paid for the advantages of feedback (they will be considered in detail in the following sections) is reduction in overall gain.   A given amplifier having an open-loop gain of A, will have a closed-loop gain

of less than A because of incorporation of negative feedback circuitry.   But the advantages of feedback are so significant that it generally pays to design our original amplifier with sufficient gain so that added feedback will not reduce gain below the desired value.   The inclusion of additional gain to offset feedback losses is a small price to pay when compared with the advantages as indicated in the paragraphs to follow.

Gain Stabilization

The gain of a feedback amplifier A_f is much lower than that of an unstabilized amplifier having open-loop gain A.   Now we investigate how stable the feedback amplifier is.   Differentiation of Eq. (4-3) yields the following expression for the fractional gain sensitivity of the feedback amplifier:

$$dA_f/A_f = [1/(1 - BA)](dA/A). \qquad (4\text{-}5)$$

Because $(1 - BA) \gg 1$, we conclude that the feedback amplifier is much less sensitive to the internal factors that cause gain changes.

Consider an amplifier with a gain of -10^4.   The fraction B is to be $1/100$. Then, from Eq. (4-3), feedback will reduce the gain to approximately -100. For a 10 percent variation in forward gain $(dA/A = 0.1)$, the overall gain of the feedback amplifier will suffer less than a 0.1 percent change (dA_f/A_f). Variations in gain will be reduced by the same amount as forward gain.

Input Impedance

The type of feedback shown in Fig. 4-1 has a strong influence upon terminal impedance levels.   From that figure, in the absence of feedback,

$$Z_i = V_g/I_i = V_i/I_i. \qquad (4\text{-}6)$$

With the feedback loop closed, Eq. (4-6) becomes

$$Z_{if} = (V_i - BV_o)/I_i. \qquad (4\text{-}7)$$

Since

$$V_o = AV_i,$$

then Eq. (4-7) becomes

$$Z_{if} = [V_i(1 - BA)]/I_i. \qquad (4\text{-}8)$$

Therefore

$$Z_{if} = Z_i(1 - BA). \qquad (4\text{-}9)$$

The presence of the $(1 - BA)$ term indicates an *increase in input impedance due to closing of the loop,* for that term is always greater than unity in systems employing negative feedback.

Output Impedance

The method used previously for determining output impedance is to connect a hypothetical generator of voltage V_o across the output terminals of a circuit or device and to measure or calculate the current I_o with the input signal source short-circuited.   Then, for the amplifier without feedback,

$$Z_o = V_o/I_o. \tag{4-10}$$

With the feedback loop closed, as in Fig. 4-2,

$$V_i = BV_o.$$

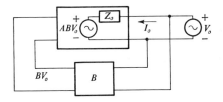

Fig. 4-2   Feedback diagram for calculation of output impedance.

The output loop current is

$$I_o = (V_o - BAV_o)/Z_o.$$

Therefore, the effective output impedance is

$$Z_{of} = V_o/I_o = Z_o/(1 - BA). \tag{4-11}$$

The conclusion that can be drawn is that *output impedance is reduced by the addition of inverse feedback*.   This reduction is not caused, in this case, by the parelleling of Z_o by the feedback network, because in the treatment given here it was assumed that no current was drawn from V_o by the feedback network.

Bandwidth Extension

To investigate the effects of feedback upon the bandwidth of a circuit, we must assign frequency dependence to the forward gain of the amplifier.   A good assumption may be that

$$A = A_o/[1 + jf/f_U], \tag{4-12}$$

where f_U, as used in Chapter 3, is the upper cutoff frequency and A_o is the low-frequency reference gain.   Substitution of Eq. (4-12) into Eq. (4-3) yields

$$A_f = A_o/[1 - BA_o + jf/f_U]. \tag{4-13}$$

Now the denominator will be of the form $K + jK$ when

$$f/f_U = 1 - BA_o. \tag{4-14}$$

Equation (4-14) indicates that the new cutoff or half-power frequency f_{Uf} occurs at

$$f_{Uf} = f_U(1 - BA_o). \tag{4-15}$$

The upper cutoff frequency has been increased, and the bandwidth extended by an appreciable amount.

To look at extension of low-frequency response, assume that low-frequency performance may be approximated by

$$A = A_o/[1 + jf_L/f], \tag{4-16}$$

where f_L is the lower cutoff frequency of the original amplifier. We may determine that the new lower cutoff frequency is

$$f_{Lf} = f_L/(1 - BA_o). \tag{4-17}$$

The lower cutoff frequency has been significantly lowered by feedback.

A word of caution is called for. The extension of the upper cutoff frequency predicted by Eq. (4-15) is *not* obtained if the gain of the forward block A contains more than one important frequency-dependent term. This is observed in the solution to Problem 4-4.

By using negative feedback in power amplifiers it is possible to obtain larger power outputs with a given percentage of amplitude distortion. The signal-to-distortion ratio with feedback is shown to be $(1 - BA)$ larger than the same circuit without feedback.

The signal-to-noise ratio is usually not improved by the use of negative feedback.

Current Feedback

The preceding discussion provides a basis for understanding the reasons why feedback is so frequently employed in electronic circuits. The discussion was limited to what is usually called " voltage " feedback; the output voltage is sampled, operated upon by the feedback network, and returned to the input circuit. It is also possible to employ "current" feedback, in which the load current is sampled and a version of that current fed back and combined with signal source current at the amplifier input terminals.

In the current feedback amplifier shown in Fig. 4-3, one may consider that the block B does not load the amplifier output, and thus $I_2 = I_o$. At the input port, a current summation takes place. Consequently,

$$I_i = I_1 - BI_o.$$

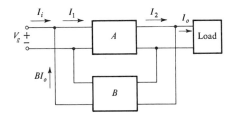

Fig. 4-3 General current-feedback diagram.

Since A_i is the ratio of I_2 to I_1, it can be shown that

$$A_{if} \equiv I_o/I_i = A_i/(1 - BA_i) \tag{4-18}$$

Normally A_i would provide phase reversal and therefore carry a negative sign. B, usually a positive quantity, represents the ratio of current fed back to load current.

It follows from this figure that

$$Z_{if} = Z_i/(1 - BA_i) \tag{4-19}$$

and

$$Z_{of} = Z_o(1 - BA_i). \tag{4-20}$$

Current feedback lowers the input impedance and increases the output impedance of an amplifying network. Otherwise, its use is equivalent to the voltage-feedback case already discussed.

4-2 INSTABILITY IN FEEDBACK SYSTEMS

The benefits derived from the use of negative feedback are great, but are achieved at the expense of gain. Additional low-level stages may usually be added to compensate for the loss of gain. Another drawback to be contended with is the possibility of self-oscillation, because forward and feedback elements are frequency-sensitive. At low and also high frequencies, the output voltage may be shifted in phase and changed in magnitude relative to the mid-frequency value. The summation of output and input voltages in a feedback circuit may, because of this additional phase shift, result in regeneration and possibly oscillation.

Self-oscillation can be visualized with the help of Eq. (4-3), which is repeated here:

$$A_f = A/(1 - BA).$$

When $BA \rightarrow +1$, $A_f \rightarrow \infty$, a condition intolerable in amplifiers that represents an output limited only by the saturation and cutoff regions of the characteristics.   Should BA approach some positive value lower than unity, regenerative operation results.

For BA to be positive, the combination of B and A must contribute the 360° phase shift required to cause a summation of in-phase signals at the amplifier input.   Although B is often a simple resistive network, wiring capacitances may cause unwanted phase shift.   A shift of 180° is normally supplied by an active circuit consisting of an odd number of common-emitter stages.   Each stage is frequency-sensitive, and with three in cascade, each need shift only an additional 60° in order for one oscillation requirement to be satisfied—namely, for BA to be positive.   If this requirement is satisfied, and, in addition, the magnitude of the loop gain (BA) is equal to or greater than unity, the network will oscillate.

In the absence of a B network, when a high-gain amplifier is operated open-loop, undesired capacitance between high- and low-level portions of the circuit may be sufficient to provide the conditions for self-oscillation. Many IC amplifiers cannot be operated open-loop.

A test for circuit stability involves plotting $-BA$ on polar coordinates. The magnitude of this quantity is examined when 180° of additional phase shift exists.   *According to the Nyquist criterion, oscillations will exist if the locus of $-BA$ encloses the $(-1,0)$ point.**

EXAMPLE.   Consider an IC amplifier with passive feedback.   The open-loop response, the ratio of BV_o to V_g, may be approximated by

$$-BA = \frac{10}{(1 + jf10^{-6})^3}.$$

We wish to investigate the stability of this amplifier.

A polar plot of this function is shown in Fig. 4-4.   The point of intersection with the negative real axis can be calculated from the equation when the imaginary part of $-BA$ is zero.   This occurs when

$$f^2 10^{-12} = 3,$$

or $f = 1.732$ MHz.   The magnitude of $-BA$ at this frequency is

$$10/(1 - 3f^2 10^{-12}) = -5/4.$$

This is larger than -1, encirclement of $-1 + j0$ exists, and the circuit is unstable.   In order to stabilize this amplifier, the gain can be reduced or a filter added to reshape the locus so that the higher frequencies are attenuated.

---

* This simplification applies to low-pass amplifiers having all-pole type transfer functions.   Circuit and system examples in this book will be of this type.

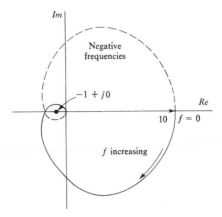

Fig. 4-4   Nyquist diagram for text example of unstable feedback amplifier.

4-3   LOCAL FEEDBACK

Feedback that is used with a single amplifying stage is termed *local feedback*. In Section 3-2, such feedback was discussed for the case of a common-emitter amplifier with a resistance connected between emitter and ground.   In that example, the resistance provided a negative feedback direct voltage to assist in stabilizing the operating point of the stage.

The most widely used local feedback for gain stability and for impedance-level modification is also a resistance in the emitter lead, as shown in Fig. 4-5(a).   Element R_e will help stabilize both gain and operating point simultaneously; however, the element values required for optimizing these two functions may be quite different.   We limit our concern in this section to signal considerations.

Equation (3-29) gives the voltage gain of a common-emitter stage if the applicable h parameters are used.   That equation does not apply to the circuit of Fig. 4-5(a) because the addition of R_e has changed the circuit.   Therefore, we develop a new relation for A_v using Fig. 4-5(b).   Note that the element h_{re} has been neglected in order to simplify results.

$$V_g = I_1(R_s + R_e) + I_2 R_e$$
$$0 = -I_1(h_{fe}/h_{oe} - R_e) + I_2(R_e + R_C + 1/h_{oe}). \tag{4-21}$$

We have used $R_s = R_G + h_{ie}$.   Note that $V_o = -I_2 R_C$.   Solution of Eqs. (4-21) yields

$$A_{vf} \equiv \frac{V_o}{V_g} = \frac{-(h_{fe} - h_{oe} R_e)R_C}{R_s(1 + h_{oe} R_e) + R_e(h_{fe} + 1)}, \tag{4-22}$$

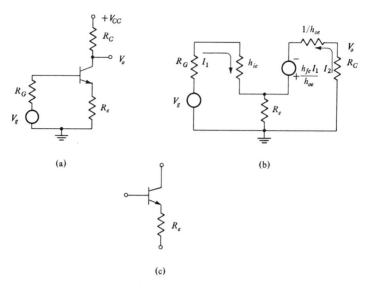

Fig. 4-5 Local emitter-leg feedback. Base biasing not shown.

provided that $h_{oe}R_C \ll 1$. It is likely also that $h_{oe}R_e \ll 1$. When this is the case,

$$A_{vf} \cong \frac{-h_{fe}R_C}{R_s + R_e(h_{fe} + 1)}. \tag{4-23}$$

Eq. (4-23) allows us to analyze easily the effect of element R_e. Since that equation is in the form $A_v = A_i R_L/R_i$, we see that the *input resistance has been significantly increased by the factor* $R_e(h_{fe} + 1)$:

$$R_i \cong h_{ie} + R_e(h_{fe} + 1). \tag{4-24}$$

The gain has been reduced, as would be expected, because of negative feedback.

Composite Parameters

Another approach to the analysis of local feedback is to derive a set of h parameters for the composite circuit including transistor and resistor R_e, as shown in Fig. 4-5(c). The location of R_e allows us to add that element to the value of each of the transistor z parameters, expressed in equivalent h values.

Using Appendix II, we can transform this new set of z parameters into h parameters.   When these steps are performed, as in Problem 4-8, we obtain:

$$h_i = \frac{h_{ie} + R_e(1 + \Delta^{he} + h_{fe} - h_{re})}{1 + R_e h_{oe}}$$

$$h_r = \frac{h_{re} + R_e h_{oe}}{1 + R_e h_{oe}}$$

$$h_f = \frac{h_{fe} - R_e h_{oe}}{1 + R_e h_{oe}}$$

$$h_o = \frac{h_{oe}}{1 + R_e h_{oe}}.$$

(4-25)

These parameters can be used directly in Eqs. (3-29) through (3-34) for gains and terminal impedances.

4-4 MULTISTAGE FEEDBACK

It will now be shown that feedback around a block of stages—multistage feedback—is more effective in reducing gain variations than repeated use of local feedback.   In Fig. 4-6(a), n identical cascaded stages stabilized by local feed-

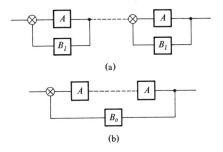

(a)

(b)

Fig. 4-6 Comparison of local and over-all feedback.

back are shown.   The forward gain of each stage is A, and the feedback factor is B_l.   Assuming no loading effects, the overall amplification is

$$A_{lf} = [A/(1 - B_l A)]^n.$$

(4-26)

An amplifier with overall feedback that also has n stages is shown in Fig. 4-6(b). The gain of this network is

$$A_{of} = A^n/(1 - B_o A^n).$$

(4-27)

The fractional gain change of the local feedback amplifier caused by changes in A is obtained by differentiating Eq. (4-26).

$$\frac{dA_{lf}}{A_{lf}} = \frac{n}{1 - B_l A}\left(\frac{dA}{A}\right). \tag{4-28}$$

The corresponding expression applicable to the overall feedback case is

$$\frac{dA_{of}}{A_{of}} = \frac{n}{1 - B_o A^n}\left(\frac{dA}{A}\right). \tag{4-29}$$

We wish to express the ratio of Eq. (4-28) to Eq. (4-29) in consistent symbols. Because $A_{lf} = A_{of}$, it follows that

$$(1 - B_l A)^n = 1 - B_o A^n.$$

Therefore

$$\frac{dA_{of}/A_{of}}{dA_{lf}/A_{lf}} = \frac{1}{(1 - B_l A)^{n-1}}. \tag{4-30}$$

For values of n greater than unity, Eq. (4-30) suggests that the *fractional gain variations of the network employing overall feedback will always be less than the corresponding quantity for a network using local feedback.* $[(1 - B_l A) > 1$ for negative feedback.] This conclusion is independent of the fractional gain variation dA/A of each forward element.

Multistage Feedback Examples

It is possible to write loop and node equations for multistage feedback amplifiers and to solve the simultaneous equations in order to analyze a given network for its terminal characteristics and amplificational properties. This turns out to be a particularly lengthy procedure, unsuitable for most purposes. Instead, it is usually more convenient to attack the problem with a block approach, using Eq. (4-3). Each stage of a feedback amplifier has associated voltage- and current-gain properties. Cascades of stages may also be treated by considering only the overall properties.

As examples, two feedback amplifiers will be considered. The active devices will provide 180° of phase shift per stage; thus we are concerned with common-emitter and common-source stages. This does not rule out the use of emitter followers or common-base or common-drain stage; those connections do not provide phase reversal, but their gains must be included in the analysis when they are employed.

The *series-shunt pair* shown in Fig. 4-7 is widely utilized for video amplifier applications. A practical example, the SA-20, is described in Section 8-3.

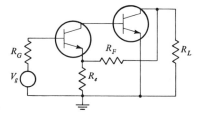

Fig. 4-7 Series-shunt amplifier with feedback elements R_l and R_F.

The primary feedback connection is from the collector of the output stage to the emitter of the first stage.   A fraction $R_e/(R_e + R_F)$ of the load voltage is added to the input loop.   In addition, R_e provides a small amount of negative local feedback in the first stage.

Voltage amplification is primarily affected in this circuit, just as in the series local feedback case discussed in the preceding section.   With two common-emitter stages in the forward path, A_v is positive.   The feedback factor B_v is negative because the voltage fedback is not of the polarity shown in the reference diagram, Fig. 4-1.   The ratio of voltage fed back to load voltage is

$$B_v \cong -R_e/(R_e + R_F). \tag{4-31}$$

The value of A_v to be used in Eq. (4-3) must include the *local* effects of R_e and R_F.   Thus $A_{v'}$ represents the gain of the two stages with R_e in place and R_F connected from the final collector to ground.   Then

$$A_{vf} = \frac{A_{v'}}{1 - B_v A_{v'}}. \tag{4-32}$$

The input resistance of the feedback circuit is

$$R_{if} = R_{i'}(1 - B_v A_{v'}). \tag{4-33}$$

Symbol $R_{i'}$ represents the input resistance of the forward gain section with the local effect of R_e included.

An approximate relation for the forward gain of the pair is

$$A_{v'} = A_{v1'} A_{v2'} \cong \frac{h_{fe1} R_{L1}}{h_{ie1} + (1 + h_{fe1})R_e} \cdot \frac{h_{fe2} R_{L2}}{h_{ie2}}, \tag{4-34}$$

with $R_{L1} = R_{C1} \parallel R_{i2}$ and $R_{L2} = R_L \parallel R_F \parallel R_{C2}$.   By using $R_F = R_e/B_v$, R_F may be eliminated from the equation.   Now R_e is the only unknown remaining. When Eq. (4-34) is differentiated with respect to R_e, and set equal to zero, it yields the value for R_e that provides the *maximum forward gain*:

$$R_e = \left[\frac{B_v h_{ie1}}{(1 + h_{fe1})(G_L + G_{C2})} \right]^{1/2}. \tag{4-35}$$

Equation (4-35) is very useful for designing amplifiers of the Fig. 4-7 type.

The second example included in this section is that of a *three-stage amplifier* with feedback from the load or final collector to the base of the first stage. The circuit is shown in Fig. 4-8. With considerable forward voltage and current gain, A_v and A_i will be large and R_F will be a high-valued resistance.

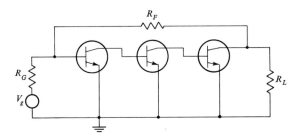

Fig. 4-8   Three-stage amplifier with feedback element R_F.

This connection will affect current gain much more than voltage gain if R_G is not large-valued because of the summation of output and input currents at the first base. The feedback factor B_i, the ratio of current feedback to load current, is

$$B_i \cong \frac{R_L}{R_F}. \qquad (4\text{-}36)$$

The current gain with feedback is

$$A_{if} = \frac{A_i}{1 - B_i A_i}, \qquad (4\text{-}37)$$

and therefore

$$A_{if} \cong -R_F/R_L.$$

B_i is positive, and A_i negative. Input resistance has been decreased by the feedback, according to Eq. (4-19), and output impedance increased.

The shunt multistage feedback circuit is widely used in operational amplifiers. At this point, we can note that when R_G is large and A_v large, the currents summed at the first base are $I_g \cong V_g/R_G$ and $I_o \cong V_o/R_F$. These currents are essentially equal in magnitude if the amplifier gain is large, so that $I_i R_i$ may be neglected. Then, the ratio of voltage feedback to load voltage is

$$B_v \cong R_G/R_F. \qquad (4\text{-}38)$$

The resulting amplifier gain with feedback is

$$A_{vf} = -R_F/R_G. \qquad (4\text{-}39)$$

Additional discussion is given in Chapter 6.

4-5   NEUTRALIZATION

The transistor contains internal feedback.   Elements $r_{b'c}$ and $C_{b'c}$ of the
hybrid-π representation provide a path for undesired negative feedback between output and input portions of a common-emitter amplifier.   At high
frequencies and under certain load conditions, this internal feedback may be
responsible for the input impedance of the stage exhibiting a negative real
part, a condition for self-oscillation.   This can be seen from observation of
Eq. (3-32):

$$Z_i = h_i - \frac{h_r h_f}{h_o + Y_L}. \tag{4-40}$$

As long as Y_L is large-valued, the negative term in this equation will not be
dominant.   When circuit conditions do not guarantee that Z_i will be positive,
external feedback may be necessary.

   Neutralization is accomplished by adding passive circuitry to provide external feedback to cancel the internal feedback that fosters instability.   Gain
of a neutralized stage is also larger than that of an unneutralized circuit.
Neutralization may be accomplished by adding an external capacitor C_n, as
shown in Fig. 4-9(a), so that $I_1 = I_2$.   When these currents are equal and

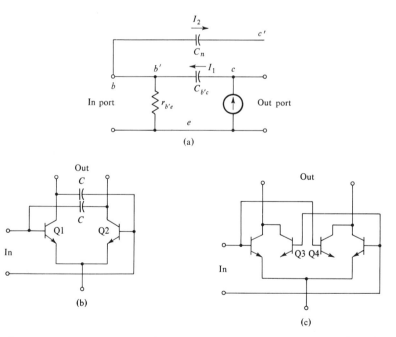

Fig. 4-9   Neutralization: (a) circuit for general discussion; (b) diff amp
with compensation; (c) practical IC diff amp.

opposite in direction, the collector is isolated from the base, for zero net current flows between those terminals. ($r_{bb'}$ has been omitted for simplicity.)

In order to practically attain a neutralized amplifier, c and c' must be connected to points where the voltage differs by 180°. In tuned amplifiers, this is accomplished with the coupling transformer used between the transistor and its load. Since the number of transformers used with IC's should be minimal, another approach is considered.

A bridge neutralization circuit, as shown in Fig. 4-9(b), can be used in differential amplifier stages. The voltages at the collectors of Q1 and Q2 are equal in magnitude but differ in phase by 180°. The capacitors labeled C provide the feedback needed to cancel internal feedback through $C_{b'c}$. A practical way of neutralizing uses the collector-base capacitances of two additional transistors Q3 and Q4 as shown in (c) of the figure. All four transistors have identical geometries and matching is thereby assured.

Problems

4-1 An amplifier with a voltage gain of $A_v = -1000$ is to be used as the forward gain block in a feedback amplifier with $A_{vf} = -10$. Determine B. If A_v is subject to ± 10 percent variation because of poor regulation in the power supply, determine the resulting variation in A_{vf}.

4-2 Consider that the amplifier described in Problem 4-1 has $Z_i = 10^4 \ \Omega$, $Z_o = 10^2 \ \Omega$, $f_u = 10^6$ Hz, $f_L = 10^2$ Hz. The feedback used to reduce the gain to -10 will result in what values for Z_{if}, Z_{of}, f_{uf}, and f_{Lf}?

4-3 Assume that an amplifier can be described by

$$A = A_o/(1 + jf/f_1)^2.$$

Feedback is added. Derive an expression for f_{uf} in terms of A_o, f_1, and B_o.

4-4 Compare f_{uf} predicted by the equation derived in the preceding problem with the result of using Eq. (4-15). Let $1 - BA_o = 10$, $f_u = f_1 = 10^6$ Hz.

4-5 Derive Eq. (4-17) for f_{Lf}.

4-6 An amplifier is described by

$$A_v = \frac{-10^4}{(1 + jf10^{-6})(1 + jf10^{-7})(1 + jf10^{-7}/3)}.$$

The feedback network B is to be a simple resistance network, and contributes no phase shift.

(a) Determine the magnitude and phase angle of $-BA$ at 10 MHz, 15 MHz, and 20 MHz.

(b) What value must B take on so that A_{vf} is barely stable?

(c) To provide a *phase margin* of 45°, determine B so that $|BA| = 1$ at $-135°$. The answers to parts (b) and (c) can be most easily estimated from a polar plot of $-BA$ as determined from the calculations in part (a).

4-7 Derive Eq. (4-22).

4-8 Derive the set of composite h parameters for local emitter-leg feedback given in Eqs. (4-25).

4-9 Consider that local feedback, in the form of a resistance between emitter and ground, is being used in a CE stage with parameters $h_{ie} = 5000$, $h_{re} = 0$, $h_{fe} = 100$, $h_{oe} = 10^{-5}$, $R_G = 0$, and $R_C = 10^4$ Ω. Determine R_i and A_v for this stage with $R_e = 0$, 10, 100, and 1000 ohms.

4-10 Derive an expression for the output impedance Z_{of} of a single CE stage with local feedback R_e in the emitter branch. Use h parameters.

4-11 It may be of value to know the effect of unbypassed emitter resistance upon the effective input capacitance of a CE stage. As a model for this study, consider that one sees $r_{bb'}$, $r_{b'e} \| C_{b'e}$, and $(h_{fe} + 1)R_e$ in series at the input port. Compare the effective shunt input capacitance with R_e in the circuit to the input condition when $R_e = 0$.

4-12 The "number of dB of feedback" is used for the ratio of amplifier gain without feedback to gain of the same amplifier with feedback, the ratio being expressed in dB. Amplifiers with open-loop gains of 10,000, 5000, and 300 have their gains reduced to 100 by feedback. Express these ratios in dB of feedback.

4-13 Derive Eq. (4-35) for maximizing forward gain.

4-14 Use Eq. (3-34) for output impedance to develop a formula for output admittance so that it has a positive and a negative part. How will neutralizing circuitry help to keep Y_o positive?

PART TWO

LINEAR
INTEGRATED
CIRCUITS
AND
SYSTEMS

5
LINEAR
IC
BUILDING
BLOCKS

The linear IC, whether it be a simple audio amplifier, a high-frequency mixer, or a high-gain operational amplifier, is composed of recognizable subcircuits. These building blocks are not revolutionary new networks; instead, they are often well-known circuits using established techniques that have gained new importance because of the limitations imposed by and the advantages gained through IC manufacturing methods.

This chapter begins with a discussion of biasing and bias-point stability using constant-current sources. The differential amplifier is considered in some detail because it is the single most important building block used in the linear IC. Also discussed is the emitter-follower, the Darlington pair, the level shifter, and IC output stages.

5-1 BIAS NETWORKS

The conventional method for biasing discrete common-emitter circuits employs a capacitance bypassing a resistance between emitter and ground to separate the paths for dc and ac within the circuit. This technique allows the use of bias-stabilizing resistance that does not disturb the overall ac performance. The large-valued capacitance required for good audio-frequency response is not practical in an IC, but the problem of bias stability remains an important consideration. One solution to the problem of operating-point stability has been referred to as *balanced biasing*.

A design philosophy to achieve a

constant Q point stems from the fact that transistors integrated on a single chip can be made with almost identical dimensions and materials, are in thermal proximity, and consequently behave almost identically.   In the circuit of Fig. 5-1(a), Q1 is connected as a diode.   The purpose of that transistor

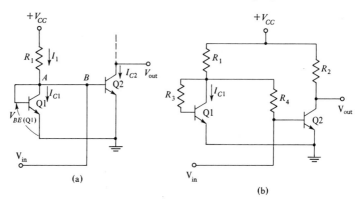

Fig. 5-1   Biasing networks.

is to bias Q2 at a particular fixed collector current.   Neglecting base currents, we find

$$I_{C1} \cong I_1 = \frac{V_{CC} - V_{BE(Q1)}}{R_1} \cong \frac{V_{CC}}{R_1}.$$ (5-1)

Q2 is matched to Q1.   The base-emitter voltages will also be equal.   As noted in Eq. (1-21), V_{BE} and I_C are related by the exponential equation. With the V_{BE} drops equal, it follows that

$$I_{C2} = I_{C1}.$$

Therefore the collector current of Q2 has been established by conditions at the Q1 transistor.   For $V_{CC} = 6$ V, $R_1 = 5$ kΩ, $I_{C2} \cong 1$ mA.   The signal may be fed to the base of Q2, and, if necessary, a resistance may be located between points A and B so that the Q1 circuit will not attract much signal current from the source connected at V_{in}.

A modification to the circuit of Fig. 5-1(a) is given in 5-1(b).   Base biasing resistances R_3 and R_4 have been added.   The base of Q2 is no longer connected across $V_{BE(Q1)}$.   Transistors Q1 and Q2 are considered to be identical and $R_3 = R_4$.   *Because the transistor bases are fed from the same point* $(+V_{CC})$ *through equal resistances, their collector currents will tend to be equal.* To determine the value of the collector current, notice that the current through R_1 is I_{C1} plus the sum of the two base currents.   The voltage equation is

$$V_{CC} = V_{BE} + I_B R_3 + (2I_B + I_{C1})R_1.$$ (5-2)

Rearranging gives

$$I_{C1} = \frac{V_{CC} - V_{BE}}{R_1} - \left(2 + \frac{R_3}{R_1}\right)I_B . \tag{5-3}$$

To simplify this equation, we note that $V_{CC} \gg V_{BE}$ and $I_C \gg I_B$. Then, for typical values like $R_1 = R_3 = 10$ kΩ, the collector currents are

$$I_{C1} = I_{C2} \cong \frac{V_{CC}}{R_1} . \tag{5-4}$$

Usually $R_2 = R_1/2$. The voltage at the Q2 collector is, therefore,

$$V_{CE2} \cong V_{CC}/2 . \tag{5-5}$$

Now we may conclude from Eq. (5-5) that the operating point is virtually independent of temperature and transistor parameters if the units are well matched. The operating point is directly dependent upon V_{CC}, but variations in V_{CC} will not cause the Q point to stray from the center of the load line.

Low-Current Source

A simple and successful constant-current source for providing current levels in the microampere range is shown in Fig. 5-2. The effectiveness of this

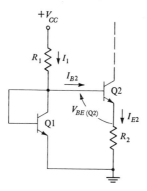

Fig. 5-2   Low-current source.

circuit is based upon the fact that the voltage $V_{BE(Q1)}$ is divided into two parts in the parallel path, the drops across R_2 and $V_{BE(Q2)}$. Therefore the base-emitter drop and the collector current of Q2 will be smaller than the corresponding voltage and current of Q1. By proper selection of R_2, the current I_{E2} can be made very small, but R_2 need not be a large resistance.

Let us determine R_2 for specified current levels. From the figure,

$$I_{E2} R_2 = V_{BE(Q1)} - V_{BE(Q2)}. \qquad (5\text{-}6)$$

The base-emitter voltages are related to the respective collector currents by the exponential equation, Eq. (1-21). Using that relation, we may solve Eq. (5-6) for R_2:

$$R_2 = (1/\Lambda I_{E2}) \ln (I_1/I_{E2}). \qquad (5\text{-}7)$$

Base currents are omitted in the derivation of this equation. The value of I_1 is simply V_{CC}/R_1, if $V_{CC} \gg V_{BE(Q1)}$. This current would generally be relatively large, for example 2 mA. Suppose that we wish I_{E2} to be one percent of that value, or 20 μA. Using $\Lambda = 40$, Eq. (5-7) suggests that the resistance R_2 be 5.7 kΩ in order to achieve the desired level for I_{E2}. This is a small resistance to provide a current level of only 20 μA.

A temperature-derived change in the base-emitter voltages will result in the right-hand side of Eq. (5-7) varying linearly with temperature (see Problem 5-5). The positive temperature coefficient of diffused resistance R_2 will tend to cancel the effects of temperature upon the operation of this circuit.

5-2 CONSTANT-CURRENT SOURCE

A source of constant direct current is customarily utilized for the transistors performing the amplification function in a linear IC. Such a circuit is referred to as a *constant-current source* or a *constant-current sink*.

A current source may be made most simply from a direct voltage source in series with a large-valued resistance and the load. Then, regardless of variations in the load, Ohm's law tells us that the current value will be determined primarily by the large resistance. In the discrete CE circuit, such a source is approximated by a bias-stabilizing resistance connected in series with the emitter, bypassed by capacitance, as has been discussed. In IC manufacture, large resistances and bypass capacitors are to be avoided; so an alternate method is sought for generating the constant current.

In the differential amplifiers and emitter-followers liberally employed in linear IC's, the constant current source does not degrade amplification.

The circuit of Fig. 5-3(a) is a typical diode-stabilized constant-current source used in IC's. As a building block, it joins a system at points Y and Z. The diodes are often transistor base-emitter junctions with collector and base joined by a short circuit as discussed in Chapter 2. *The current to be held constant is I_O.* For analysis, N diodes are connected in series to provide temperature stabilization, with the voltage drop across each when forward-

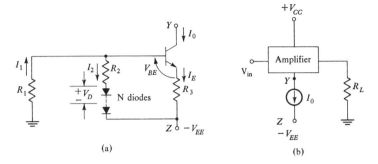

Fig. 5-3   Diode-stabilized constant-current source and application.

biased given as V_D. We write two loop equations and one current equation to find I_E:

$$V_{EE} = I_1 R_1 + I_2 R_2 + N V_D$$
$$V_{EE} = I_1 R_1 + V_{BE} + I_E R_3 \qquad (5\text{-}8)$$
$$I_E/(\beta_{dc} + 1) = I_1 - I_2.$$

Solution of Eqs. (5-8) for I_E yields

$$I_E = (\beta_{dc} + 1) \left[\frac{R_2 V_{EE} + R_1 N V_D - (R_1 + R_2) V_{BE}}{(\beta_{dc} + 1)(R_3)(R_1 + R_2) + R_1 R_2} \right]. \qquad (5\text{-}9)$$

A form of Eq. (5-9) more convenient for study is

$$I_E = \frac{\Theta[V_{EE} + (R_1/R_2)N V_D - (1/\Theta)V_{BE}]}{R_3 + (1 - \alpha_{dc})R_1 \Theta} \equiv \frac{V_T}{R_T}, \qquad (5\text{-}10)$$

with $\Theta = R_2/(R_1 + R_2)$. The required constant current I_O is equal to $\alpha_{dc} I_E$ and since α_{dc} is approximately unity, I_O and I_E are essentially equal.

The main reason for current variations is the effect of temperature excursions. Terms such as Θ and R_1/R_2 will not be affected, for all resistances are formed alike in the manufacturing process and their ratios will be more-or-less temperature independent.

It is possible to achieve a zero temperature coefficient for the numerator of the I_E expression, Eq. (5-10), by using the V_D term to cancel the V_{BE} term. We assume $V_D = V_{BE}$. Then the required number of diodes N is

$$N = (1 + R_2/R_1). \qquad (5\text{-}11)$$

Often R_2/R_1 is chosen as unity so the required number of diodes is two.

The denominator of Eq. (5-10) will have a positive temperature coefficient. To minimize this effect it can be matched against the load supplied by the constant-current source. I_O normally forms the emitter current for another stage shown as the "amplifier" in Fig. 5-3(b).   The amplifier gain is usually proportional to load resistance R_L and its transconductance g_m.   It was noted in Chapter 3 that $g_m = \Lambda I_C$, and $I_C = \alpha_{dc} I_E$.   Therefore, the amplifier gain A_v may be expressed by

$$A_v = K\alpha_{dc} I_0 R_L.$$

K represents a constant more or less independent of operating point.   Substitution of Eq. (5-10) into this gain expression gives

$$A_v = \frac{K\alpha_{dc}^2 R_L V_T}{R_T}. \tag{5-12}$$

For A_v to be invariant, the temperature variation in $\alpha_{dc}^2 R_L$ can be matched with the variation in R_T, and the term V_T chosen to have zero coefficient, as given by Eq. (5-11).   Additional discussion of this network is given in the literature[8].

The output impedance of the current-source transistor is of concern.   For common-emitter stages, it is known that a high level of output impedance may be accomplished if the base-terminating resistance is low-valued, and thus it is desirable to have R_1 and R_2 small.   A disadvantage of small resistance values is their increased power dissipation.   A limit can be put on the sum of R_1 and R_2, for if this sum is greater than 7 kΩ, their self-dissipation will usually be less than 5 mW.

It is possible for the output impedance of a current-source stage to approach 500 kΩ at a low reference frequency, such as 10 kHz, for base termination resistance of 1000 ohms, and R_3 of 500 ohms.   However, for frequencies in the neighborhood of 100 MHz, one finds that this output impedance has deteriorated to perhaps 1000 ohms.   This rather substantial reduction has not necessarily caused significant performance degradation, because the magnitude of the load impedance fed by the constant-current source in all likelihood has also decreased over this range of frequency.

Because of the several compromises necessary, a satisfactory circuit may be designed with values such as the following:

$$R_1 = 5000 \ \Omega \qquad N = 2$$
$$R_2 = 5000 \ \Omega \qquad R_3 = 3000 \ \Omega.$$

These values are somewhat dependent upon the load fed by the constant-current source.

5-3 DIFFERENTIAL AMPLIFIER

The workhorse of the linear IC is the *differential amplifier*, or *diff amp*, circuit.   The simplest differential amplifier is shown in Fig. 5-4.   Because

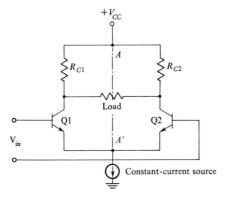

Fig. 5-4   Simple differential amplifier with transistors connected common-emitter.

this circuit is symmetrical about a line such as *A-A'* drawn down its center, and because the load is usually connected between collectors, the differential amplifier is insensitive to many influences that would disturb the performance of other circuits.   A properly designed differential amplifier pair with Q1 matched to Q2 and R_{C1} matched to R_{C2} will provide stable amplification down to zero frequency, even when the circuit is subject to temperature excursions.   Any change in gain or leakage in the Q1 signal channel will be balanced out by an equal change in the Q2 channel.

AC Analysis

To show the differential action of this type of circuit, either a dc or an ac analysis may be made.   The circuit of Fig. 5-5 will be used here with a small-signal low-frequency hybrid-π model employed to describe the transistors. The model has been modified; $r_{b'c}$ and r_{ce} have been omitted, and a current generator of βI_b is used instead of $g_m V_{b'e}$.

Because of the proximity of the transistors and the matching possible with IC manufacturing techniques, the analysis will assume transistor parameters to be identical.   Parameter $r_{bb'}$ has been lumped with the source resistance R_G, and the sum termed R_S.   Equations for the Fig. 5-5(b) circuit are

$$V_1 = I_{b1}(R_S + r_{b'e}) + I_x R_x$$
$$V_2 = I_{b2}(R_S + r_{b'e}) + I_x R_x \qquad (5\text{-}13)$$
$$I_x = I_{e1} + I_{e2}$$
$$I_{e1} = (\beta + 1)I_{b1} \text{ and } I_{e2} = (\beta + 1)I_{b2}.$$

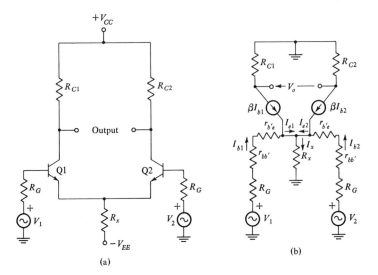

Fig. 5-5 Differential amplifier: (a) with signal source and constant-current source; (b) equivalent circuit.

For $R_x(\beta + 1) \gg r_{b'e}$ and $\gg R_S$, solution of Eqs. (5-13) yields

$$I_{b1} \cong \frac{V_1 - V_2}{2[R_S + r_{b'e}]} \tag{5-14a}$$

and

$$I_{b2} \cong \frac{V_2 - V_1}{2[R_S + r_{b'e}]}. \tag{5-14b}$$

The voltage between collectors is

$$V_o = \beta R_C I_{b1} - \beta R_C I_{b2}. \tag{5-15}$$

We may substitute Eqs. (5-14) into Eq. (5-15). In terms of circuit elements, the differential output voltage is

$$V_o = \frac{(V_1 - V_2)\beta R_C}{R_S + r_{b'e}}. \tag{5-16}$$

Eq. (5-16) gives the output in terms of input voltages. Normally V_1 and V_2 are not equals, for, as the equation suggests, *zero output voltage would result*. Rather, it is more probable that in a practical application V_1 and V_2 will be 180° out of phase, and therefore $V_1 = -V_2$. Then the circuit has satisfactory amplifying properties.

Other conclusions can be drawn from this discussion:

(1) *R_x does not appear in the expression for V_o.* When V_1 and V_2 are oppositely phased, the currents through R_x contributed by those generators will also be oppositely phased and will cancel, resulting in zero net current in that branch.

(2) *Input impedance does not depend upon R_x if R_x is large.* Consider channel 1 with $V_2 = 0$. From Eq. (5-14a),

$$Z_i = \frac{V_1}{I_{b1}} = 2(R_S + r_{b'e})$$

This condition results because the easiest path for current will be around the loop containing generators V_1 and V_2.

(3) *If R_S is reasonably small, gain is independent of β.* Beta may be partially cancelled from numerator and denominator (recall that $r_{b'e} = \beta/g_m$).

(4) *For single-sided operation, with the output taken between either collector and ground, the effective gain is one-half of the differential output gain:*

$$V_o(\text{collector to ground}) = \frac{(V_1 - V_2)\beta R_C}{2(R_S + r_{b'e})} \tag{5-17}$$

(5) *If a channel has no path for signal base current such as I_{b2} to flow, we no longer have differential amplifier operation*; the input impedance and output become respectively

$$Z_i \cong r_{bb'} + r_{b'e} + (\beta + 1)R_x \tag{5-18}$$

and

$$V_o(\text{collector to ground}) = I_{c1}R_{C1} \cong \frac{\beta V_1 R_{C1}}{R_S + r_{b'e} + (\beta + 1)R_x}. \tag{5-19}$$

Common-mode Operation

Common-mode signals, as mentioned in Chapter 1, are signals applied to the input terminals of a diff amp that are in phase with each other. The diff amp behaves differently from Eq. (5-15) when V_1 and V_2 are in phase. No longer is the contribution of each emitter signal current effectively cancelled in R_x, the constant-current source portion of the circuit. Rather, each half of the diff amp looks like the representation in Fig. 5-6.

We can determine the common-mode gain of this circuit by adaptation of Eq. (4-22). Substitution of $2R_x$ for R_e yields

$$A_v(\text{common}) = \frac{V_o}{V_g} = -\frac{(h_{fe} - 2R_x h_{oe})R_C}{(R_S + r_{b'e})(2R_x h_{oe} + 1) + 2R_x(h_{fe} + 1)}. \tag{5-20}$$

To arrive at Eq. (5-20) it has been assumed that $h_{oe}R_C \ll 1$.

The common-mode gain will be quite small if R_x is large. Suppose

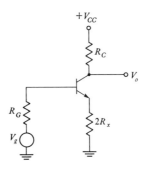

Fig. 5-6   Circuit for Eq. (5-20).

$h_{fe} = 100$, $R_S + r_{b'e} = 5$ kΩ, $h_{oe} = 0$, and $R_C = 5$ kΩ. Eq. (5-20) predicts that $A_v(\text{common}) = 0.25$ for $R_x = 10$ kΩ, and about 0.025 for $R_x = 100$ kΩ.

For the differential mode, the expression for gain equivalent to Eq. (5-20) is available from Eq. (5-17):

$$A_v(\text{diff}) = \frac{V_o}{V_1 - V_2} = \frac{-h_{fe}R_C}{2(R_S + r_{b'e})}.$$

The *common-mode rejection ratio* (CMRR) is a measure of departure from the ideal case where $A_v(\text{common}) = 0$. The CMRR is defined as the ratio of differential to common-mode gains:

$$\text{CMRR} = \frac{A_v(\text{diff})}{A_v(\text{common})}. \tag{5-21}$$

Before substitution of Eqs. (5-20) and (5-17) into this defining equation, we must require that those expressions reflect the same transistor model. Eq. (5-17) omitted h_{oe}, and Eq. (5-20) employed that parameter. We simplify Eq. (5-20) by assuming $h_{oe} = 0$. Then

$$\text{CMRR} = \frac{R_S + r_{b'e} + 2R_x(h_{fe} + 1)}{2(R_S + r_{b'e})}. \tag{5-22}$$

For $h_{fe} = 100$, $r_{b'e} = R_S = 5$ kΩ, and $R_x = 10$ kΩ, CMRR $= 101$. If the value of R_x can be increased to 100 kΩ, CMRR $= 1010$. This latter value expressed in dB, as commonly done with CMRR, equals 60 dB.

5-4   TRANSFER CHARACTERISTICS

The relation between the input and output quantities of an electrical circuit is the transfer characteristic. This relation may be given graphically, or it may take the form of an equation. We will look at both means of description for the diff amp in this section.

To arrive at a mathematical expression for the transfer characteristic of the diff amp, a development based upon the large-signal properties of bipolar transistors will be made.   Again, the devices will be considered to be identical.   Then, for each transistor of the differential pair shown in Fig. 5-7, it is known that the direct currents are

$$I_C = \alpha_{dc} I_E \qquad \text{and} \qquad I_B = (1 - \alpha_{dc}) I_E. \tag{5-23}$$

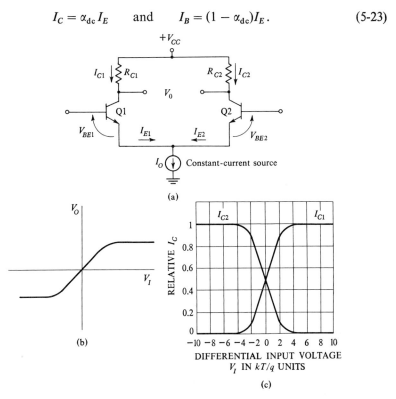

Fig. 5-7   Transfer characteristic derivation: (a) differential amplifier; (b) graphical transfer characteristic; (c) collector current variation with V_I.

The relation between input junction voltage and current is given by the familiar diode equation, with I_{ES} the saturated value of reverse current,

$$I_E = I_{ES}(\epsilon^{\Lambda V_{BE}} - 1), \tag{5-24a}$$

where $\Lambda = q/kT$, as noted in Chapter 1.   For practical values, the -1 term may be omitted.   Then

$$I_E \cong I_{ES} \epsilon^{\Lambda V_{BE}}. \tag{5-24b}$$

Kirchhoff's current law allows us to sum the currents at the common emitter terminal:

$$I_O = I_{E1} + I_{E2}$$

or

$$I_O = I_{ES}\,\epsilon^{\Lambda V_{BE2}}(1 + \epsilon^{\Lambda(V_{BE1} - V_{BE2})}).\qquad(5\text{-}25)$$

Eq. (5-25) follows from Eq. (5-24b).

For convenience we let $V_I \equiv V_{BE1} - V_{BE2}$. We may now solve Eq. (5-25) for I_{E2} in terms of I_O and V_I:.

$$I_{E2} = \frac{I_O}{1 + \epsilon^{\Lambda V_I}}.\qquad(5\text{-}26)$$

and therefore

$$I_{C1} = \frac{\alpha_{dc}\,I_O}{1 + \epsilon^{-\Lambda V_I}} \quad\text{and}\quad I_{C2} = \frac{\alpha_{dc}\,I_O}{1 + \epsilon^{\Lambda V_I}}.\qquad(5\text{-}27)$$

The transfer characteristic will be obtained from

$$V_O = I_{C1}R_{C1} - I_{C2}R_{C2}.\qquad(5\text{-}28)$$

Thus, for $R_{C1} = R_{C2} = R_C$,

$$V_O = \frac{\alpha_{dc}\,I_O R_C(\epsilon^{\Lambda V_I} - \epsilon^{-\Lambda V_I})}{2 + \epsilon^{\Lambda V_I} + \epsilon^{-\Lambda V_I}}.\qquad(5\text{-}29)$$

Mathematical manipulation can yield

$$V_O = \frac{\alpha_{dc}\,I_O R_C \sinh \Lambda V_I}{1 + \cosh \Lambda V_I}.\qquad(5\text{-}30)$$

This relation between output and input is the voltage transfer characteristic, and is plotted in Fig. 5-7(b). To be noted are the saturation regions that exist at large positive and negative values of V_I. The circuit has natural limiting capabilities. Small-signal operation as an amplifier should take place in the linear region near the center of the curve close to $V_I = 0$.

A plot of Eqs. (5-27), collector currents versus V_I, is also given. If we consider V_{BE2} equal to zero, then when V_{BE1} is also zero the collector current of each is $\alpha_{dc}\,I_O/2$. I_{C1} varies from 0 when V_{BE1} is large and negative to $\alpha_{dc}\,I_O$ when V_{BE1} is large and positive.

Transconductance

A parameter used by some to describe differential amplifier performance is the *transconductance*. Recall that transconductance has been used for vacuum-tube, transistor, and FET devices, and is universally defined as the ratio of an incremental change in output current to the increment in input voltage causing the change, under short-circuit load conditions.   In mathematical language,

$$g_m \equiv \lim_{\Delta v_{IN} \to 0} \left. \frac{\Delta i_{OUT}}{\Delta v_{IN}} \right|_{v_{OUT} = \text{constant}} \tag{5-31}$$

We now consider the use of the transconductance parameter for a circuit rather than for a device.   For the diff amp shown in Fig. 5-8(a),

$$g_m(\text{d.a.}) \equiv \left. \frac{\partial I_{CC}}{\partial V_I} \right|_{V_{CE} = \text{constant}} \tag{5-32}$$

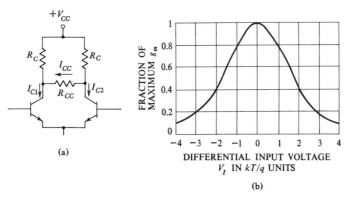

(a)

(b)

Fig. 5-8   (a) Circuit for determination of g_m; (b) behavior of g_m with V_I.

It can be shown that as $V_{CE1} \to V_{CE2}$, a short-circuit R_{CC},

$$I_{CC} = (I_{C1} - I_{C2})/2, \tag{5-33}$$

and, therefore, using Eq. (5-27), we may obtain

$$I_{CC} = \frac{\alpha_{dc} I_O}{2} \left[\frac{1}{1 + \epsilon^{-\Lambda V_I}} - \frac{1}{1 + \epsilon^{\Lambda V_I}} \right]. \tag{5-34}$$

It follows from Eq. (5-32) that

$$g_m(\text{d.a.}) = \frac{\alpha_{dc} I_O \Lambda}{(1 + \epsilon^{-\Lambda V_I})(1 + \epsilon^{\Lambda V_I})} \tag{5-35a}$$

or

$$g_m(\text{d.a.}) = \frac{\alpha_{dc} I_O \Lambda}{2(1 + \cosh \Lambda V_I)}. \tag{5-35b}$$

This behavior is shown in Fig. 5-8(b).   The maximum value for $g_m(\text{d.a.})$ can be obtained by setting the first derivative of Eq. (5-35) to zero to find V_I, and then using that value of V_I in Eq. (5-35).   However, the mathematics can be circumvented by observation.   From the last equation, the maximum value of this function exists at $V_I = 0$ and is given by

$$g_m(\text{d.a.}) = \frac{\alpha_{\text{dc}} I_0 \Lambda}{4} \qquad (maximum) \qquad (5\text{-}36)$$

As an example, for $\Lambda = 40$, $I_0 = 1$ mA, $\alpha_{\text{dc}} \cong 1$, it follows that $g_m(\text{d.a.})$ is 10,000 μmhos.

5-5   DIFF AMP WITH EMITTER DEGENERATION

The curve of transconductance versus differential input voltage shown in Fig. 5-8(b) would be improved for some applications if it were flattened particularly in the neighborhood of $V_I = 0$.   Such a modification would show up as straightening the transfer characteristic, the plot of V_O versus V_I, in the center or "linear" region.

By flattening the $g_m(\text{d.a.})$ curve, the user of the diff amp would be able to apply larger input signals to the circuit without becoming overly concerned about distortion in the output waveform.   Nonconstant transconductance, as shown in Fig. 5-8(b), has the effect of amplifying portions of a signal near $V_I = 0$ greater than any other portions, and this gives rise to distortion.

Let us investigate the addition of resistances R_{e1} and R_{e2} between each emitter and the constant-current terminal, as shown in Fig. 5-9.   Negative

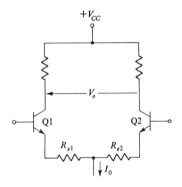

Fig. 5-9   Diff amp with emitter degeneration.

local feedback of this type will tend to straighten the input diode characteristic because the series addition of a linear element with a nonlinear element tends to dilute the nonlinearity.

By writing voltage equations of the Eq. (5-13) type and including $I_e R_e$ drops in each loop, expressions for I_{b1} and I_{b2} can be obtained.   In most practical circuits R_x will be much greater than the other resistances.   Upon solution for V_o, we obtain

$$V_o = \frac{(V_1 - V_2)\beta R_C}{R_S + r_{b'e} + R_e(\beta + 1)}.$$ (5-37)

The transconductance of this circuit is $\beta/[R_S + r_{b'e} + R_e(\beta + 1)]$, and this is considerably smaller than g_m for the non-feedback case.   For $\beta = 100$, $r_{b'e} = 5000\ \Omega$, $R_S = 0$, $R_e = 50\ \Omega$, the effective transconductance is reduced to less than half of its original value.   However, the linear range of operation has been extended.

Diff Amp Gain

Several different diff amps have been discussed in this chapter.   For convenience, a listing of the most widely accepted circuits is given in Table 5-1 along with performance equations.   It can be noted that g_m(d.a.) is equal in effective value to one half the individual transistor transconductance, g_m. All equations assume that the channels have identical signal sources, loads, and transistors.

5-6   OFFSETS IN THE DIFF AMP

An *ouput voltage offset* (V_{OO}) in a differential amplifier is defined as a potential difference between transistor collectors in the absence of signals at the input terminals of the transistors.   A perfect differential amplifier would exhibit zero potential difference between collectors.   When discussing offset, the input terminals are generally grounded.

When an *input voltage offset* (V_{IO}) is discussed, the term refers to the amount of input voltage that must be supplied to one transistor in order to reduce the output voltage offset to zero.   Therefore the input voltage offset, if multiplied by the gain of the circuit, would be a measure of the output voltage offset.

An *input current offset* (I_{IO}) is defined as the current that must be supplied to one input in order for the output voltage offset to be reduced to zero.

Offsets can be attributed to: (1) geometry variations; (2) impurity concentration variations resulting from the diffusion process; (3) variations in the thermal resistance paths and thus the temperature of components.   These variations result in differences in transistor input characteristics (particularly V_{BE}), transistor parameters (particularly β), and circuit resistances.

In order to determine the relative effects of unbalances in the various transistor parameters and in the resistances of an integrated differential amplifier, we shall derive an equation for offset voltage.   Using Fig. 5-10,

TABLE 5-1   DIFF AMP DIFFERENTIAL GAINS

Circuit	$A_v = V_o/V_i$	A_v for $R_S = R_G + r_{bb'} = 0$	$A_{v'} = V_o'/V_i$
	$\dfrac{\beta R_C}{R_S + r_{b'e}}$	$g_m R_C$ or $g_m(\text{d.a.})R_C$	$\dfrac{\beta R_C}{2(R_S + r_{b'e})}$
	$\dfrac{\beta R_L}{R_S + r_{b'e}}$ where $R_L = \dfrac{R_C R_{CC}}{R_{CC} + 2R_C}$	$g_m R_L$ or $g_m(\text{d.a.})R_L$	—
	$\dfrac{\beta R_C}{2(R_S + r_{b'e})}$	$\dfrac{g_m R_C}{2}$	$\dfrac{\beta R_C}{2(R_S + r_{b'e})}$
	$\dfrac{\beta R_C}{R_S + r_{b'e} + (\beta + 1)R_e}$	$\dfrac{\beta R_C}{r_{b'e} + (\beta + 1)R_e}$	$\dfrac{\beta R_C}{2[R_S + r_{b'e} + (\beta + 1)R_e]}$

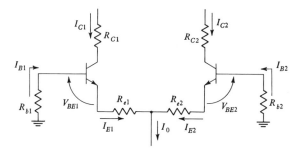

Fig. 5-10   Portion of differential amplifier used to study offsets.

it is possible to write a Kirchhoff direct voltage equation for the path that includes both bases and emitters. Symbols R_{b1} and R_{b2} represent the sums of external resistances in each base circuit. Therefore

$$I_{B1}R_{b1} + V_{BE1} + I_{E1}R_{e1} - I_{E2}R_{e2} - V_{BE2} - I_{B2}R_{b2} = 0. \quad (5\text{-}38)$$

Using $I_B = I_C/\beta$ and $I_E = [(\beta + 1)/\beta]I_C$, Eq. (5-38) can be solved for I_{C1}:*

$$I_{C1} = \frac{(\beta_1/\beta_2)[R_{b2} + (\beta_2 + 1)R_{e2}]}{R_{b1} + (\beta_1 + 1)R_{e1}} I_{C2} + \frac{\beta_1(V_{BE2} - V_{BE1})}{R_{b1} + (\beta_1 + 1)R_{e1}}. \quad (5\text{-}39)$$

An analogous equation may be written for I_{C2}. The offset voltage is then

$$V_{O1} - V_{O2} = I_{C1}R_{C1} - I_{C2}R_{C2}. \quad (5\text{-}40)$$

Using Eqs. (5-39) and (5-40), it is possible to perform a number of studies to determine the relative effects of the various parameters. Here we will present one example of such a study.

Consider $\beta_1 = \beta_2 = 100$, $R_{b1} = R_{b2} = 100$ ohms, $R_{C1} = R_{C2} = 1000$ ohms, $R_{e1} = R_{e2} = 5$ ohms. We wish to find the offset output voltage caused by a 1-mV difference in $V_{BE1} - V_{BE2}$ ($\equiv V_I$). Observe that V_I is not a signal imposed upon the circuit but simply the difference in voltages of biased transistors caused by physical differences in the units. It follows from Eq. (5-39) that

$$I_{C1} - I_{C2} = \frac{-\beta_1 V_I}{R_{b1} + (\beta_1 + 1)R_{e1}} = -0.165 V_I. \quad (5\text{-}41)$$

Eq. (5-41) follows, since beta's and resistances for the transistors are equal.

---

* For simplicity the symbol β will often be used where technically β_{dc} is the correct parameter. This tends to simplify certain equations by eliminating subscripts. Likewise, α and α_{dc} are sometimes interchanged.

From Eq. (5-40)

$$V_{o1} - V_{o2} = (0.165)(1000) = 165V_I.$$

Thus, for a one millivolt difference in the base-to-emitter voltages, we would observe an output offset voltage of 165 mV. This is indeed a sensitive situation.

The assumption used in the preceding example that $R_e = 5$ ohms implied absence of external resistance in the emitter leads. Let us now assume $R_{e1} = R_{e2} = 50$ ohms. Then

$$I_{C1} - I_{C2} = -0.019V_I.$$

The offset voltage now is $19V_I$, and a one-millivolt difference causes only 19 mV output offset. It is clear that emitter degeneration desensitizes the circuit to parameter changes.

To examine the case of an input current offset, assume that the 1 mV difference cited in the preceding paragraph applies ($V_I = 0.001$). With all parameters equal, we find from Eq. (5-39):

$$I_{C1} = I_{C2} + \frac{-100(0.001)}{100 + (101)50}.$$

It follows that I_{C1} will be about 20 μA smaller than I_{C2}. The input current offset, found by dividing this number by beta, is of the order of 0.2 μA.

Input Bias Current

If the input terminals of a diff amp are directly connected to the bases of the input stage transistors, *a path for direct base current is necessary* to assure biasing at the design level. By connecting the input terminals to ground, that path is established, and the base current flowing through the connection is primarily determined by power supplies, the characteristics of the constant-current source, and the h_{FE} of the transistors.

Input bias current is, therefore, a necessary evil in certain circuits. It is not a factor with FET inputs. Usually this current is discussed as a parameter; as such it is defined as one-half of the sum of the separate currents entering the two input terminals of a balanced amplifier.

5-7   PRACTICAL DIFF-AMP CIRCUITS

To achieve more gain than is available from a simple differential amplifier pair, another diff amp can be added in cascade. This connection is shown in Fig. 5-11. The bases of Q3 and Q4 are fed from the collectors of the input pair.

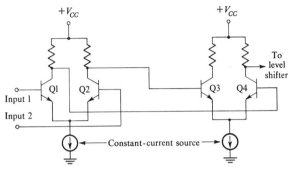

Fig. 5-11   Cascaded differential amplifiers.

The effective load between collectors on the first stage pair to small ac signals is the sum of the input impedances of Q3 and Q4 in series.

From the figure, it is clear that the single output at the collector of Q4 will be at a voltage level considerably more positive than the input lines to Q1 and Q2. It is usually necessary to follow such amplifying stages by "level shifting" circuits, the main purpose of which is to bring the quiescent direct-voltage level to the vicinity of zero, or at least to prepare the level so that it is compatible with the requirements of an emitter-follower, the commonly used output stage.

A differential amplifier packaged as an IC is the Fairchild $\mu A730$ shown in Fig. 5-12(a). Upon examination of the schematic, one can note that Q5 is the constant-current source, Q1 and Q2 are the differential pair, and Q3 and Q4 are emitter-followers. The purpose of Q3 and Q4 is to provide a lower output impedance than available from the differential pair. If a higher output impedance is desirable, the user can connect his load to the collectors of Q1 and Q2, designated Hi Z output in the figure.

This amplifier operates from a single power supply of $+15$ volts or less. Because of this, there is a finite quiescent direct voltage with respect to ground at input and output terminals. Base currents for Q1 and Q2 are each about 5 μA, and input offsets are typically 1 mV and 0.5 μA.

This circuit will be used for an example of a method of analysis of dc conditions in an IC. Consider that $V_{CC} = +12$ V. First we find the value of the constant current source, $I_{o(Q5)}$. The voltage at the base of Q5 is $(12)(1.3 \text{ k})/(6.9 \text{ k}) = 2.3$ V. Subtraction of 0.7 volt for the base-emitter drop for Q5 leaves 1.6 V across the 2.1 $k\Omega$ resistance. This corresponds to an emitter current of 0.76 mA. One half of this current flows through Q1 and one half through Q2.

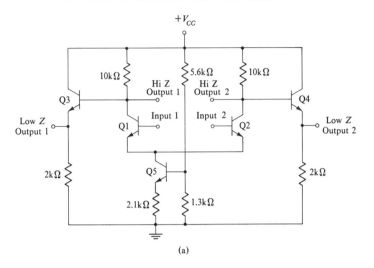

(a)

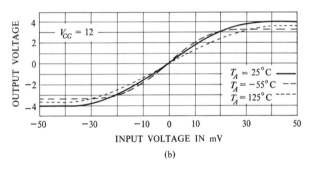

(b)

Fig. 5-12   IC differential amplifier: (a) complete schematic; (b) transfer characteristic at various temperatures.

At the collectors of Q1 and Q2 the voltage is $12 - (0.38 \times 10^{-3})(10 \times 10^3)$ $= 8.2$ V.   The drop from base to emitter for Q3 and Q4 will be about 0.7 V. Thus the direct voltage level at the low-Z outputs is 7.5 V.   The quiescent currents for Q3 and Q4 are $7.5/2k = 3.75$ mA.   The total current drain from V_{CC} is determined to be 10 mA from addition of the four branch currents.

Voltage gain for a differential pair is approximately $g_m R_C$. At 0.38 mA, g_m is 0.015 mho.   For $R_C = 10$ kΩ, the gain is 150.   This will be reduced by emitter-followers Q3 and Q4 to an overall A_v of perhaps 140 to 145.   The gain generally declines with temperature by about 0.16 percent/°C.   Input impedance is of the order of 20 kΩ.   Output impedance is less than 100 ohms.

The voltage transfer characteristic at three ambient temperatures is shown in Fig. 5-12(b).

5-8   THE EMITTER-FOLLOWER

Emitter-follower is a familiar name for the transistor common-collector configuration.   This circuit is used in almost all linear IC's and in many digital
IC's to perform a function commonly called *isolation*.   An emitter-follower
basically exhibits a high input impedance and a low output impedance.
Thus it does not draw much signal current from preceding stages (does not
"load them down"), and, because of its low output impedance, it does not
adversely affect the operation of succeeding stages.

The desirable impedance levels noted are the result of the large amount
of negative feedback inherent in the emitter-follower stage.   As shown in
Fig. 5-13(a), the collector is ac grounded through the V_{CC} supply, but the
emitter path to ground contains the load resistance R_e.   A signal voltage
applied between base and ground is opposed by the entire output voltage
across R_e.   All of the output is present in the input loop; consequently the
input must be slightly larger than V_o.   It follows that the voltage gain of this
stage is less than unity.   Current gain does exist, however, so that the
emitter-follower will provide power amplification.

From Table 3-1, approximate relations between common-collector and
common-emitter hybrid parameters are available.   Thus

$$h_{ic} = h_{ie} \qquad h_{fc} = -(h_{fe} + 1)$$
$$h_{rc} = 1 \qquad h_{oc} = h_{oe}.$$

Therefore $\Delta^{hc} = h_{fe} + 1$.   We may use the general formulas for the hybrid
equivalent to predict the behavior of an emitter-follower.   Using typical
values of $h_{ie} = 2000$, $h_{re} = 10^{-4}$, $h_{fe} = 100$, $h_{oe} = 10^{-4}$, $R_e = R_L = 5000\ \Omega$,
and $R_G = 5000$, this sample stage would provide:

$$A_v = 0.99 \qquad R_i = 338\ \text{k}\Omega$$
$$A_i = -67.5 \qquad R_o = 69\ \Omega.$$

(Common-emitter parameters are changed to common-collector before using
in the equations of Table 3-3).   From these results, we clearly see the special
characteristics of the emitter follower.

In the circuit of Fig. 5-13(a), biasing is accomplished from a single V_{CC}
supply, and base current is set by resistance R_B.   Direct voltage values are
shown.   A direct collector current of 0.5 mA with $h_{FE} = 100$ requires that
$R_B \cong 2.8/(5 \times 10^{-6}) = 560$ kΩ.   From the standpoint of the signal source
V_g, R_B will reduce the total input resistance because R_B appears in parallel
with R_i.   Also R_B will reduce the overall current gain because the fraction
$R_i/(R_i + R_B)$ of the total signal current will reach ground through R_B instead
of R_i.   And further deterioration will be evident in A_v owing to the additional drop across R_G that results because the total input impedance has been

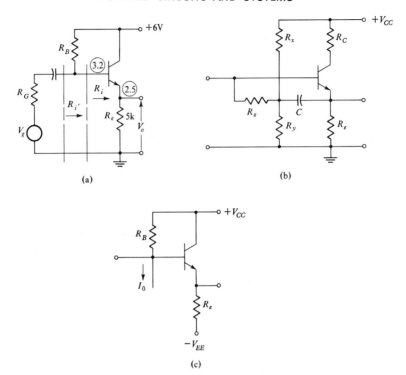

Fig. 5-13   Emitter-follower stages: (a) basic circuit; (b) boot-strapped circuit; (c) circuit operating with two power supplies.

lowered.   On the plus side, the circuit Z_o has been slightly decreased because source resistance is now $R_B \| R_G$, a value smaller than R_G alone.

To minimize the adverse effects of base-biasing resistance, "bootstrapping" may be used, as in the circuit of Fig. 5-13(b).   Element R_C is included for protection of the collector-base junction.   When an extremely large signal is applied to the base, the transistor saturates and R_C limits the forward current of that diode.   One hundred ohms is usually satisfactory.

Elements R_x, R_y, and R_z provide for the base current.   Element C blocks dc but allows the time-varying signal at the emitter to be fed back and made available at the junction of the three resistances.   Because the signal at the emitter is approximately of the same magnitude and phase as the incoming signal, there exists across R_z essentially *zero signal-potential difference*.   As a consequence, the base-biasing elements look like an open circuit, and input impedance is determined simply from Eq. (3-32).   A slight modification to that equation can be made, for R_x and R_y appear in parallel with R_e.   Because of the blocking capacitor required, this circuit is more suitable for discrete rather than integrated applications.

Many IC's are designed to work with both positive and negative power supplies. A typical emitter-follower stage is shown in Fig. 5-13(c). With two power supplies a base current need not be fixed, but of course a path for direct base current is always necessary. In the circuit a constant-current source I_O feeds R_B, causing the base terminal to be at $V_{CC} - I_O R_B$ volts. The base current is *not* being set, but a path through R_B to V_{CC} is available. In this circuit, the static emitter current is set. Suppose, for simplicity, that the base is at ground potential. Then

$$I_E = \frac{-V_{EE} + V_{BE}}{R_E}. \tag{5-42}$$

Since $|V_{BE}| \ll |V_{EE}|$, it is apparent that I_E is dependent only upon V_{EE} and R_E when the base is near ground.

Instability

It is possible under certain conditions for a capacitively loaded emitter-follower to exhibit a negative input admittance, and, if the stage is fed from an inductive source, the entire network could oscillate. In order to examine this premise, consider the circuit shown in Fig. 5-14. A hybrid-π model with r_{ce} and $r_{b'c}$ omitted is being used to represent the transistor; R_L and C_L comprise the load. It is most convenient in the analysis of this circuit to consider the admittance Y_i looking to the right from node b'. This admittance may be determined to be

$$Y_i = \frac{as^2 + bs + 1}{ds + e}, \tag{5-43}$$

with

$$a = T_E T_L \qquad d = (C_{b'e} + C_L) r_{b'e} R_L$$
$$b = T_E + T_L \qquad e = r_{b'e} + R_L (1 + g_m r_{b'e}),$$

where

$$T_E = r_{b'e} C_{b'e}$$
$$T_L = R_L C_L.$$

Let us study the Real part of Y_i. In order that the circuit not oscillate, Re $Y_i > 0$. In Problem 5-22, it is shown that the sign of Re Y_i is the same as Re $Y_{i'}$. Because the expressions are simpler, Y_i will be used in this analysis. For Re Y_i to be positive, it is necessary for

$$T_E T_L < \frac{(T_E + T_L)(T_L r_{b'e} + T_E R_L)}{r_{b'e} + R_L (1 + g_m r_{b'e})}. \tag{5-44}$$

The condition for a positive Re Y_i will be most useful if expressed in terms

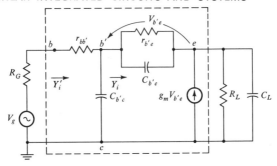

Fig. 5-14   Emitter-follower feeding load composed of $R_L \| C_L$.

of the direct collector current I_C.   Note that $g_m = h_{fe}/r_{b'e}$ and $g_m = \Lambda I_C$. To insure stability, we conclude the following from study of Eq. (5-44):

$$g_m < \frac{h_{fe} C_{b'e}{}^2}{R_L C_L (h_{fe} C_{b'e} - C_L)}. \tag{5-45}$$

This equation predicts a level of I_C for biasing the emitter-follower.   If $C_L = 0$, it follows from Eq. (5-45) that $g_m < \infty$ or I_C can take on any finite value to assure stability.

Equation (5-45) is of value if h_{fe} and $C_{b'e}$ are not dependent upon I_C.   If we consider $C_{b'e} = K\Lambda I_C$ and h_{fe} constant, it follows that

$$I_C > \frac{C_L}{K\Lambda h_{fe}}, \tag{5-46}$$

provided that $R_L C_L \gg K$.   The constant K has a value of perhaps 10^{-12}.

5-9   DARLINGTON PAIR

The *Darlington pair*, or *compound connection*, is a useful subcircuit both in integrated and in discrete form.   The main feature of the connection, as seen in Fig. 5-15, lies in the fact that the emitter lead of Q1 is tied directly

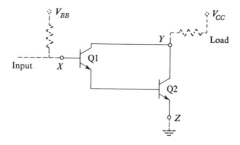

Fig. 5-15   Darlington pair with terminals X, Y, and Z. External connections are shown dashed.

to the base of Q2.   It can also be observed that the collectors are connected together.

When both transistors are connected common-emitter, and therefore share a common load as shown in the figure, both collector currents flow through that load.   Because of the current gain of Q2, that transistor will contribute most heavily to the total load current.

Analysis of the circuit, with each transistor represented by two parameters h_{ie} and h_{fe}, yields the following expression for short-circuit current gain:

$$h_f = h_{fe1} + (1 + h_{fe1})h_{fe2}.$$ (5-47)

This gain may be approximated by $h_{fe1}h_{fe2}$.

Q2 is in the common leg of Q1; it will provide negative feedback that will serve to gain-stabilize the composite circuit as well as raise the input impedance seen at terminal X to a level considerably higher than that of a single common-emitter–connected transistor.

Again using a simplified two-parameter model for each transistor, the input impedance parameter for the pair is

$$h_i = h_{ie1} + (1 + h_{fe1})h_{ie2},$$ (5-48)

as expected.

Voltage amplification in a general form is given by

$$A_v = \frac{A_i R_L}{R_i}.$$ (5-49)

Suppose that we use Eq. (5-47) for A_i and Eq. (5-48) to approximate R_i. For the Darlington pair it follows that

$$A_v \cong \frac{h_{fe2} R_L}{h_{ie2}}.$$ (5-50a)

and, if $h_{ie2} = r_{b'e2}$, this equation can be written

$$A_v \cong g_{m2} R_L.$$ (5-50b)

If R_L is large, the value used in these equations must include the paralleling effect of transistor output resistance.

A dual Darlington array, with four transistors sharing a common collector terminal, is available in IC form.   As is the case in the Fig. 5-15 circuit, load and biasing circuitry must be provided by the user external to the IC.   One application for the array is as a stereo phonograph amplifier.

Modifications to the basic Darlington circuit are considered in the end-of-chapter problems.

5-10   DC LEVEL SHIFTING

It has been mentioned that when direct-coupled stages or building blocks are cascaded there is a tendency for the magnitude of the quiescent collector voltages to increase as one progresses through the circuit toward the load. It therefore becomes necessary in such circuits to lower or *shift* the direct voltage level.   In direct-coupled discrete circuit design, the remedy is to alternate complementary transistor types in cascade.   This approach is sometimes used in integrated circuits, but the bandwidth is limited if lateral *pnp*'s are used.

A level shifting network is shown in Fig. 5-16(a).   The signal to be amplified is impressed upon the base of Q1, which is functioning as an emitter-follower.   Q3 is acting as a constant-current source, with its base fed from a constant-voltage source.   The output transistor Q2 is also operating as an emitter-follower, with output taken from its emitter terminal.

The level shift that takes place is approximately equal to $I_{E1}R_1$ volts. To be more precise, the level at the output is $V_{BE(Q1)} + I_{E1}R_1 + V_{BE(Q2)}$ volts more negative than the level at the In terminal.

A variation in the level shifting network of Fig. 5-16(a) is shown in Fig. 5-16(b).   Resistance R_4 is connected in series with the emitter of the output device.   This feedback connection is further discussed later in this section.

When considering ac operation, the level shifter does have a limiting effect upon bandwidth.   This is primarily caused by the input capacitance of Q2 and the wiring and stray capacitances; in Fig. 5-16(c), these effects are lumped into the element C_{in} shown dashed.   In conjunction with R_1, the resulting time constant $(R_1 C_{in})$ may be the significant bandwidth limiting factor.   In order to broadband the level shifter, the addition of a capacitance has been proposed.   The effect of the feedback capacitance C_{fb} is to *add* at point X a current of magnitude equal to that which goes to C_{in}.   This may be accomplished if $C_{fb} = C_{in}$ because Q2 has a voltage gain of about unity. The result is that the C_{in} current is cancelled and frequency response extended.

Another variation to the basic level shifter circuit places a capacitor across R_1.   This has the effect of extending the bandwidth of the circuit by assisting in the cancellation of falloff caused by C_{in}.

Positive Feedback

We return to the circuit of Fig. 5-16(b) for further analysis.   The signal path is simply two cascaded emitter-followers, Q1 and Q2.   The voltage to be fed back appears across R_4, and is amplified by Q3 acting as a common-base stage.   The collector current of Q3 then adds in phase with the signal from Q1.

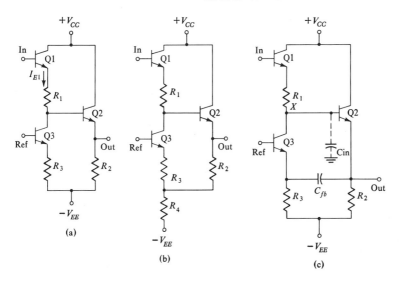

Fig. 5-16   DC level shifter: (a) basic circuit; (b) modification; (c) capacitance added to improve frequency response.

Voltage gain of this subsystem between Out and In terminals is

$$A_v = \frac{K_1}{K_2 - (r_{e1} + R_1)[R_4 - (r_{e3} + R_3 + R_4)/h_{fe2}]}, \tag{5-51}$$

where $K_1 = (r_{e3} + R_3 + R_4)(R_2 + R_4) - R_4{}^2$

$K_2 = K_1 + r_{e2}(r_{e3} + R_3 + R_4)$.

Symbols r_{e1} and so on represent the reciprocal of transconductance $(1/g_m)$ for the respective transistors.   The equation uses a simple model composed of $r_{b'e}$ and h_{fe} for each transistor.

The negative term in Eq. (5-51), if large enough, would result in a net zero value for the denominator, representing oscillation.

A sample stage, with $R_1 = 3$ kΩ, $R_2 = 1.47$ kΩ, $R_3 = 140$ Ω, $R_4 = 60$ Ω, $r_{e1} = r_{e3} = 29$ Ω, $r_{e2} = 8$ Ω and $h_{fe} = 100$ provides a voltage gain of 1.9. Element values could be selected to provide gains of 15 or greater.   Output impedance tends to be increased with positive feedback, bandwidth is reduced, and input impedence becomes negative.

5-11   OUTPUT STAGES

The final stage in a linear IC is referred to as the output stage.   This stage feeds the external load which may be another transistor, a passive network, or a form of energy conversion device such as a loudspeaker.

Characteristics of output stages vary because all linear IC's are not designed to meet the same specifications.   In general, we expect the output stage to be able to handle the *highest power, current,* or *voltage levels* in the IC.   As mentioned in Chapter 1, a *low output impedance is desired.*   The designer of the output stage is faced with the question of allowable *self-dissipation* because of the high signal level.   He must decide upon the merits of Class-A versus Class-B operation.   Does the application require *differential output,* or is a single-ended output satisfactory?   Often it is required that the output terminal be at *zero direct voltage* in the absence of signal.

The simplest form of output stage is a single emitter-follower shown in Fig. 5-17(a).   This circuit has a fairly high self-dissipation because one-half of the total available supply potential is lost across R_e.   If a differential output is needed, two emitter-followers could be used with their bases connected to the collectors of a diff amp driver.

The circuit of Fig. 5-17(b) has emitter-follower Q2 as its output stage. In addition, this circuit employs a small amount of positive feedback.   Transistor Q1 is an emitter-follower driver, and Q3 functions as a constant-direct–current source.   The signal emitter current of Q2 flowing through R_4 causes a voltage drop that is transferred to the base of Q2 in a direction such as to reinforce the signal there.   This use of positive feedback has the advantage of increasing the gain of the Q1-Q2 pair from its normal value of less than unity to perhaps 1.5 or more.

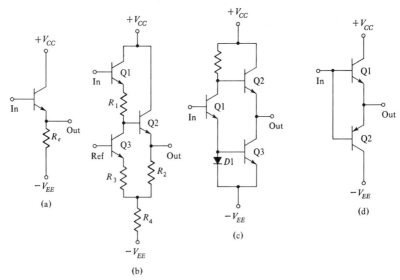

Fig. 5-17   Output stages: (a) simple emitter follower; (b) emitter follower with positive feedback; (c) push-pull Class-A output pair; (c) push-pull Class-B pair.

To achieve highest power capability, one seeks to do away with the resistance inherent in the emitter-follower connection, for that resistance results in a waste of signal as well as quiescent power.   In the circuit of Fig. 5-17(c), Q2 and Q3 are operating Class A, and Q1 serves as a *phase inverter*.   The phase inverter provides a positive-going signal to Q2 at the same time it supplies a negative-going signal to Q3.   Thus a load connected between the junction of the output transistors and ground is fed simultaneously from Q2 acting as an emitter-follower and Q3 acting as a common-emitter stage.   Diode $D1$ is for biasing purposes.

The *npn-pnp* pair shown in Fig. 5-17(d) operates Class B.   A positive-going input signal turns Q1 ON, but Q2 remains OFF for that half-cycle. On the alternate half-cycle, Q2 is ON and Q1 OFF.   Both transistors have their collectors grounded.   The crossover distortion can be significantly reduced by feedback.

Problems

5-1   For a balanced-bias circuit of the Fig. 5-1(b) type, consider the transistors to be identical, $V_{CC} = 6$ V, $V_{BE} = 0.7$ V, $R_1 = R_3 = R_4 = 10$ kΩ and $R_2 = 5$ kΩ. Determine the operating-point coordinates (I_C and V_{CE}) of Q1 and Q2 if $I_B = I_C/50$.

5-2   For the balanced-bias network of the preceding problem, consider that uneven temperature distribution causes $V_{BE(Q2)}$ to be 0.75 V, with all other conditions as previously given.   This effect will result in $I_{B2} \neq I_{B1}$.   Find the operating point coordinates for each transistor.

5-3   Derive Eq. (5-7) for R_2 of the low-current source.

5-4   Design a 10 μA current source of the Fig. 5-2 type.   Consider V_{CC} to be $+9$ V, and that R_1 and R_2 should be limited to 10 kΩ or less.

5-5   Use Eq. (1-19) to show that $I_{E2} R_2$ in the low-current source circuit varies directly with temperature.

5-6   Verify Eqs. (5-9) and (5-10).

5-7   A diode-stabilized constant-current source is to be designed with $R_2 = 0$ and $N = 1$. I_O is to be 2 mA.   We wish to determine values for R_1 and R_3.
(a) Can we use Eq. (5-9) to find these resistances?
(b) If $N = 2$, determine R_1 and R_3 for conditions given.

5-8   Verify Eq. (5-16), starting with Eqs. (5-13).

5-9   For a diff amp with $R_C = 5$ kΩ, $R_x = 10$ kΩ, $R_s = 2$ kΩ, $h_{fe} = 100$, $h_{oe} = 10^{-4}$ mho, $g_m = 0.02$ mho, determine:
(a) A_v(diff)
(b) A_v(common)
(c) CMRR

5-10   Derive Eq. (5-30), by starting with Eqs. (5-27).

5-11   Show that for a differential amplifier feeding a resistance R_{CC} connected between collectors the effective differential load on the stage is $R_{CC}/2$. Consider that R_C elements are very large resistances, and that the current generators approach ideal.

5-12   Explain why Eq. (5-33) is true.   Use mathematics if necessary.

5-13   Show that the use of emitter degeneration in a transistor differential amplifier reduces the sharpness of the saturation "knee" of the transfer characteristic.

5-14   Derive Eq. (5-39).

5-15   Determine A_v, A_i, R_i, and R_o of a simple emitter-follower of the Fig. 5-13(a) type with the following parameters: $h_{ie} = 5000$, $h_{fe} = 200$, $h_{re} = 10^{-4}$, $h_{oe} = 200 \times 10^{-6}$.   Consider $R_e = 10^3 \ \Omega$ and $R_G = 10^3 \ \Omega$.

5-16   In the bootstrapped emitter-follower of Fig. 5-13(b), R_z does not appear to be infinite because V_o is not exactly equal to V_i.   Show that the effective value of that resistance is

$$R_z(\text{effective}) = \frac{R_z}{1 - A_v}$$

5-17   The formula given in Problem 5-16 applies only when the reactance of C is negligible.   At low frequencies that formula is in error.   Modify the equation for $R_z(\text{effective})$ so that it includes C, f, R_x, and R_y.   You can assume that the transistor draws no base current and that its output impedance is zero.

5-18   Design a bootstrapped emitter-follower subject to the following: $V_{CC} = 10$ V, $h_{FE} = 200$, $R_e = 10^3 \ \Omega$, $I_{C(Q)} = 4$ mA.

5-19   In the circuit of Fig. 5-13(c), the supplies are ± 6 V, $R_e = 10$ kΩ, and $V_{B(Q)} = +4$ V.   Determine values for I_o, R_B, I_E, and $V_{E(Q)}$.

5-20   Derive Eq. (5-43), for emitter-follower input admittance.

5-21   Derive Eq. (5-44), for emitter-follower stability.

5-22   Show that the algebraic sign of Re Y_i is the same as Re $Y_{i'}$ in Fig. 5-14.

5-23   An emitter-follower stage has $C_{b'e} = 10$ pF, $C_L = 20$ pF, $h_{fe} = 80$, $R_L = 10^4 \ \Omega$.   Use both criteria, Eqs. (5-45) and (5-46), to comment upon the stability of the stage with respect to $I_{C(Q)}$.

5-24   The emitter-follower shown in the figure is used in certain linear IC's. To analyze this circuit, consider that the transistor may be represented by a modified hybrid-π with $r_{b'e}$, $r_{b'c}$, and $g_m V_{b'e}$ elements, and R parallels $r_{b'e}$.   Show that the voltage gain V_o/V_g for this circuit will not exceed unity regardless of the value of R.

5-25   Assume that the transistors in the Darlington pair shown in Fig. 5-15 can be represented by merely h_{fe1}, h_{ie1} and h_{fe2}, h_{ie2}.   Draw an equivalent circuit for the pair and determine the h_f and h_i parameters of the pair in terms of those parameters of the individual transistors.

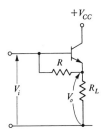

Problem 5-24

5-26   A Darlington pair can be used as a compound emitter-follower as shown in the figure.   The parameters of the stages usually do not have identical values.

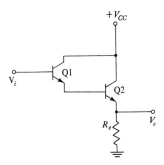

Problem 5-26

For this connection, show

(a) $A_i \cong \dfrac{(1 + h_{fe1})(1 + h_{fe2})}{1 + h_{oe1}h_{fe2} R_e}$

(b) $R_i \cong A_i R_e$

(c) $A_v \cong 1 - h_{ie1}/(1 + h_{fe2})R_e$.

For each part, list the assumptions being made to arrive at the equations given.

5-27   A bootstrapped Darlington pair is shown in the figure.   Element R_C has been added to remove the collector of Q1 from ac ground, and C provides a path for ac feedback from V_o to the Q1 collector.   Without these elements, the input resistance of the pair is limited to about $1/h_{ob}$.

(a) Using common-emitter h parameters, draw a complete equivalent circuit for the pair;

(b) From (a), show that

$$R_i \cong h_{fe1}h_{fe2} R_e.$$

What assumptions must be made to arrive at this equation?

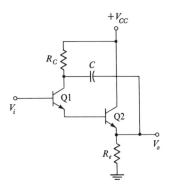

Problem 5-27

5-28   A modified Darlington connection is shown in figure. Transistors Q1 and Q2 form the Darlington pair. Since R and " diode " Q3 are connected from base-to-emitter of Q2 it follows that

$$I_3 R = V_{BE(Q2)} - V_{BE(Q3)}.$$

The base-to-emitter voltages are related exponentially to the respective collector currents. Show that

$$I_3 R \cong \frac{1}{\Lambda} \ln \frac{I_2}{I_3}.$$

In the typical circuit, current I_2 would be determined by power supplies and other external elements.   If $R = 2$ kΩ and $I_2 = 2$ mA, use the above equation to find I_3.

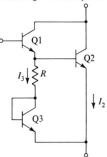

Problem 5-28

5-29   Consider the practical IC diff amp shown in the figure, with terminal 8 grounded and signal applied between 3 and 9.

(a) Show that by connecting terminals 4 with 6 and 5 with 7, the effective resistance between emitter of Q3 and V_{EE} is 3.37 kΩ.

(b) Determine the Q3 emitter current.   Assume $V_{BE} = 0.7$ V and $I_B = I_E/100$. Do not neglect current through $D1$.

(c) Calculate the differential voltage gain of the stage.
(d) Calculate the input resistance between terminals 3 and 9. Consider that $r_{bb'} = 0$.

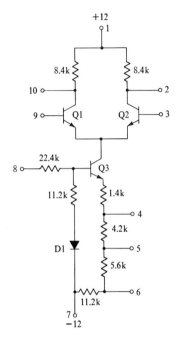

Problem 5-29

5-30 Explain why R_4 in Fig. 5-17(b) is responsible for positive feedback.

6
THE OPERATIONAL AMPLIFIER

The preceding chapter discussed many of the foundations or building blocks that can be used in the design of high performance linear IC's. These building blocks are so valuable that some of them, such as the diff amp and the Darlington pair, have been made available in IC form for use by system designers.

Let us now assemble the building blocks to form the linear IC. Because of its flexibility, the *operational amplifier*, or *op amp*, has been the goal of IC manufacturers since the beginnings of linear integrated circuits. We shall discuss the overall characteristics and some applications of this extremely useful circuit in this chapter; Chapter 7 is devoted to circuit analysis of practical op amps.

6-1 OP AMP DESCRIPTION

An operational amplifier is generally considered a high gain, high input impedance amplifier capable of amplifying dc as well as ac signals. To be widely acceptable, it must also have good noise properties, low dc offset and drift, wide bandwidth, low output impedance, and a large dynamic range.

The operational amplifier is a "workhorse." Originally it was conceived to perform the mathematical operations of summing, integration, differentiation, and multiplication by a constant as required in analog computers. It soon became apparent that such amplifiers could satisfy other needs, as in wave shapers, equalizers, filters, signal and control amplifiers, nonlinear function generators, voltage comparators, and so on.

When IC technology had advanced to the point where analog circuits could be fabricated, the op amp was considered because its stages are direct coupled, and therefore it does not require coupling capacitors.   The fact that all elements of a monolithic IC are fabricated under identical conditions on and within the same wafer meant that the previously objectionable aspects of op amps made by assembling discrete elements, particularly temperature sensitivity, would be minimized.

The op amp relies upon a differential amplifier first stage.   This may be followed by other diff amp stages—sometimes there are single-ended stages. A liberal sprinkling of emitter-followers is common for isolation.   A level translator is usually included.   The output stage is often an emitter-follower. Fig. 6-1 depicts the functions employed in a single dc IC amplifier for op amp service.   Feedback paths are not included in the figure.

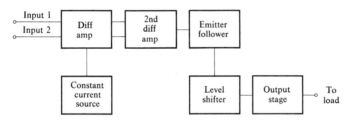

Fig. 6-1   Block diagram of a typical operational amplifier.

The typical operational amplifier utilizes external negative feedback to improve its functional performance.   As has been noted, if the forward gain is sufficiently large, a feedback amplifier system is primarily dependent only upon the properties of the feedback elements.   Thus the system behavior may be more or less independent of variations in the active devices in the forward path, and if the feedback elements are stable, dependable, passive devices, the entire amplifier will be just as stable and dependable.

It has also been shown that other benefits may result from the use of negative feedback: increased input impedance; reduced output impedance; improved gain stability; extended bandwidth.   The price paid for these benefits is twofold: gain reduction always results, and a tendency toward self-oscillation can exist.

The functional blocks as shown in Fig. 6-1 are, in IC form, referred to as op amp.   When external feedback elements are added, the composite is also likely to be referred to as an op amp even though the characteristics of the feedback amplifier are vastly different.   The reader is cautioned to note the double use of the term op amp.   Where necessary for clarity, the Fig. 6-1 assemblage will be referred to as the *open-loop* circuit; with feedback, it may be called a *closed-loop* amplifier.

6-2   MATHEMATICAL OPERATIONS

As previously mentioned, the original purpose of the operational amplifier was to perform mathematical operations in analog computing applications. Here we shall discuss how the op amp contributes to the electronic computation processes of summation, integration, differentiation, and how it can be used to multiply variables.

Summation

A dc amplifier with no load and high input resistance is depicted in Fig. 6-2 with two input voltages, v_1 and v_2, available.   More than two inputs are of

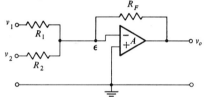

Fig. 6-2   Summing amplifier.

course possible.   The amplifier open loop voltage gain is given by A.   It is desired that the output voltage v_o be proportional to the sum of v_1 and v_2. Thus, the relation required is

$$v_o = K(v_1 + v_2), \tag{6-1}$$

where K is a proportionality or gain constant.   A summation of currents at the significant node yields

$$\frac{v_1 - \varepsilon}{R_1} + \frac{v_2 - \varepsilon}{R_2} = \frac{\varepsilon - v_o}{R_F}. \tag{6-2}$$

Eq. (6-2) assumes that there is no current flow into A from ε.   For the amplifier alone, including the assumed 180° phase inversion,

$$A = -v_o/\varepsilon. \tag{6-3}$$

Eq. (6-3) is used to eliminate ε from Eq. (6-2).   Solving for v_o, with A very large, we obtain

$$v_o \cong -R_F\left[\frac{v_1}{R_1} + \frac{v_2}{R_2}\right]. \tag{6-4}$$

If all resistances are of equal magnitude,

$$v_o = -(v_1 + v_2). \qquad (6\text{-}5)$$

Addition of the inputs has been accomplished.   It can be easily proved that if finite amplifier input resistance had been considered, input resistance would have acted as if it were parallel with R_1 and R_2.   However, a large A would result in an expression identical with Eq. (6-4).

By feeding the input into the $(-)$ terminal as shown in the figures of this section, we are using the *inverting* input terminal of the op amp.   The terminal with the $(+)$ sign is referred to as the *noninverting* terminal.   These symbols designate the sign of an input voltage to cause a positive-going output voltage.   Further discussion is given in Section 6-5.

For a single input to the circuit of Fig. 6-2, the expression for output voltage including forward gain is

$$v_o = -v_1 A R_F / [R_F + (1 + A)R_1]. \qquad (6\text{-}6)$$

Equation (6-6) represents a standard feedback circuit that, for large value of A, exhibits a gain of

$$v_o/v_1 \cong -R_F/R_1. \qquad (6\text{-}7)$$

Since this ratio is the overall gain of the network, *gain control* is achieved by selection of the resistances R_F and R_1.   If R_F is made larger than R_1, the amplifier performs the function of *multiplication by a constant* greater than unity.   When R_1 is the larger, the constant is less than unity.   While either R_F or R_1 could be adjustable resistors for performing this multiplication in analog computation systems, it is usually more acceptable to precede the amplifier by a potentiometer if gain control is required.

Consider the amplifier with single input channel shown in Fig. 6-3. Impedances Z_F and Z_I will determine the operation of the overall circuit. Because this circuit is very similar to the case just discussed, the output voltage, for large A, is given by a modification of Eq. (6-7) applicable to the steady-state:

$$V_o/V_i = -(Z_F/Z_I). \qquad (6\text{-}8)$$

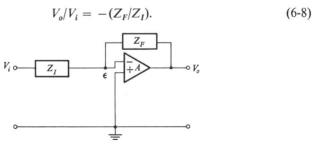

Fig. 6-3   Computing amplifier.

Here we have the most important equation in op amp analysis. It indicates that the voltage transfer ratio is dependent only upon the feedback elements Z_F and Z_I. With correct choice of those elements, the circuit may be useful in a great many applications. Some will be discussed here and in Section 6-3.

Integration and Differentiation

We use Eq. (6-8) to show that the circuit is capable of mathematical *integration*. Let Z_I be resistive and equal R, and let Z_F be capacitive. In Laplace notation, $Z_F = 1/sC$, with $s = j\omega$ for the sinusoidal steady state. Equation (6-8) can be written

$$V_o(s) = -\frac{V_i(s)}{RCs}.$$ (6-9)

In the time domain, the corresponding expression is

$$v_o = -\frac{1}{RC}\int v_i\, dt.$$ (6-10)

Therefore the circuit will integrate the input voltage.

To *differentiate* a waveform, the connections of capacitance and resistance in Fig. 6-3 are reversed, so that $Z_I = 1/sC$ and $Z_F = R$. Then we obtain, from Eq. (6-8),

$$V_o(s) = -V_i(s)RCs.$$ (6-11)

In the time domain, this corresponds to

$$v_o = -RC\frac{dv_i}{dt}.$$ (6-12)

The input voltage is differentiated if the amplifier gain is large. Differentiation is to be avoided if the v_i waveform contains electrical noise, for the derivative of noise pulses and spikes usually contains pulses and spikes of even greater amplitude.

To include an *initial condition* in the integration process, a circuit such as the one shown in Fig. 6-4 can be used. The capacitor C is initially charged to the voltage V_C with the switch in the position shown. (Because ε is very small, the input terminal of the dc amplifier is ground, for practical purposes.) When the switch is thrown, v_I is applied to the input R_1 line, V_C is grounded through 10 kΩ, and the result is

$$v_o = -\frac{1}{R_1 C}\int_0^t v_I\, dt + V_C.$$ (6-13)

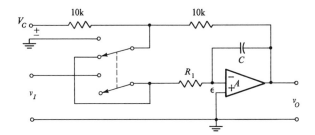

Fig. 6-4   Integrator with provision for initial condition.

Should v_I be a constant direct voltage V_I, the output will rise with time according to

$$v_O = -\frac{V_I t}{R_1 C} + V_C \qquad (6\text{-}14)$$

until amplifier saturation occurs.

Logarithmic Operation

If feedback element Z_F in Fig. 6-3 is a transistor, it is possible to construct a system in which $V_O = K \log V_I$. Such behavior can be visualized with the help of Fig. 6-5(a) and the following discussion. The collector current of the feedback transistor is essentially equal to the input current:

$$I_C \cong \frac{V_I}{R_I}. \qquad (6\text{-}15)$$

For most transistors I_C and V_{BE} are related exponentially, as given by Eq. (1-21):

$$I_C = K_1 e^{\Lambda V_{BE}} \qquad (6\text{-}16)$$

It may be necessary in this application to select a transistor behaving as much like Eq. (6-16) as possible. The constant K_1 represents the reverse current of the base-emitter junction and the transistor gain.

By equating (6-15) with (6-16), and substituting V_O for V_{BE}, we obtain

$$V_O = \frac{1}{\Lambda} \ln \frac{V_I}{R_I K_1}. \qquad (6\text{-}17)$$

Empirical results have been reported for

$$V_O = 0.062 \log_{10} V_I + 0.450,$$

and adherence to this equation has been extremely good over about six decades of V_I.[9]

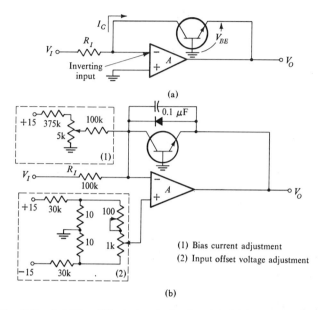

(a)

(1) Bias current adjustment
(2) Input offset voltage adjustment

(b)

Fig. 6-5 (a) Simplified circuit for log function; (b) details.

In a practical implementation of a logarithmic amplifier, it is necessary to compensate for the input offset voltage and the input bias current of the op amp in order to achieve a wide range of input values over which the desired relation holds. The compensating networks are shown in Fig. 6-5(b). It may also be noted that a 0.1 μF capacitor and a protection diode are connected across the transistor. Since a common-base stage can provide considerable ac voltage gain that is not wanted in this application, the capacitor effectively kills that gain.

Multiplication

In analog computation, the multiplication of two voltages representing variables is required. Several means exist for accomplishing this operation. The method discussed here uses operational amplifiers. Another method is introduced in Problem 8-19.

We wish to form $Z = XY$. This can be done with logs. The functional equation used is

$$Z = \log^{-1} (\log X + \log Y). \tag{6-18}$$

This implies three steps: (1) take log of X and of Y; (2) sum; (3) take antilog of the sum.

The system for multiplication of voltages representing X and Y is shown in Fig. 6-6. Constants associated with the log operations can be removed in the summing operation by adding a direct voltage as shown as $-B_3$. To perform the $\log^{-1}$ function a suitably biased common-base transistor is employed in an op amp circuit.

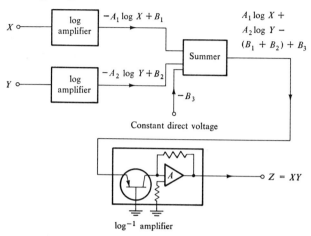

Fig. 6-6   System for multiplication.

6-3   TRANSFER FUNCTION SYNTHESIS

The relation given for the circuit of Fig. 6-3, namely,

$$\frac{V_o}{V_i} = -\frac{Z_F}{Z_I}, \tag{6-8}$$

provides the basis for the synthesis of many frequency-dependent transfer functions by selecting impedances Z_F and Z_I so that their quotient equals the required function. As we have seen, integration and differentiation are functions easily achieved.

A simple low-pass filter can be developed if Z_F is the parallel combination of R_F with C_F, and Z_I is a resistance R_I. Then the voltage transfer function, according to Eq. (6-8), is

$$\frac{V_o}{V_i} = -\frac{R_F/R_I}{R_F C_F s + 1}. \tag{6-19}$$

This circuit, shown in Fig. 6-7, has a dc and low-frequency gain of $-R_F/R_I$. The gain may be greater than unity. At a frequency where $R_F C_F \omega = 1$, the gain has declined to 0.707 of its low-frequency value, and falls off with frequency at the rate of -6 dB per octave. (An octave is a doubling of frequency.)

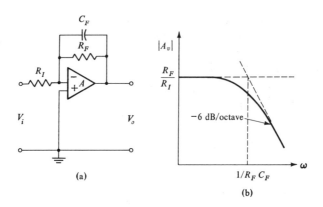

Fig. 6-7   Simple low-pass filter: (a) circuit; (b) frequency response.

Impedances Z_F and Z_I must be two-terminal networks, and, in order to be physically realizable, the corresponding impedance functions represented as polynomials in s cannot differ in the order of numerator and denominator by more than one power in s. This limitation seriously restricts the design of circuits of the type represented by Eq. (6-8).

In Fig. 6-8, we add networks I and F to the dc amplifier designated by A. This connection of two-port networks allows the generation of voltage

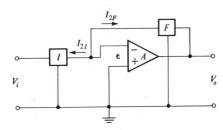

Fig. 6-8   Transfer function synthesis with two-port passive networks.

transfer functions not possible using the Fig. 6-3 scheme. The output terminals of networks I and F feed the *virtual ground* ε. We consider that voltage to be zero. Consequently, in the y-parameter representation for the networks, the non-zero terms remaining are

$$I_{2I} = y_{21I} V_i$$

and

$$I_{2F} = y_{21F} V_o.$$

It can be noted that $I_{2I} = -I_{2F}$ because A draws little or no input current.

Therefore, combining these equations yields the voltage-transfer function

$$\frac{V_o}{V_i} = -\frac{y_{21I}}{y_{21F}}.\qquad (6\text{-}20)$$

It remains to select passive networks having transfer admittance functions that satisfy the requirements of a particular design problem.

One important application of the system of Fig. 6-8 is for filtering. Since, at low frequencies, inductors are to be avoided because of their size, weight, and cost, *RC* networks are sought to perform their function. The *bridged-T* and the *twin-T* networks shown in Fig. 6-9 have proved very successful. It is found, for the bridged-*T*,

$$y_{21} = -\frac{T_1 T_2 s^2 + C_1(R_1 + R_2)s + 1}{R_1 T_2 s + (R_1 + R_2)},\qquad (6\text{-}21)$$

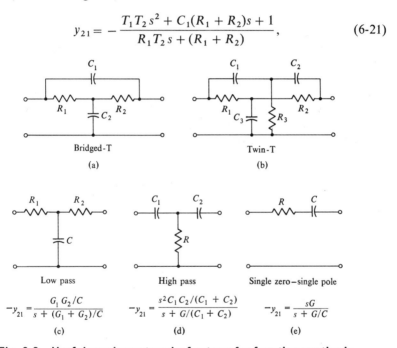

Fig. 6-9   Useful passive networks for transfer function synthesis.

with $T_1 = R_1 C_1$ and $T_2 = R_2 C_2$. For the twin-*T* of Fig. 6-9(a),

$$y_{21} = -\frac{T_1 T_2 T_3 s^3 + C_1 C_2 R_3(R_1 + R_2)s^2 + R_3(C_1 + C_2)s + 1}{(C_3 R_1 R_2 s + R_1 + R_2)(R_3(C_1 + C_2)s + 1)},\qquad (6\text{-}22)$$

with $T_1 = R_1 C_1$, $T_2 = R_2 C_2$, $T_3 = R_3 C_3$. Three other RC networks are also shown in the figure. They are useful to perform the function of impedance *I*.

To realize a *low-pass* filter having a Butterworth type of response of the normalized form

$$\frac{V_o}{V_i} = \frac{-K}{s^2 + \alpha s + 1}, \tag{6-23}$$

we may use a bridged-T for network F and the low-pass circuit of Fig. 6-9(c) for network I. It is necessary that the poles of networks F and I cancel. α is chosen to be $\sqrt{2}$ for Butterworth response.

A *high-pass* filter with transfer function

$$\frac{V_o}{V_i} = \frac{-Ks^2}{s^2 + \alpha s + 1} \tag{6-24}$$

can be achieved with the high-pass circuit of Fig. 6-9(d) as network I, and the bridged-T used again in the feedback path.

As an example of a *band-pass* filter, consider the requirement to provide a resonant frequency of 100 Hz, maximum gain of 10 and effective Q of 10. (Q is resonant frequency divided by bandwidth.)   The twin-T and the single-zero, single-pole networks will be used.   The final design with element values given is shown in Fig. 6-10.   To arrive at the band-pass form

$$\frac{V_o}{V_i} = \frac{-H\omega_o s}{s^2 + \alpha\omega_o s + \omega_o^2}, \tag{6-25}$$

design equations are available in the literature.[10]

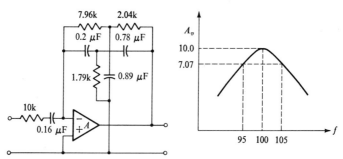

Fig. 6-10   Bandpass filter with 100Hz center frequency and band width of 10Hz.

6-4   INDUCTANCE SIMULATION

At low frequencies, inductors and transformers are bulky and costly; they have the additional disadvantage of not being integratable.   The synthesis of voltage transfer functions discussed in the preceding section allows us to realize inductorless filter networks. Another approach to filter design is to simulate inductance using operational amplifiers and RC networks.[11]

Consider the network shown in Fig. 6-11. The op amp is connected as a

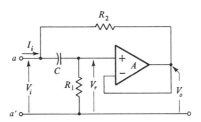

Fig. 6-11   Inductance is simulated at terminals a-a'.

voltage-follower (See Problem 6-6). The entire output voltage V_o is fed back to the inverting input terminal marked with $(-)$. This is the terminal that has been used in our study up to this point. The voltage V_r is fed to the *noninverting* input. This voltage-follower has the following characteristics: $A_{vf} \equiv V_o/V_r \cong 1$, Z_i is high, Z_o is low.

We determine the input impedance at terminals a-a':

$$I_i = (V_i - V_r)j\omega C + \frac{V_i - V_o}{R_2}. \qquad (6\text{-}26)$$

By voltage division,

$$V_r = V_i\left(\frac{R_1}{R_1 + 1/j\omega C}\right), \qquad (6\text{-}27)$$

and it is clear that $V_o = V_r$. When these equations are solved for Z_i, we obtain

$$Z_i \equiv \frac{V_i}{I_i} = \frac{R_2(R_1 R_2 C^2 \omega^2 + 1)}{1 + \omega^2 R_2{}^2 C^2} + \frac{j\omega R_2 C(R_1 - R_2)}{1 + \omega^2 R_2{}^2 C^2}. \qquad (6\text{-}28)$$

The first term is the effective series resistance, and the second term is inductive if $R_1 > R_2$. The quality factor Q of the R-L circuit is

$$Q \equiv \frac{\omega L}{R} = \frac{\omega C(R_1 - R_2)}{R_1 R_2 C^2 \omega^2 + 1}. \qquad (6\text{-}29)$$

For satisfactory L and Q values, we select $R_1 \gg R_2$. The maximum Q equals $(\frac{1}{2})(R_1/R_2)^{1/2}$, and occurs at $\omega = 1/[C(R_1 R_2)^{1/2}]$. For $C = 0.1\ \mu F$, $R_1 = 100\ k\Omega$, $R_2 = 100\ \Omega$, and $L = 1\ H$, $Q_{max} = 15.8$ at 505 Hz.

By paralleling terminals a-a' with a capacitance, the result is a tuned circuit.

6-5 SPECIFICATIONS

The performance capabilties of a general purpose class of practical operational amplifiers are summarized in Table 6-1.   It is important to note that

TABLE 6-1   TYPICAL SPECIFICATIONS FOR IC OP AMPS

Open-loop dc voltage gain	10^3–10^5
Gain-bandwidth product	1–100 MHz
Slewing-rate limit	0.1–30 V/μs
Initial input voltage offset	0.5–5 mV
Initial input current offset	5–500 mA
Drift with temperature	3–50 μV/°C
Open-loop differential input impedance*	10k–50 kΩ
Maximum differential input voltage	1–8 V
Maximum common-mode input voltage	1–11 V
Output voltage	3–12 V
Output load current	1–10 mA
Common-mode rejection ratio	60–120 dB
Open-loop output impedance*	10–5 kΩ

* Highly dependent upon input and output stages.With FET, $Z_i = 10^{10}$–10^{12} Ω

significant strides are being made in the design and fabrication of IC's, and therefore the numbers listed in the table are useful only as a reference level with which to compare new developments.

Certain of the entries in Table 6-1 deserve comment.

Op amps are seldom used open-loop, but, as noted in the feedback discussion, a large open-loop gain is necessary to assure that the performance of a feedback amplifier will be independent of variations in the forward path. *Gain-bandwidth product* is the result of multiplying the low-frequency gain by the bandwidth of the circuit.   It is a measure of the gross capability of an amplifier.

Slewing rate, or *slew rate*, is the time rate of change of a closed-loop amplifier output voltage under large-signal conditions.   For this definition, a large signal is the maximum ac input voltage for which the amplifier performance remains linear.   Stabilization networks affect the slew rate and are included in its measurement.

Slew rate is simply a measure of the ability of an op amp to follow a transient. This parameter may be measured using a fast-rising pulse generator at the amplifier input. If the amplifier output voltage changes ΔV volts in Δt seconds, the slew rate is equal to $\Delta V/\Delta t$.

Transient response of an op amp is depicted in Fig. 6-12. At time t_o, an input pulse is applied. The slew rate determines the rise time. Another

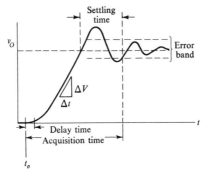

Fig. 6-12 Pictorial definitions of op amp transient response terms.

condition, *settling time*, is of importance. Settling time is the time required to reach an error band such as $\pm 5\%$ of the final steady-state value of v_O. The *acquisition time*, as shown, is the sum of delay and slewing and settling times.

Input voltage offset and input current offset were discussed in Chapter 5 with relation to the diff amp. *Drift* refers to the variation in output caused by variations in the amplifier due to temperature or aging. It is a change in the output, with no input variation causing it. Drift will be further discussed in Section 6-9.

Common-mode rejection ratio (CMRR) was discussed in Sections 1-6 and 5-3; it is the ratio of differential-mode amplifier gain to common-mode amplifier gain. The need for a high value of CMRR is determined primarily by the application of the amplifier.

As an example, consider an op amp with 60 dB voltage gain, and a low CMRR, say 50 dB. The amplifier is being used as a comparator, with a one-volt signal on one input line being compared against a one-volt reference on the other line. The differential mode input is therefore zero volts. However, since there is one volt on each input line, this constitutes a common-mode voltage level of one volt. Perfect rejection is not achieved; rather, the net common-mode voltage is -50 dB from one volt, or 3.2 mV. When amplified 1000 times, the undesired output is 3.2 volts.

As a rule of thumb, the CMRR must be at least 20 dB greater than the differential gain for successful operation.

As another example of an extreme, consider that a signal of 10 mV is to be amplified with 0.1 percent accuracy, where inaccuracy is defined as the result of noninfinite CMRR.   Suppose that the interference or common-mode voltage present is 10 volts.   The rejection factor must be 120 dB to meet the accuracy specification.

6-6   INVERTING FEEDBACK

Operational amplifiers can be single-input or differential-input.   The single-input system allows only the *inverting* type of feedback, but when differential inputs are employed, *noninverting* feedback may also be used, thus extending the versatility of the basic circuit.   *Inverting feedback is the more common, and represents the case where the instantaneous output voltage is opposite in polarity to the instantaneous input signal.*

Consider the system shown in Fig. 6-13, an operational amplifier with load and source terminations and external feedback elements.   Element

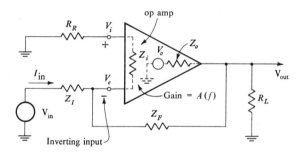

Fig. 6-13   Operational amplifier for discussion of inverting feedback.

Z_I includes both source impedance and elements purposely added for operational purposes.   Symbols Z_i and Z_o stand for the intrinsic or internal input and output impedances of the basic operational amplifier.   The so-called *open-loop* gain, the differential voltage gain of the basic amplifier, is symbolized by $A(f)$ and is a function of frequency.

Resistance R_R should be equal to the net resistance to ground represented by $Z_I \| (Z_F + Z_o \| R_L)$ in order that the dc offset voltage be minimized.   This is considered in Section 7-4.   The effects of R_R upon amplifier gain and impedance levels will be clearly obtainable from the equations presented in this section.

The circuit load R_L will be assumed to be infinitely large in the derivations that follow for closed-loop gain, A_{vf}, and Z_{if}.   This assumption is valid, for in all practical cases the intrinsic output impedance Z_o will be low-valued, and when feedback is added, the resulting output impedance Z_{of} generally will be very small. (To include the effect of finite R_L upon voltage-transfer function, we define A_{vf*} as the finite-load-resistance gain. Then

$$A_{vf*} = A_{vf}\left(\frac{R_L}{R_L + Z_{of}}\right)$$

Symbols A_{vf} and Z_{of} include feedback.)

Plus (+) *and minus* (−) *symbols at input terminals refer to noninverting and inverting inputs, respectively.* The V_o generator represents the open-circuit output voltage of the amplifier, and Z_o is the intrinsic amplifier output impedance. Differential amplification is provided; thus we may relate output to input:

$$V_o = -A(f)(V_e - V_i). \tag{6-30}$$

$A(f)$ serves to indicate that the amplifier gain is frequency-dependent. The negative sign modifying the right-hand side of Eq. (6-30) represents the phase reversal between V_o and V_e that is characteristic of the inverting input. A positive-going signal at V_e, with $V_i = 0$, results in V_o negative. The equation shows this, provided we consider $A(f)$ as a positive quantity in this discussion.

Let us determine the closed-loop performance. First, note a relation between two voltages:

$$V_i = V_e R_R/(Z_i + R_R). \tag{6-31}$$

By writing current equations at the V_e and V_{out} nodes, and by using Eqs. (6-30) and (6-31), the expression for the closed-loop gain is obtained:

$$A_{vf} \equiv \frac{V_{\text{out}}}{V_{\text{in}}} = \frac{Z_o(Z_i + R_R) - A(f)Z_i Z_F}{(Z_F + Z_o)(Z_i + R_R) + Z_I(Z_F + Z_o + Z_i + R_R) + A(f)Z_i Z_I}. \tag{6-32}$$

Eq. (6-32) applies for $R_L \to \infty$. It may be simplified by noting that Z_i is often very large and Z_o very small. Therefore, under those conditions we have

$$A_{vf} \cong \frac{-A(f)Z_F}{Z_F + Z_I + A(f)Z_I} \cong \frac{-A(f)}{1 + A(f)Z_I/Z_F}. \tag{6-33}$$

To make an insensitive amplifier, the forward gain $A(f)$ must be maximized. When this is possible, Eq. (6-33) becomes

$$A_{vf} \cong -Z_F/Z_I. \tag{6-34}$$

This equation represents the ideal operational amplifier, for overall gain is independent of the forward amplification properties.

The feedback used in Fig. 6-13 has altered the input impedance.   It is no longer simply Z_i.   Solution of the Kirchhoff equations gives

$$Z_{if} \equiv \frac{V_{in}}{I_{in}} = Z_I + \frac{Z_o Z_i + Z_o R_R + Z_F R_R + Z_F Z_i}{Z_F + Z_o + Z_i + R_R + A(f)Z_i}.\tag{6-35}$$

Again using the assumption that $A(f)$ is very large, we conlude that

$$Z_{if} \cong Z_I.\tag{6-36}$$

The large forward gain assumed for the derivation of Eqs. (6-34) and (6-36) implies that the signal voltage at the point noted as V_e is very small.   Consequently, that node is referred to as a *virtual ground*.   The consequence of considering V_e as ground is that I_{in} and the feedback current through Z_F are equal.

The output impedance of any network is the Thevenin equivalent impedance looking back from the load terminals.   An expression for output impedance can be derived by straightforward circuit analysis, or on occasion it is convenient to use the relation

$$Z_o = \frac{V_{out}(R_L \rightarrow \infty)}{I_{out}(R_L \rightarrow 0)}.\tag{6-37}$$

For the inverting feedback circuit of Fig. 6-13, the complete expression for Z_{of} is

$$Z_{of} = Z_o - Z_o \frac{(Z_I + Z_i + R_R)Z_o + A(f)Z_i Z_I}{(Z_i + R_R)(Z_o + Z_f) + Z_I(Z_o + Z_F + Z_i + R_R + A(f)Z_i)}.\tag{6-38}$$

When Z_i is the largest of the resistances, and is therefore dominant, Eq. (6-38) may be reduced to

$$Z_{of} \cong Z_o \left[\frac{1}{1 + (Z_I/Z_F)A(f)} \right].\tag{6-39}$$

Use of Eq. (6-39) is dependent upon the actual level of Z_i.

6-7   NONINVERTING FEEDBACK

The noninverting feedback connection is shown in Fig. 6-14.   This connection differs from the inverting case in that Z_I is connected directly to ground and the signal V_{in} is applied through R_R to the $(+)$ or noninverting input.

The voltage transfer function for this connection as $R_L \rightarrow \infty$ is

$$A_{vf} \equiv \frac{V_{out}}{V_{in}} = \frac{Z_I Z_o + A(f)Z_i(Z_I + Z_F)}{(Z_i + R_R)(Z_I + Z_F + Z_o) + Z_I(Z_F + Z_o) + A(f)Z_i Z_I}\tag{6-40}$$

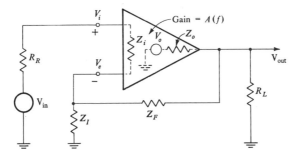

Fig. 6-14   Operational amplifier for discussion of noninverting feedback.

This relation is proved in Problem 6-13.   If Z_o is assumed low-valued, and $Z_i \to \infty$, Eq. (6-40) becomes

$$A_{vf} \cong \frac{A(f)}{1 + A(f)/(1 + Z_F/Z_I)}. \tag{6-41}$$

For a very large forward gain, we arrive at

$$A_{vf} \cong 1 + Z_F/Z_I. \tag{6-42}$$

Eq. (6-42) does *not* contain the minus sign indicative of phase inversion.
Input impedance for this connection may be derived:

$$Z_{if} = Z_i + R_R + Z_I \frac{Z_F + Z_o + A(f)Z_i}{Z_I + Z_F + Z_o} \tag{6-43}$$

For the normal case of large $A(f)$ and Z_i and small Z_o, we may obtain

$$Z_{if} \cong \frac{A(f)Z_i}{1 + Z_F/Z_I}. \tag{6-44}$$

The input impedance has been raised in value over the non-feedback case.
Output impedance is given by

$$Z_{of} = Z_o \left[\frac{Z_I Z_F + (Z_I + Z_F)(R_R + Z_i)}{(R_R + Z_i)(Z_o + Z_F + Z_I) + Z_I(Z_o + Z_F + A(f)Z_i)} \right]. \tag{6-45}$$

Under the assumptions that $A(f)$ and Z_i are very large, we obtain

$$Z_{of} \cong Z_o \frac{1 + Z_F/Z_I}{A(f)}. \tag{6-46}$$

6-8   FREQUENCY COMPENSATION

The frequency behavior of the open-loop amplification and output phase-angle change for a typical 60-dB (1000) operational amplifier is shown in Fig. 6-15.   No compensation techniques are employed to obtain the data shown, and no feedback is included.

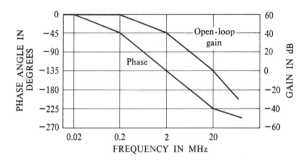

Fig. 6-15   Gain characteristic of amplifier given in Eq. (6-47).

The behavior shown in the figure can be approximated by the following expression:

$$A(f) = \frac{1000}{(1 + jf/0.2 \times 10^6)(1 + jf/2 \times 10^6)(1 + jf/20 \times 10^6)}. \quad (6\text{-}47)$$

To appreciate that Eq. (6-47) does represent the response shown in the figure, consider that f increases to the first "break" frequency, 0.2 MHz. The first denominator term becomes $1 + j1$, corresponding to an angle of $45°$. Other denominator terms are still predominately real at this frequency. The figure shows a $-45°$ phase angle, as expected. Open-loop gain is shown as remaining at 60 dB because the gain curve is approximate; the exact gain curve would be $1/\sqrt{2}$ or 0.707 of the value shown. The 0.707 level corresponds to -3 dB, so that the true gain behavior is 57 dB at $f = 0.2$ MHz. Similarly, at the other break frequencies, the gain curve is 3 dB high.

The reasons for the gain falloff predicted by Eq. (6-47) will now be investigated. We consider as an example an op amp composed of two stages of differential amplification, a level shifter stage, and an emitter-follower output stage. A model for this op amp is given in Fig. 6-16. The input circuit cutoff frequency is determined in part by the resistance of the signal source and the input capacitance of the initial transistor stage. For convenience, we omit this cause of high-frequency gain falloff and concern ourselves with reactive loads on the diff amps and the level shifter.

The load on the first diff amp is $R_C \| C_S \| r_{b'e} \| C_{b'e}$, as shown in the figure. The transfer impedance, V_2/I_1 is of the form

$$\frac{K_1}{1 + sR_1C_1},$$

where $R_1 = R_C \| r_{b'e}$ and $C_1 = C_{b'e} \| C_S$.

The second diff amp feeds an emitter-follower level shifter. The high input resistance of the level shifter will be neglected, but the input capacitance

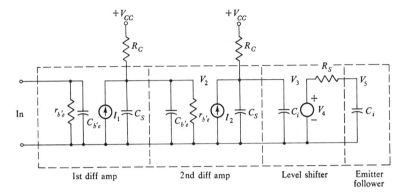

Fig. 6-16   Op amp model for high frequency study.

C_i will be retained.   The transfer impedance V_3/I_2 for this portion of the circuit is

$$\frac{K_2}{1 + sR_2 C_2},$$

with $R_2 = R_C$ and $C_2 = C_i \parallel C_S$.

The final emitter follower is being represented by its input capacitance C_i. We assume that its load is purely resistive.   The level shifter is represented by a voltage source because its output impedance is low-valued.   Then V_5/V_4 is given by

$$\frac{K_3}{1 + sR_S C_i}$$

R_S is a series element used to drop the direct voltage level.

The product of the three transfer impedances can be put in the following form:

$$\frac{K}{(1 + jf/f_1)(1 + jf/f_2)(1 + jf/f_3)},$$

with

$$f_1 = \frac{1}{2\pi R_1 C_1}, f_2 = \frac{1}{2\pi R_2 C_2}, f_3 = \frac{1}{2\pi R_S C_i}.$$

For $R_1 = 5\text{ k}\Omega$, $C_1 = 160\text{ pF}$, $f_1 = 200{,}000\text{ Hz}$.   For $R_2 = 10\text{ k}\Omega$, $C_2 = 8$ pF, $f_2 = 2 \times 10^6\text{ Hz}$.   For $R_S = 5\text{ k}\Omega$, $C_i = 1.6\text{ pF}$, $f_3 = 20 \times 10^6\text{ Hz}$. These values for the break frequencies are consistent with the breaks shown in Fig. 6-15.

Stability

With all feedback systems, the possibility of undesired self-oscillations exists. A test such as the Nyquist criterion may be used to establish specific conditions for freedom from oscillations.   As noted in Chapter 4, it is necessary for the loop gain $-BA$ to have a magnitude of less than unity at the frequency where the phase angle of that term equals 180°.   We consider only the inverting connection here—the noninverting case is similar:

$$BA = -A(f)Z_I/Z_F, \qquad (6\text{-}48)$$

as seen from Eq. (6-33).   Several cases will be considered to illustrate stability considerations.

Z_F and Z_I resistive.   No phase angle is contributed by the denominator of Eq. (6-48).   Therefore, to insure stability, it is necessary that

$$\left| \frac{A(f_{180})Z_I}{Z_F} \right| < 1, \qquad (6\text{-}49)$$

where $A(f_{180})$ is the gain at the frequency where the *added* phase shift is 180°. Obviously, for an amplifier with that gain equal to 10, it is necessary for Z_F/Z_I to exceed 10 to insure stability.

Pole cancellation by external feedback.   Eq. (6-47) represents the unaltered open-circuit gain.   Since this equation will be multiplied by Z_I/Z_F to form BA, it would be possible using reactive elements in the feedback blocks to cancel one of the poles of $A(f)$.   The poles of $A(f)$ in this example are frequencies of 0.2, 2, and 20 MHz.   If Z_F/Z_I has the form

$$\frac{K}{1 + jf/0.2 \times 10^6}, \qquad (6\text{-}50)$$

the amplifier could not be unstable.   The resulting function

$$\frac{1000/K}{(1 + jf/2 \times 10^6)(1 + jf/20 \times 10^6)} \qquad (6\text{-}51)$$

has a low frequency gain of $1000/K$, usually significantly less than 1000, and a maximum phase shift of 180° that occurs as $f \to \infty$.   For all finite frequencies, the shift is less than the critical 180°, and thus stability is assured.

Modification of Forward Transfer Function

The operation of an op amp may be modified by a change in circuitry affecting the signal transfer within the forward path.   As we have seen, the forward gain of a cascaded amplifier is of the general form

$$A(f) = \frac{K}{(1 + jf/f_1)(1 + jf/f_2)(1 + jf/f_3) \cdots}. \qquad (6\text{-}52)$$

A compensating network composed of C_ϕ and R_ϕ is shown between stages in Fig. 6-17(a).   Analysis of the effect of this circuit follows.

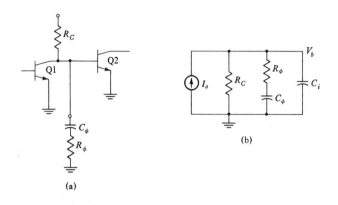

(a)

(b)

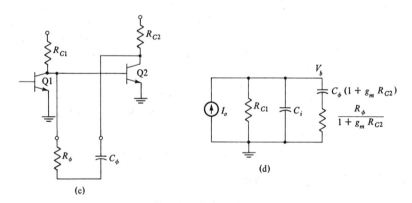

(c)

(d)

Fig. 6-17 (a) (b) Load compensation and equivalent circuit; (c) (d) Miller effect compensation and equivalent circuit.

Transistor Q1 is represented by a current-generator I_o, in the model shown in Fig. 6-17(b), and Q2 is represented by its input capacitance, C_i.   *Without* the compensating network, the transfer impedance is

$$\frac{V_b}{I_o} = \frac{R_C}{1 + j\omega C_i R_C}.$$ (6-53)

With the network, Eq. (6-53) becomes

$$\frac{V_b}{I_o} = \frac{R_C(1 + j\omega C_\phi R_\phi)}{(1 + j\omega C_\phi R_\phi)(1 + j\omega C_i R_C) + j\omega C_\phi R_C}.$$ (6-54)

The denominator will be approximated by $C_\phi R_C$ because that product is the largest of the three. Thus

$$\frac{V_b}{I_o} \cong \frac{R_C(1 + j\omega C_\phi R_\phi)}{1 + j\omega C_\phi R_C}. \qquad (6\text{-}55)$$

Now the denominator of this expression will take the place of the denominator in Eq. (6-53), resulting in a new $f_1' = 1/2\pi C_\phi R_C$. If the circuit designer makes the numerator term of Eq. (6-55) *cancel* the f_2 term in Eq. (6-52), the resulting compensated transfer function becomes

$$A'(f) = \frac{K}{(1 + jf/f_1')(1 + jf/f_3)}. \qquad (6\text{-}56)$$

The compensation network can be included in the IC; more usually, it is added externally. The results of such compensation naturally include gain falloff lower than f_1, namely at f_1', so that the high frequency gain is lowered. However, the cancellation of the f_2 pole has stabilized the network.

Similar results are possible if an R-C compensating network is connected between the collector and the base of a common-emitter connected transistor. Connecting R_ϕ and C_ϕ as shown in Fig. 6-17(c) can be analyzed using the equivalent circuit shown in Fig. 6-17(d). Elements C_ϕ and R_ϕ are treated as the $C_{b'c}$ element in the discussion of Miller effect in Section 3-9. For this arrangement, we find

$$\frac{V_b}{I_o} = \frac{R_{C1}(1 + j\omega C_\phi R_\phi)}{(1 + j\omega C_\phi R_\phi)(1 + j\omega C_i R_{C1}) + j\omega C_\phi R_{C1}(1 + g_m R_{C2})}. \qquad (6\text{-}57)$$

To simplify this expression, let

$$C_\phi R_{C1}(1 + g_m R_{C2}) \gg C_\phi R_\phi + C_i R_{C1}.$$

Then Eq. (6-57) may be written

$$\frac{V_b}{I_o} \cong \frac{R_{C1}(1 + j\omega C_\phi R_\phi)}{1 + j\omega C_\phi R_{C1}(1 + g_m R_{C2})}. \qquad (6\text{-}58)$$

Again, a new break frequency has been established, at $f_1' = 1/2\pi C_\phi R_{C1} \times (1 + g_m R_{C2})$ instead of $f_1 = 1/2\pi C_i R_{C1}$. Cancellation of the f_2 break by the numerator of Eq. (6-58) is usually desired by the system designer.

6-9   TRANSIENT RESPONSE

Compensation networks designed to improve the steady-state frequency response of an op amp also affect the transient response of the system. Transient study data, as illustrated in Fig. 6-18, may be used for the selection of values for external compensating elements.

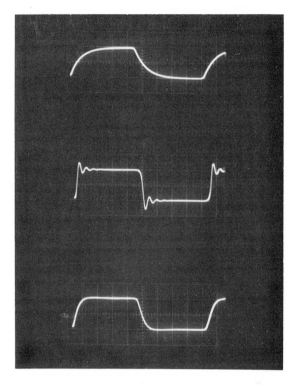

Fig. 6-18   Op amp response overdamped (top), underdamped (center) and critically damped (bottom).

The oscillograms were taken at the output of a type CA3015 op amp connected as a unity-gain inverter.   Compensation was achieved with series RC elements connected between each collector and base of the second diff amp stage.   This connection has been previously referred to as Miller-effect compensation.   A square wave generator operating at 100 kHz supplied the input signal of one volt, peak-to-peak.   The top photo shows an overdamped or overcompensated curve for $C = 470$ pF, $R = 226$ Ω.   The middle photo is representative of undercompensation with $C = 20$ pF, $R = 226$ Ω.   At the bottom is the more satisfactory critically damped response obtained with $C = 200$ pF, $R = 226$ Ω.

Mathematical analysis of the transient performance of a typical op amp often can be accomplished by considering the feedback amplifier as a *second-order system*.   In the preceding section, we noted three break frequencies in the denominator of the voltage transfer function (see Eq. [6-52]).   No serious error is introduced if the highest break frequency is neglected; the resulting second-order system is easily analyzed.

The differential equation of the system is to be determined, and that equation will be solved for time response. We first determine the closed-loop transfer function. We retain two denominator terms in Eq. (6-52) and substitute that expression for $A(f)$ into the general feedback relation

$$A_{vf} = \frac{A(f)}{1 - BA(f)}. \tag{6-59}$$

The result is

$$A_{vf}(s) = \frac{K}{s^2/\omega_1\omega_2 + s(1/\omega_1 + 1/\omega_2) + 1 - BK}, \tag{6-60}$$

where s, the Laplace variable, is being used to represent $j\omega$. Since $A_{vf}(s) = V_o(s)/V_i(s)$, we may rearrange Eq. (6-60) to obtain

$$[s^2/\omega_1\omega_2 + s(1/\omega_1 + 1/\omega_2) + 1 - BK]V_o(s) = KV_i(s). \tag{6-61}$$

Eq. (6-61) represents the differential equation

$$\left(\frac{1}{\omega_1\omega_2}\right)\frac{d^2v_o(t)}{dt^2} + \left(\frac{1}{\omega_1} + \frac{1}{\omega_2}\right)\frac{dv_o(t)}{dt} + (1 - BK)v_o(t) = Kv_i(t) \tag{6-62}$$

because $s \equiv d/dt$, $s^2 \equiv d^2/dt^2$, and so on.

To determine the transient response of the system to a step input of amplitude V_I, ($v_i(t) = V_I u(t)$ or $V_i(s) = V_I/s$), we solve Eq. (6-61) for $v_o(t)$. The solution may be written for $\delta < 1$

$$v_o(t) = \frac{KV_I}{1 - BK}\left[1 - \frac{\epsilon^{-\delta\omega_n t}}{\sqrt{1 - \delta^2}}\sin(\omega_n\sqrt{1 - \delta^2}t + \phi)\right] \tag{6-63}$$

$\phi = \cos^{-1}\delta$, $\omega_n^2 = \omega_1\omega_2(1 - BK)$ and $\delta = (\omega_1 + \omega_2)/2\omega_n$. The parameters ω_n and δ are referred to as the "undamped natural frequency" and the "damping ratio," respectively.

The complicated expression for $v_o(t)$, Eq. (6-63), predicts all forms of response shown in Fig. 6-18. Naturally, values for δ and ω_n determine the shape of the response curve.

When $\delta = 0$, the system is undamped with large overshoot and continued oscillation in $v_o(t)$. Critical damping exists for $\delta \cong 0.7$. The natural frequency ω_n is the angular frequency of oscillations for $\delta = 0$.

Equation (6-63) may be familiar to the reader, for it is encountered in the study of electrical circuit analysis and of control systems. Parameters δ and ω_n are also used in those disciplines.

6-10   DC AMPLIFIERS

Amplification of direct quantities can be accomplished in several ways that differ from the fundamental op amp shown in Fig. 6-1. These methods in

one way or another seek to circumvent limitations present in the specifications given in Table 6-1.

Chopper Amplifier

The operational amplifiers previously described are subject to the drifts inherent in dc amplifiers even though a large amount of feedback may be
employed.   Chopper amplifiers tend to have lower drifts, but they are
severely limited in their bandwidth.

The principle of operation of a chopper amplifier relies upon using a
" chopper " or " modulator," a switching scheme, to cause the incoming dc
to vary with time.   This is often accomplished by periodically grounding the
dc input line.   The resulting time-varying or ac voltage is then amplified and
finally " demodulated " by using a phase-sensitive discriminator circuit to
reconstruct dc at the load.   These operations are depicted in Fig. 6-19.   The

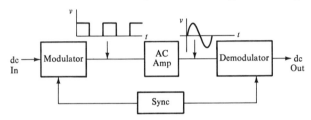

Fig. 6-19   Chopper amplifier block diagram.

block marked " sync " represents the need to feed both modulator and demodulator with a synchronized chopper drive waveform so that the proper
phase relations are not lost.   It is of course required that a positive input
signal be supplied to the load as a positive voltage and vice versa.

The rate at which chopping takes place is most important, for it is generally
considered unwise to try to amplify any waveform containing frequencies
greater than one-tenth of the chopping frequency.

Chopper-Stabilized Amplifier

Automatic balancing circuits based upon chopper stabilization have proved
very successful because the freedom from drift of the chopper amplifier and the
superior high-frequency response of a conventional dc amplifier are both
realized.

In the circuit of Fig. 6-20(a), D represents a dc amplifier having a wide
bandwidth, but drift, denoted by d, will be present and can be referred to
the circuit input. D has a differential amplifier input stage of high input

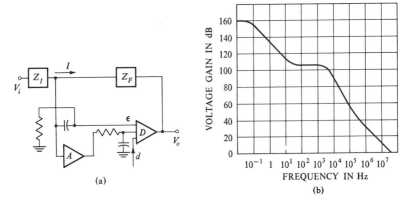

Fig. 6-20   Chopper-stabilized dc amplifier: (a) block diagram; (b) frequency response.

resistance.   The amplifier A is a chopper circuit that will pass all low frequencies.   At the output of A is a low-pass filter, while in the path from Z_I to D a filter is used to block dc and pass the highs.   Low-frequency amplification is the product of A and D; high frequencies will exhibit a gain of D.   Both lows and highs are modified by the appropriate networks.

It is desired to eliminate or at least greatly reduce the amount of d that is present at the output.   For simplicity, we neglect the effects of the filters.   A voltage equation yields

$$\varepsilon = V_i - IZ_I = V_i - \frac{(V_i - V_o)}{Z_I + Z_F} Z_I.$$

Output voltage is

$$V_o = -AD\varepsilon - Dd - D\varepsilon.$$

Solution of these equations results in

$$V_o \cong -V_i\left(\frac{Z_F}{Z_I}\right) - \frac{d}{A}\frac{(Z_I + Z_F)}{Z_I}. \tag{6-64}$$

It can be seen that the unwanted signal d has been decreased by a factor close to $1/A$.

Figure 6-20(b) shows gain behavior with frequency for a typical chopper-stabilized amplifier.

Chopper-stabilized amplifiers have been built with initial input offsets less than $\pm 20~\mu V$ and ± 20 pA and drift less than 0.5 $\mu V/°C$ and 1 $\mu V/week$.   The chopper-stabilized amplifier is unsurpassed for long-term stability.

Temperature-Stabilized Substrate

An approach to IC circuit design with promise of solving the drift problem is the incorporation of a substrate temperature regulator in order that all components of the circuit be kept at constant temperature.   The active regulator is included on the chip.   This approach gives drift performance approaching that of the best chopper-stabilized amplifiers, with the further advantage of small size, low cost, moderate power consumption, and good reliability.

A preamplifier featuring two differential amplifier stages, emitter-followers at both input and output, and regulating circuitry, Fairchild μA727, is available with the following characteristics:

Voltage gain = 100	Input offset voltage drift = 0.6 μV/°C
Input impedance = 300 kΩ	Input offset current drift = 2 pA/°C
Input offset voltage = 2 mV	Input bias current drift = 15 pA/°C
Input offset current = 2.5 nA	Long term drift = 5 μV/week
Input bias current = 12 nA	

A simplified schematic of the temperature regulating circuit is shown in Fig. 6-21.   Transistors Q1 and Q2 are used to sense the substrate temperature.   A

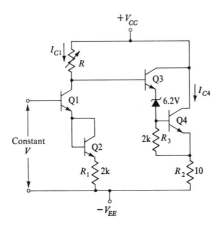

Fig. 6-21   Simplified schematic of temperature regulator.

change in that temperature changes the collector current of those transistors. Consequently the IR drop across the external resistance R also changes.   This voltage change is fed to the base of the large power transistor Q4 through Q3 and the Zener diode.   The collector current of Q4 is thereby altered in order to keep chip temperature constant.   Electric power is transformed into heat in the 10-ohm load and at the collector junction. It has been reported that the chip reaches its temperature steady state in a few seconds after turn-on. Power consumption of the preamplifier is about 800 mW at $-60°C$ but only 50 mW at $+120°C$.

Analysis of the operation can proceed as follows. All changes in quantities are symbolized by Δ, and we consider the changes to be temperature-derived. Thus, since the base of Q1 is held at a constant voltage,

$$\Delta I_{E2} = -\frac{2 \, \Delta V_{BE}}{R_1} \cong \Delta I_{C1}. \tag{6-65}$$

Both Q1 and Q2 are sensing the temperature. The negative sign appears because temperature increases cause ΔV_{BE} to be negative, and this results in an increase in the value of I_{C1}. Temperature change information is transferred to Q3-Q4 by the drop across R, and the resulting changes in the Q4 branch are

$$-\Delta I_{C1} R \cong \Delta I_{C4} R_2 + \Delta V_{BE(Q4)} \tag{6-66}$$

Changes in I_C are related to V_{BE} changes by the exponential equation, Eq. (1-21):

$$\Delta V_{BE(Q4)} = \frac{kT_C}{q} \ln \frac{(I_{C4} + \Delta I_{C4})}{I_{C4}}. \tag{6-67}$$

where T_C is the chip temperature. This equation and Eq. (6-65) may be incorporated into Eq. (6-66), with the following result:

$$\frac{2R \, \Delta V_{BE}}{R_1} = \Delta I_{C4} R_2 + \frac{kT_C}{q} \ln \left(1 + \frac{\Delta I_{C4}}{I_{C4}}\right). \tag{6-68}$$

Power dissipation in the Q4 branch is changed into heat. The temperature of the chip T_C in the immediate vicinity of the energy conversion (collector-base junction for a transistor) is related to the ambient temperature T_A and power being dissipated P_T according to

$$T_C = T_A + \Theta_T P_T \quad °C. \tag{6-69}$$

Θ_T is the thermal resistance of the chip-to-ambient thermal path, and is measured in $°C/W$. According to Eq. (6-69), the required relation for control without change in T_C is

$$\Delta I_{C4}(V_{CC} + V_{EE})\Theta_T = -\Delta T_A \tag{6-70}$$

The chip temperature change is found from Eq. (6-68) by writing ΔV_{BE} as $\left(\dfrac{\Delta V_{BE}}{\Delta T_C}\right) \Delta T_C$:

$$\Delta T_C = -\frac{R_1}{2R(\Delta V_{BE}/\Delta T_C)} \left[\frac{R_2 \, \Delta T_A}{\Theta_T(V_{CC} + V_{EE})} - \frac{kT_C}{q} \ln\left(1 - \frac{\Delta T_A}{I_{C4} \, \Theta_T(V_{CC} + V_{EE})}\right)\right]. \tag{6-71}$$

This behavior is described in Fig. 6-22. It is desired that the chip temperature remain constant, and the Figure indicates very satisfactory performance.

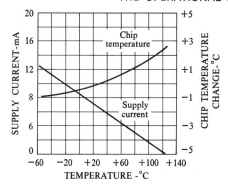

Fig. 6-22 Chip temperature variation versus ambient temperature for circuit with temperature regulated substrate.[12]

Problems

6-1 Derive an expression similar to Eq. (6-7) but include A in the expression as well as Z_i, the input impedance of A.

6-2 Design a four-channel adder of the Fig. 6-2 type using $R_F = 10^6\ \Omega$. The channel gains are to be 10, 5, 2, and 1. Consider that A is very large.

6-3 The circuit shown is an *adder-subtractor*. Express V_o in terms of the six input voltages and the resistance values. Consider that A is very large.

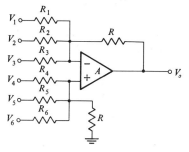

Problem 6-3

6-4 The circuit shown is a *differential input/output* amplifier. Explain why the differential gain is approximately equal to R_2/R_1.

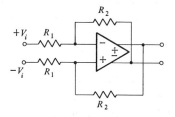

Problem 6-4

6-5   A DIDO system using two op amps is shown in the figure.   DIDO stands for *differential input, differential output*.   For minimizing offset and matching of parameters, the two op amps are fabricated on the same chip.   Use $V_{oA} = A_{vA} V_a$ and $V_{oB} = A_{vB} V_b$.   Prove, for this system,

(a) $A_v(\text{diff}) = \dfrac{V_{oA} - V_{oB}}{V_i} = \dfrac{A_{vA}}{1 + A_{vA}[(R_2 + R_4)/(R_1 + R_2 + R_3 + R_4)]}$

(b) $Z_i = \dfrac{V_i}{I_i} = 2R_{iA}(1 + A_{vA}K),$

where $R_{iA} = R_{iB}$, $A_{vA} = A_{vB}$ and $K = (R_2 + R_4)/(R_1 + R_2 + R_3 + R_4)$.

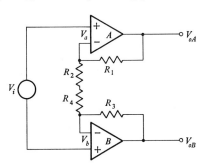

Problem 6-5

6-6   The connection shown in the figure is referred to as a *voltage-follower or noninverting buffer*.   Discuss the reasons why $V_O = V_I$, and comment upon the terminal impedance levels.

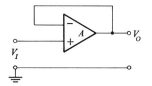

Problem 6-6

6-7   Show that when two parallel networks a and b are feeding ϵ in Fig. 6-8 from voltages V_{ia} and V_{ib}, the relation between output and input is

$$V_o = -\left[\frac{y_{21a}}{y_{21F}} V_{ia} + \frac{y_{21b}}{y_{21F}} V_{ib}\right].$$

6-8   Prove Eq. (6-21) for the bridged-T network.

6-9   Prove Eq. (6-22) for the twin-T network.

6-10   Prove Eq. (6-32) for inverting feedback.

6-11   Prove Eq. (6-35) for inverting feedback.

6-12   Prove that the complete expression for output impedance of an operational amplifier with inverting feedback is given by Eq. (6-38).

6-13   Prove Eq. (6-40) for noninverting feedback.

6-14   Prove Eq. (6-43) for noninverting feedback.

6-15   Prove Eq. (6-45) for noninverting feedback.

6-16   An op amp of the Fig. 6-13 type uses a diff amp input stage.   Use an ac analysis for V_o of the diff amp (similar to the analysis given in Section 5-3) to derive an expression for V_o of the stage when the resistance R_R differs from the resistance of Z_I between V_e and ground.

6-17   Derive Eq. (6-57).

6-18   Use Eq. (6-71) to determine the change in chip temperature for a regulator having the following characteristics: $R = 100$ k Ω, $R_1 = 2$ kΩ, $\Delta V_{BE}/\Delta T_C = -0.002$ V/°C, $R_2 = 10$ Ω, $\Theta_T = 80$ °C/W, $V_{CC} + V_{EE} = 20$, $I_{C4} = 0.04$ A.   Consider $\Delta T_A = 20$, 40, and 60°C.

7
OPERATIONAL AMPLIFIER CIRCUITS

Several examples of practical IC operational amplifiers are given in this chapter. As will be noted, the circuits make use of many of the concepts presented in earlier chapters.

The examples presented are, in a sense, "case studies." While it is not possible to consider all variables important to the design of these circuits, such as economics and manufacturing methods, study of these cases provides insight into certain of the limitations involved in the design and in the application of IC op amps.

The practical monolithic op amp circuits given in Sections 7-1 through 7-3 have been selected because of their educational value. The type MC1530 unit is a straightforward, easily understood circuit. The type 709 device is a "first-generation" or early design, but includes some rather sophisticated circuitry. It has been an industry standard. The type CA3033A is a "second-generation" design with higher power-handling capacity.

In addition to the analysis of op amp circuits, this chapter discusses bias currents and the importance of offset, operation from a single-polarity power supply, and discrete power stages that can be useful when the amplifier is feeding a high power load. A sample amplifier design is given in Section 7-7.

7-1 OP AMP ANALYSIS—I

As an initial example, consider the Motorola MC1530 op amp shown in

Fig. 7-1.   The figure has been partitioned into three sections; input differential amplifier, second differential amplifier, and level translator with output stage.   A companion circuit, the MC1531, differs from the circuit shown by having Darlington pairs in place of transistors Q1 and Q2.   Naturally, a higher input impedance (2 MΩ) results.

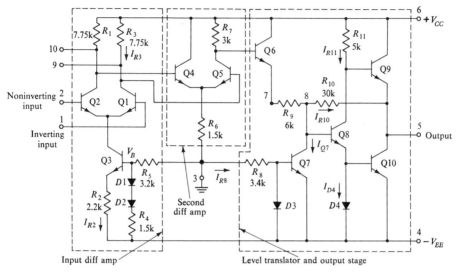

Fig. 7-1   Motorola MC1530 op amp circuit diagram.[13]

Some of the more important operating characteristics of the MC1530 are listed in Table 7-1.   Bandwidth is not listed because it is highly dependent upon the compensation network and amount of feedback that is used.   The widest bandwidth is obtained with no compensation and is in excess of 1 MHz.

TABLE 7-1   MC1530 TYPICAL CHARACTERISTICS

Open-loop voltage gain	5000
Output impedance	25 Ω
Input impedance	20 kΩ
Common-mode rejection ratio	75 dB
Input bias current	3 μA
Input offset current	0.2 μA
Input offset voltage	1 mV
Dc power dissipation (±6 V)	110 mW
Slewing rate ($A_{vf} = 10$, 10% overshoot)	4.5 V/μs
Input noise voltage	
(50Ω source, bandwidth = 5 MHz)	10 μV(rms)
Power supply sensitivity (+ or −)	100 μV/V
Drift with temperature	3.8 μV/°C

DC Analysis

The Q3 network is naturally a current source. Calculations will yield its value. Between V_{EE} and ground are diodes D1 and D2 and resistances R_4 and R_5. Both V_{CC} and V_{EE} will be considered as 6-volt supplies. It follows that the base-to-ground voltage for Q3 is

$$[-6 + 2(0.7)]R_5/(R_4 + R_5) = -3.13 \text{ V},$$

where we have assumed diode and base-emitter drops to be 0.7 volt. Then

$$I_{R2} = \frac{-3.13 - 0.7 - (-6)}{2.2(10^3)} = 1 \text{ mA}$$

in the direction shown. Assuming Q1 and Q2 identical, each collector current will be 0.5 mA, and the collector-to-ground voltages are about $+2.1$ volts. This is also the level of base-to-ground voltages for Q4 and Q5.

Allowing 0.7 volt for base-emitter junction of Q4, the total emitter current through R_6 must be

$$\frac{2.1 - 0.7}{1.5(10^3)} = 0.93 \text{ mA}.$$

Resistance R_6 is helping to stabilize this current by acting as a pseudo-constant current source. Since the bases of these transistors are treated equally, it is reasonable to assume that the collector currents are equal. Therefore, the collector-to-ground voltage of Q5 and thus the base-to-ground voltage of Q6 is

$$6 - 0.47(10^{-3})(3000) = 4.6 \text{ V}.$$

The Q6 emitter voltage is then

$$4.6 - 0.7 = 3.9 \text{ V}.$$

Transistor Q7 is acting as a constant-current source. By assuming the drop across D3 and the emitter junction of Q7 to be 0.7 V, the current through R_8 is

$$\frac{-6 + 0.7}{3.4(10^3)} = -1.56 \text{ mA}.$$

This current will flow through D3. A current of equal value will result as the emitter current of Q7, and most of it will become the collector current of that transistor. As before, base currents are considered negligible compared to collector currents.

The emitter current of Q6 may be calculated. Assume the voltage at the output terminal 5 to be zero.   A summation of voltages from Q6 emitter to output yields:

$$3.9 - (I_{Q7} + I_{R10})(6 \text{ k}) - I_{R10}(30 \text{ k}) = 0.$$

Since I_{Q7} is known, we may solve for I_{R10}.   The value of I_{R10} is found to be negative; therefore it flows away from terminal 5, with a magnitude of 0.15 mA.   The voltage at the base of Q8 is

$$3.9 - (1.41 \times 10^{-3})(6 \times 10^3) = -4.5 \text{ V}.$$

For the output to be at zero potential, we may calculate the current through R_{11}:

$$\frac{6.0 - 0.7}{5(10^3)} = 1.06 \text{ mA}.$$

This is essentially the collector current of Q8 and is the diode D4 current. The Q10 collector current will also be this value if D4 and Q10 have identical dimensions.   However, the manufacturer has purposely made the area of Q10 three times that of D4, so that the direct current in the Q9-Q10 leg is of the order of 3.18 mA.

Calculation of the dc conditions in the amplifier of Fig. 7-1 is now complete.   These conditions are summarized in Table 7-2.

TABLE 7-2   DC CONDITIONS IN MC1530 OP AMP

Transistor	Voltages to Ground			I_C in mA
	Base	Emitter	Collector	
Q1	0	-0.7	2.1	0.5
Q2	0	-0.7	2.1	0.5
Q3	-3.1	-3.8	-0.7	1.00
Q4	2.1	1.4	6	0.47
Q5	2.1	1.4	4.6	0.47
Q6	4.6	3.9	6	1.39
Q7	-5.3	-6	-4.5	1.56
Q8	-4.5	-5.3	0.7	1.06
Q9	0.7	0	6	3.18
Q10	-5.3	-6	0	3.18

Gain Analysis

According to Eq. (5-16), the differential voltage gain of a pair with negligible source and base-spreading resistances is

$$A_v(\text{diff}) = \frac{V_o}{V_1 - V_2} = \frac{\beta R_C}{r_{b'e}} \cong g_m R_C. \tag{7-1}$$

R_C is the collector-return resistance from either transistor to V_{CC}.

When feeding a second diff amp, the additional collector-to-collector load resistance is two series $r_{b'e}$ elements. This load, designated by $2r_{b'e(2)}$, appears effectively in parallel with $2R_C$ between collectors of the first diff amp. For compensation of the frequency response, a capacitance, C, is sometimes connected *between* collectors; when it is used, the entire load on the first diff amp for use in place of R_C in Eq. (7-1) is

$$Z_L = \cfrac{1}{\cfrac{1}{R_C} + \cfrac{1}{r_{b'e(2)}} + 2Cs}. \qquad (7\text{-}2)$$

We wish to determine the open-loop gain of the op amp shown in Fig. 7-1. The differential gain of the first pair may be calculated from knowledge of the load on that stage. Consider that β for all transistors is 50. First-stage transconductance $= \Lambda I_{C(Q1)} = 0.02$ mho, and $r_{b'e(2)} = 50/\Lambda\,(0.47 \times 10^{-3}) = 2675\ \Omega$ for each of the second stage devices. The effective load on the first pair is $7750 \,\|\, 2675 = 1990\ \Omega$. It follows from Eq. (7-1) that the first stage gain is

$$A_v(\text{diff}) \cong (0.02)(1990) = 40.$$

Single-ended gain of the second pair is

$$A_v \cong (0.019)(3000)/2 = 28.$$

The emitter-follower Q6 input impedance has been considered to be much greater than R_7.

Taking the level translator and output stage as a single entity, that portion of the network can be considered to be a sort of miniature op amp with R_{10} and R_9 as feedback and input resistances. A discussion of output stages is given in Section 5-11. We conclude from Eq. (6-8) that this structure should provide a gain of approximately R_{10}/R_9, or 5. If we assume that emitter-follower Q6 provides voltage gain of about 0.9, the total amplification for the entire circuit is $(40)(28)(5)(0.9) \cong 5000$.

Frequency Compensation

Provision is made for a frequency-compensation capacitance C to be connected between the collectors of Q1 and Q2 at terminals 9 and 10. Equation (7-2) gives the form of the total load impedance on the input pair. After rearrangement, that equation becomes

$$Z_L = \cfrac{1}{2C\left[s + \cfrac{R_C + r_{b'e(2)}}{2r_{b'e(2)}R_C\,C}\right]}. \qquad (7\text{-}3)$$

Consider that the capacitance is sufficiently large so that it will constitute the major reason for gain falloff. To find an expression for the frequency-dependent gain of the entire amplifier, Eq. (7-3) can be multiplied by g_m of the first stage and by the gains of the emitter-follower and output stage. Such an expression will not be given here because we are concerned only with the upper cutoff angular frequency, ω_U. This can be obtained after substitution of $j\omega_U$ for s in the Eq. (7-3) and solution for ω_U by equating real and imaginary parts of the denominator. The result is

$$\omega_U = \frac{R_C + r_{b'e(2)}}{2r_{b'e(2)}R_C C}.\tag{7-4}$$

For values given in this Section,

$$\omega_U = \frac{0.25 \times 10^{-3}}{C} \quad \text{or} \quad f_U = \frac{40 \times 10^{-6}}{C}.\tag{7-5}$$

A value for C of 0.1 μF would yield an upper cutoff frequency of 400 Hz.

A form of compensation called *lead compensation* results when a capacitance C is connected between terminals 7 and 8 across resistance R_9. When this modification is made, the gain of the level translator and output stage becomes

$$\frac{R_{10}}{R_9}(R_9 Cs + 1),\tag{7-6}$$

instead of R_{10}/R_9, as in the uncompensated case.

The bandwidth of the amplifier given by Eq. (7-5) is very narrow, as evidenced by the example with $C = 0.1$ μF. It must be remembered, however, that op amps are seldom used open-loop, and, in a circuit employing negative feedback, the open-loop bandwidth can be extended or effectively multiplied by the factor $1 - BA$. To show the bandwidth-widening effect of negative feedback upon this amplifier, we select a numerical example. The open-loop gain will be considered to be 5000. The circuit is shown in Fig. 7-2.

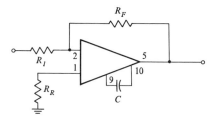

Fig. 7-2   Frequency compensation for MC1530. Feedback elements are also shown.

EXAMPLE.   A capacitance of 1800 pF is connected between terminals 9 and 10, and $R_F = 100$ kΩ, $R_I = R_R = 1$ kΩ.

Using Eq. (7-5) the open-loop upper cutoff frequency is found to be 22 kHz.   From Eq. (6-8) we determine the closed-loop gain as 100.   The factor $1 - BA$, as discussed in Section 4-1, is

$$1 - BA = \frac{A}{A_f} = \frac{5000}{100} = 50.$$

It follows that the closed-loop bandwidth is

$$f_{Uf} = (1 - BA)f_U = 1.1 \text{ MHz.}$$

It is seen that a significant extension of the upper-cutoff frequency is predicted.

7-2   OP AMP ANALYSIS—II

The type 709 op amp pioneered by Fairchild Semiconductor forms our second circuit example.   This circuit is in some ways similar to the MC1530 previously discussed.   It represents the thinking of different designers, however, and is therefore of value in our study of the linear IC.

The 709 is constructed on a 55 mil-square silicon die.   Typical characteristics are given in Table 7-3.   Voltage gain, CMRR, and terminal impedance

TABLE 7-3   709 TYPICAL CHARACTERISTICS

Open-loop voltage gain	45,000
Output impedance	150 Ω
Input impedance	400 kΩ
Common-mode rejection ratio	90 dB
Input bias current	200 nA
Input offset current	50 nA
Input offset voltage	2 mV
Dc power dissipation (± 15 V)	80 mW
Slewing rate ($A_{vf} = 10$)	3 V/μs
Input noise voltage (100-Ω source, bandwidth $= 100$ kHz)	3 μV(rms)
Power supply sensitivity	25 μV/V

levels are higher than given for the MC1530; typical power supply voltages are ± 9, or ± 15 V.

Refer to Fig. 7-3.   Input lines feed the bases of Q1 and Q2.   Those transistors are supplied from a starved current source of only 20 μA (Q11 is the current source and Q10 is the stabilizing diode). R_1 and R_2 are the load resistors, and Q7 is an emitter-follower.

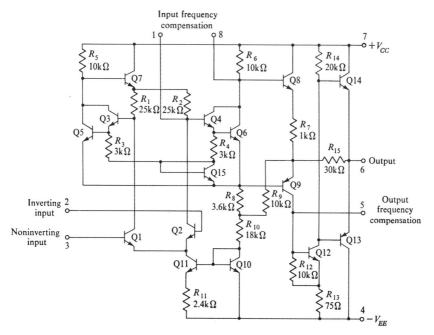

Fig. 7-3   Circuit diagram for type 709 op amp.

The second stage is also differential, and uses the modified Darlington circuit.   Q3 and Q5 form one channel, and Q4 and Q6 the other. R_3 and R_4 and diode Q15 are also part of the Darlington circuit, as discussed in Section 5-9.

A transition from differential to single-ended operation is made in the second stage.   Q5 is the main component in a balanced-biasing network. The simplified schematic shown in Fig. 7-4 bears a high degree of similarity to Fig. 5-1(b).

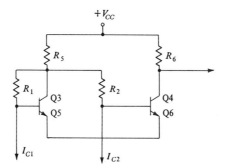

Fig. 7-4   Simplified schematic of load on first stage transistor.

The entire differential output voltage of the Q1-Q2 diff amp appears across R_1 and R_2. All of this voltage is made available to Q4, and hence Q6, because the base of Q6 is connected to R_2 and the other end of R_1 is connected through Q3-Q5 to that same transistor, Q6.

The signal at the collector of Q6 is available to the emitter-follower, Q8. The remaining transistors form a feedback amplifier, with R_{15} performing the function of Z_F and R_7 of Z_I. The voltage gain of this section is about 30.

Transistor Q9, a *pnp* unit, acts as a level-shifter, with most of V_{EE} appearing across its collector-base junction.

The driver stage is Q12. That transistor feeds *npn* transistor Q14 and *pnp* transistor Q13. These two output transistors are biased to operate Class-B. When the collector signal voltage at Q12 is positive-going, only Q14 will conduct; when negative-going, only Q13 will conduct. Each output transistor is completely turned off before the other is allowed to conduct. So-called crossover distortion usually results, but the R_7-R_{15} feedback circuit reduces this distortion and assures a low level of output resistance. An advantage of Class-B operation is that the quiescent supply current drain is minimized.

The high input impedance quoted in Table 7-3 for this amplifier is the result of operating the input stage at low quiescent collector current levels. It was determined in Section 3-6 that the input resistance of a transistor stage is

$$R_i \cong h_{fe}/g_m = h_{fe}/\Lambda I_C. \tag{7-7}$$

If typical values are assumed such as $h_{fe} = 100$, and $\Lambda = 39.6$, and if $I_C = 10$ μA, Eq. (7-7) predicts $R_i = 250$ kΩ. Notice that this value is highly dependent upon h_{fe} and I_C, and would be susceptible to manufacturing variations.

Compensation

It is necessary to compensate most op amp circuits for offset. The external resistance R_a and the potentiometer R_b in Fig. 7-5 are used for this purpose.

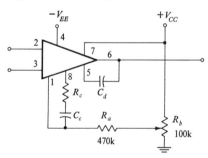

Fig. 7-5   Compensation elements for 709 op amp.

Current can flow in either direction through R_a, depending upon the fraction of V_{CC} available at the adjustable tap on R_b. Thus current can either be added to or subtracted from the Q2 collector branch. Adjustment of that potentiometer alters the collector current of Q2 so that no differential voltage unbalance exists between the bases of Q3 and Q4. This adjustment results in zero output offset.

Frequency compensation is a must, for the 709 will generally oscillate when operated open-loop. The R_c-C_c network shown in Fig. 7-5 provides negative feedback from collector to base of Q4, and assures a satisfactory gain rolloff with frequency. Capacitance C_d connected between output terminal and base of Q12 compensates for the effect of parasitic capacitance at the base of Q14. Element values recommended by the manufacturer result in open-loop performance as shown in Fig. 7-6.

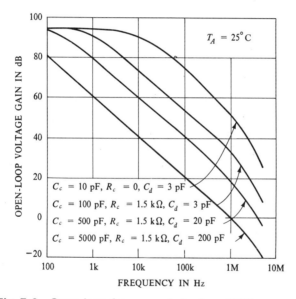

Fig. 7-6 Open-loop frequency behavior of 709 op amp.[12]

7-3 OP AMP ANALYSIS—III

The RCA CA3033A medium power op amp shown in Fig. 7-7 can supply a 500 Ω external load with 255 mW of signal power when operating from ±18 V power supplies. The user can select the pair of input terminals to meet the needs of his application. Terminals 9 and 12 are directly connected to the bases of the first diff amp pair Q3 and Q4. If input resistance of approximately 1 MΩ is desired, terminals 10 and 11 connect to the bases of

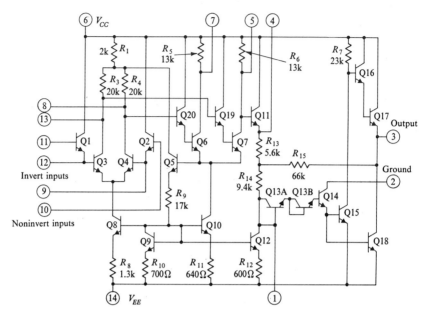

Fig. 7-7   Circuit diagram of RCA CA3033A medium-power op amp.

emitter-followers Q1 and Q2.   Using the emitter followers also has the advantage of reducing the input capacitance of the amplifier.

A summary of typical characteristics for the amplifier is given in Table 7-4. It is apparent that this amplifier is a high-gain, high-CMRR circuit.

TABLE 7-4   CA3033A TYPICAL CHARACTERISTICS

Open-loop voltage gain	63,000
Input impedance	1 MΩ
Common-mode rejection ratio	108 dB
Input bias current	103 nA
Input offset current	9 nA
Input offset voltage	2.9 mV
Dc power dissipation ($\pm$18 V)	250 mW
Slewing rate ($A_{vf} = 10$)	2.5 V/μs
Power supply sensitivity (+ or −)	200 μV/V
Drift with temperature	6.6 μV/°C

Terminals 8 and 13 are used for frequency compensation.   Terminals 1, 4, 5, and 7 are available for monitoring or compensation purposes, but are not normally used.

DC Conditions

Operating as a diode, Q9 provides temperature stabilization for constant-current sources Q8, Q10, and Q12.   However, the stabilization is somewhat different from cases previously considered.   Since no ground is apparent in the low-signal level portions of the IC, we seek a new method to determine the important direct currents.

Observe that $R_8 : R_{10} = 1300 : 700 = 1.86$.   This will be the approximate ratio of $I_{E(Q9)}$ to $I_{E(Q8)}$.   A voltage equation that contains these quantities can be written in order to evaluate the currents:

$$V_{CC} = (I_{C(Q8)} + I_{C(Q5)})R_1 + I_{C(Q4)}R_4 + V_{BE(Q20)} + V_{BE(Q6)}$$
$$+ V_{BE(Q5)} + I_{E(Q5)}R_9 + V_{BE(Q9)} + I_{E(Q9)}R_{10} - V_{EE}. \qquad (7\text{-}8)$$

We assume all base-emitter drops equal 0.7 V, and neglect all base currents. For ± 18 V supplies, Eq. (7-8) yields $I_{E(Q9)} = 1.3$ mA, and therefore $I_{E(Q8)} = 0.7$ mA.   Using the resistance values given for R_{11} and R_{12}, nominal values for the other currents are calculated: $I_{E(Q10)} = 1.4\,\text{mA}$, $I_{E(Q12)} = 1.5\,\text{mA}$.

Transistor Q5 provides a feedback voltage to help compensate for operating-point changes caused by power-supply variations.   Suppose that such a variation results in a rise in the voltage at the emitters of Q6 and Q7.   This point is also the base of Q5, and therefore emitter and collector currents of that transistor will increase.   A larger voltage drop will appear across resistance R_1 because $I_{C(Q5)}$ flows through it.   This increased drop will tend to compensate for a rise in V_{CC} as far as the collector voltages of Q3 and Q4 are concerned.

The increase in $I_{E(Q5)}$, flowing through R_{10}, will also result in an increase in the Q8 (and Q10) current-source values, causing increased drops across R_1, R_3, and R_4 to help maintain the collector voltages of Q3 and Q4 constant in the presence of power-supply variations.

We have noted that $I_{E(Q10)}$ is about 1.4 mA.   The voltage at the collector of Q7 is therefore $18 - (0.7 \times 10^{-3})(13\text{k}) = 9$ V.   At the emitter of Q11, we find $9 - 0.7 = 8.3$ V.   Since $I_{E(Q11)} = I_{E(Q12)} = 1.5$ mA, the voltage at the junction of R_{13}-R_{14} is just about zero, and at the collector of Q12 we find -14 V.   Diodes Q13A and Q13B provide further level shifting, yielding V_{CE} for Q14 of 14.7 V.

The bases of Q15 and Q18 are tied together, and those transistors are matched.   Therefore their collector currents will be equal.   About 0.7 mA is typical for this amplifier.

AC Conditions

Transistors Q19 and Q20 are emitter-followers that serve to reduce the loading effect of the second differential pair (Q6 and Q7) upon the first diff amp

(Q3 and Q4).   Transistor Q11 is also an emitter-follower, and R_{13} forms the input resistance (Z_I) of the "internal" feedback amplifier composed of Q14, Q15, Q16, Q17, and Q18.   Resistance R_{15} is the feedback (Z_F) resistance. The ratio of these two values is 11.8.   Element R_{14} has been discussed as providing dc level shifting; because the output resistance of Q12 is high, there is little loss of signal caused by the high value of R_{14}.   A low driving impedance for the high signal level stages is assured by emitter-follower Q14. The output pair Q17-Q18 is essentially Class-B; both are *npn* transistors, but Q17 is an emitter-follower and Q18 is a common-emitter.   This connection requires that phase reversal be provided in the Q17 channel.   This is accomplished by *CE* Q15; *CC* Q16 reduces the signal-source impedance for Q17.

The voltage gain of the overall amplifier is given as typically 96 dB or 63,000.   The first diff amp stage provides the highest gain because of the 20 kΩ load resistances.

Transistor Q5 provides improved ac performance in addition to its role in maintaining operating point stability discussed earlier.   Normally, signal voltage is absent at the base of Q5.   However, when a common-mode voltage *is* present in the Q3-Q4 pair, there will be a voltage at the emitter terminals of the second diff amp.   The common-mode voltage is amplified by Q5 and is fed to the Q3-Q4 emitters.

Thus Q5 provides negative feedback for common-mode signals.   The emitter current of that transistor contains the unwanted signal, and because that current flows through Q9 and R_{10}, it is available to the base of constant-current source Q8 with a phase such as to further decrease the common mode voltage.

Compensation

In the diagram of Fig. 7-8, the op amp is providing a closed-loop gain of 1000 or 60 dB, because $Z_F = 100$ kΩ and $Z_I = 100$ Ω.   The 2-μF capacitance is

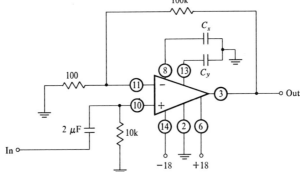

Fig. 7-8   CA3033A op amp with feedback and compensating elements.

coupling the circuit to a signal source. Between terminals 8 and 13 and ground are the compensating capacitors C_x and C_y. These terminals correspond to the collectors of the first diff amp transistors.

For the 60 dB amplifier, $C_x = C_y = 0.001$ μF will assure a stable circuit with a closed-loop bandwidth of close to 400 kHz. The additional phase shift at that frequency will be about 150°, providing a satisfactory phase margin. Wider bandwidths are possible with lower gain; larger capacitance values must be used.

7-4 BIAS CURRENT, OFFSETS, AND LATCH-UP

The model of Fig. 7-9 will be used to determine the importance of offset voltages and currents. Elements R_I, R_F, and R_R are external to the IC, and are used for gain setting. Output impedance is assumed negligible, and $-AV_X$ is the output signal; V_X is the portion of the input V_I that reaches R_i, the input resistance. Note that only direct currents and voltages are used in this analysis.

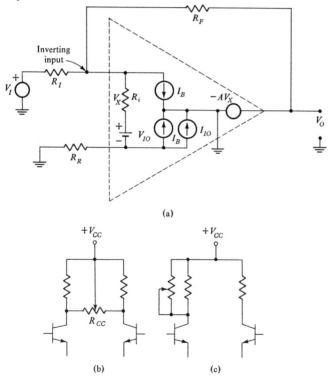

(a)

(b)   (c)

Fig. 7-9   Model of op amp for study of offset; (b) (c) methods for control of offset.

Currents I_B and I_{IO} represent average input bias current and input offset current, respectively. V_{IO} represents the input offset voltage necessary to cause V_O to equal zero when input signal is not present.

Analysis of the circuit yields

$$V_O = -\frac{KV_I}{D} + \frac{(K+1)V_{IO}}{D} + \frac{[KR_I - (K+1)R_R]I_B}{D}$$

$$-\frac{(K+1)R_R I_{IO}}{D}, \qquad (7\text{-}9)$$

with $K = R_F/R_I$, the closed-loop gain of ideal amplifier, and

$$D = 1 + \frac{K}{A}\left[\frac{1}{R_i}(R_I + R_R(K+1)/K) + (K+1)/K\right].$$

The factor D reverts to unity if A, the forward gain, is large. The final three terms on the right-hand side of Eq. (7-9) clearly show the effects of V_{IO}, I_B and I_{IO}. When R_I, R_F, and R_R are not pure resistances, their dc resistances would be used in evaluating Eq. (7-9).

Equation (7-9) is of considerable value because it explains a method for the minimization of bias current contribution to output offset. The I_B term can be eliminated if $R_R = R_I R_F/(R_I + R_F)$. The resulting expression for V_O becomes

$$V_O = -\frac{KV_I}{D} + \frac{(K+1)V_{IO}}{D} - \frac{R_F I_{IO}}{D}. \qquad (7\text{-}10)$$

Thus we conclude that the contribution of average bias current to the output voltage can be minimized by proper selection of R_R. This does not eliminate the effects of either V_{IO} or I_{IO}.

The offsets that have been mentioned in op amp specs and in Eq. (7-9) demand attention. A 1-mV input voltage offset in an amplifier with an open-loop gain of 1000 causes 1 volt at the output terminal in the absence of signal. Closing the feedback loop reduces the magnitude of the output offset voltage, but signal gain is also reduced.

Two circuits have thus far been presented to control offset. In the log amplifier of Fig. 6-5, an adjustment is included in one diff amp base circuit. As has been noted, the combination of R_a and R_b in the 709 circuit of Fig. 7-5 compensates first diff amp collector current for offset. Two other methods for offset elimination are also possible. In Fig. 7-9(b), an external high resistance pot has been connected between collectors of the first diff amp. The collector currents of those transistors can be modified by adjustment of the variable resistance. In Fig. 7-9(c), an external pot is connected across one of the collector-return resistors in the first diff amp stage, thereby altering the collector current and reducing offset.

Certain of the methods of offset control have the disadvantage of upsetting the circuit balance, causing gain imbalance that may adversely affect dynamic operation.

Latch-up

A condition referred to as *latch-up* is prevalent with certain types of op amps. A transistor connected in the common-emitter mode is an inverter because conditions at its collector differ in phase by 180° from the base conditions. When the transistor is saturated by a large driving signal, a direct connection is effectively made between base and collector because that junction becomes forward-biased.   Thus the overdriven transistor no longer provides phase inversion.   This may result in the feedback amplifier changing from inverting to noninverting operation.   If it is then possible for the output voltage to hold the input stage in saturation because the feedback has changed to positive, the condition is referred to as latch-up.

7-5   SINGLE SUPPLY-POLARITY OPERATION

Under certain circumstances it is impractical to provide both positive and negative power supply polarities for an op amp application.   It is possible to attain good ac performance using one supply polarity—in this discussion we consider that the positive supply is available.

The method used in Fig. 7-10 is referred to as the split zener method. The supply is $+12$ V to ground.   We consider that normally the op amp

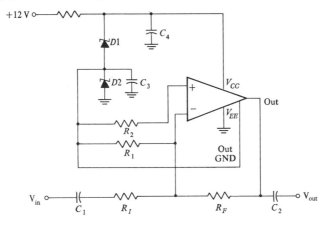

Fig. 7-10   Single-polarity supply used for op amp biasing.

works from $+6$, ground, and -6.   Therefore, the supply must be partitioned into three levels ($+12$, $+6$, 0).   Six-volt zener diodes D1 and D2 accomplish this, with capacitances C_3 and C_4 providing decoupling to minimize supply impedance.

Resistances R_I and R_F are the gain-determining elements, and C_1 and C_2 are serving to block dc from source and load, if necessary.

Signal input terminals marked $(+)$ and $(-)$ are normally near ground potential.   However, with the single polarity supply, they are now at $+6$ V. The output is also raised above ground by that same amount.

Inverting feedback is being used in the circuit of Fig. 7-10.   Resistance R_2 is performing the function of R_R discussed in the preceding chapter, and, as previously noted, R_R is usually selected to equal the parallel combination of R_F and R_I. R_1 must be large enough so as not seriously to degrade the inverting terminal input impedance.

The CA3033A op amp discussed as example III is shown in Fig. 7-11 operating from a single supply of $+36$ V.   The power supply is bisected by

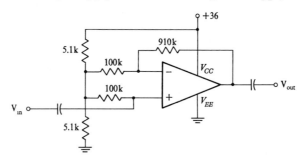

Fig. 7-11   Single-polarity operation for CA3033A op amp.

the two 5.1 kΩ resistances, and the V_{EE} terminal is grounded.   This amplifier is functioning in the noninverting mode, for the input is fed directly into the $(+)$ terminal. The overall gain is given by Eq. (6-42), and, with $Z_F = 910$ kΩ, $Z_R = 100$ kΩ, may be calculated to be 10.1. Although not shown in the figure, terminal 2, the output ground, can be conveniently connected to the V_{CC} supply line.

7-6   POWER STAGES

All IC op amps are limited in their power-handling capability, and for this reason loads that require power levels above 0.5 W cannot be fed directly from the output of a monolithic op amp.   Therefore discrete elements are often used as output stages, and the op amp acts as a pre-amplifier.

In the circuit of Fig. 7-12(a), all resistances and Q1 and Q2 are discrete.

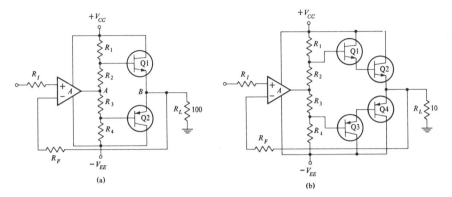

Fig. 7-12   Discrete elements used with an op amp to feed high-power loads.

The op amp, symbolized by A, is connected in a noninverting manner, with R_F and R_I providing feedback from the load, R_L.   Transistors Q1 and Q2 are matched complements, functioning as Class-B emitter-followers.   A load of perhaps 1 W is symbolized by the 100 Ω resistance.

Points A and B are at ground potential in the absence of signal.   The R_1-R_2 and R_3-R_4 voltage dividers are designed to provide a small forward base bias for the transistors in order to minimize crossover distortion when a signal is present.   This bias may result in some collector-current flow, but point B remains at ground level.

When a time-varying signal is to be amplified, the positive half-cycle is processed by Q1 and the negative by Q2.   The overall feedback helps to minimize distortion.   The emitter-followers do not provide voltage gain greater than unity, so their function can be described as current amplification.

The circuit of Fig. 7-12(b) provides additional current gain by using complementary Darlington pairs.   Higher power levels are therefore possible.

7-7 OP AMP DESIGN EXAMPLE

The thoughts contained in the preceding sections will now be composed in order to create a new circuit to meet predetermined performance specifications.   This design example is presented in order to indicate to the reader some of the decisions and compromises required in the creative process.   The ammunition for this design is contained in Chapters 5 through 7.

Other design problems are given at the end of this chapter.   It is not possible to consider all the design constraints applicable to a practical design simply because they involve economic and manufacturing limitations peculiar to a particular plant or operation.

The solution to this problem is not unique. Alternate solutions may be more appealing to the reader.

Specifications for Op Amp Design Example

1. Voltage gain: 1000 minimum.
2. Differential input, single-ended output.
3. Power supplies: ± 6 V.
4. Input impedance: 10 kΩ minimum.
5. Resistances limited to 10 kΩ.
6. Transistors: $h_{fe} = 100$ (nominal)
7. Standby power: 100 mW maximum.

Diff Amps

The circuit will of course use a diff amp at its input. Conventional level shifters and output stages provide voltage gain of less than unity; let us conservatively ascribe 0.5 to the collective voltage gain of those blocks. Then the first important question that arises is whether a single diff amp can provide a gain of 2000. We use the relation

$$A_v(\text{diff}) = \Lambda I_C R_L.$$

At $I_C = 1$ mA, the requirement would be for the effective load of 50 kΩ. It is clear that the entire gain requirement cannot be obtained from one diff amp pair. Two pairs are necessary. These are shown in Fig. 7-13.

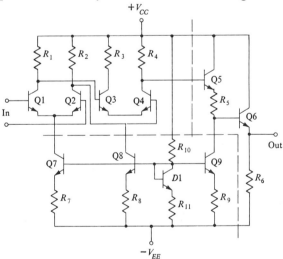

Fig. 7-13.   Circuit for op amp design example.

It is not possible to arbitrarily select the quiescent currents for Q1 and Q2. The direct voltage drop across the collector-return resistances must be less than 6 V in order to insure that the collector-to-ground voltage will be positive. We would like the bases of those transistors to be very nearly at dc ground. Let us consider that the drop across R_C will be 4 V. The resulting value for $I_{C(Q1)}$ and $I_{C(Q2)}$ is 0.4 mA if each R_C is chosen to be 10 kΩ.

With $\beta = 100$, 0.4 mA collector current corresponds to an input resistance per transistor of 6,250 Ω; this assures that the input resistance will exceed the listed requirement.

The bases of Q3 and Q4 are at +2 V because they are directly connected to the collectors of Q1 and Q2. Making $I_{C(Q3)} = I_{C(Q4)} = 0.4$ mA and $R_3 = R_4 = 5$ kΩ would result in $V_{CE(Q3)} = 2.7$ V. We will consider that acceptable.

Now that a decision has been made regarding the second pair, we are able to calculate the anticipated gain of the first pair. The load is 10 kΩ in parallel with $r_{b'e} = 6250$ Ω, the input resistance of Q3-Q4. The equivalent load is therefore 3.84 kΩ. The differential gain of Q1-Q2 is

$$A_v(\text{diff}) = (40)(0.0004)(3840) = 61.4.$$

The second pair will feed a high impedance emitter-follower. Therefore R_3 and R_4 are most important. The single-sided gain of this pair is given by Eq. (5-17) as

$$(1/2)[(40)(0.0004)(5000)] = 40.$$

The product of the gains of the two diff amps is $(61.4)(40) = 2460$. This will be sufficient to meet the required minimum.

While the quiescent collector currents of both diff amps have been chosen equal, the values of V_{CE} differ, with the larger V_{CE} selected for Q3-Q4 to allow larger signals to be handled by that pair. Also, to minimize the noise figure of cascaded stages, high gain is required in the lowest level stage. This has been accomplished.

Level Shifter and Output

Level-shifter transistor Q5 and output transistor Q6 will be emitter-followers. The direct voltage at the base of Q5 is +4 V. Since the output line should be at 0 V, the lower terminal of R_5 will be +0.7. The drop required across R_5 is therefore 2.6 V. An arbitrary selection of 2600 Ω for R_5 and 1 mA for the level-shifter emitter current will assure this drop. If Q6 is operated at 2 mA, R_6 must be 3000 Ω.

Constant-Current Sources

The current through D1 is selected to be 1 mA. The major portion of the available 12 V will be dropped across R_{10}. A solution is $R_{10} = 10$ kΩ and

R_{11} is 1300 Ω.   Element R_9 will also equal 1300 Ω since the currents through D1 and Q9 are to be equal.   For R_7,

$$R_7/R_{11} = 1 \text{ mA}/0.8 \text{ mA}.$$

So $R_7 = 1624 \ \Omega = R_8$.   The standby power requirement is determined by the product of 12 V and total current drain, and is 67 mW.

7-8   SUPER-GAIN TRANSISTORS

The op amp input stage must have high gain to minimize noise, and must have low offset and bias currents to minimize undesired voltages at the output port.   These requirements are not always compatible.   However, a recent technological breakthrough, the technique of producing *super-gain* or *super-beta* transistors, can be used to improve the characteristics of input stages.

The super-gain transistor has a beta in the 1000 to 10,000 range, at collector current levels as low as 1 μA.   It has the disadvantage of collector-to-emitter voltage breakdown of less than 5 volts.

It has been noted that narrow base width results in high beta transistors. We know too that a narrow base will not withstand high reverse junction voltage, because the depletion layer widens with reverse bias; if too wide, that depletion layer reaches the emitter junction and results in *punch-through*, virtually a short circuit between collector and emitter.

If the emitter diffusion is deep but stopped in time, super-gain transistors can be practically made.   By using a second emitter diffusion for super-gain transistors, they can be fabricated on the same chip as regular high voltage *npn* devices.

Super-gain transistors have been shown to have worst-case bias currents of 3 nA, and offset of 400 pA.   Matched pairs have typical offset voltages of 1 mV, and drift of 2 μV/$^\circ$C; in these specifications, they are superior to both the JFET and the MOSFET.   The super-gain transistor is superior to the Darlington pair in the matters of noise and temperature sensitivity.

When super-gain transistors are used in the first stage of an op amp, the collector and emitter terminals are usually shunted by a forward-biased diode junction to assure that the voltage breakdown limit will not be exceeded.

7-9   THE OP AMP IN ELECTRONIC SYSTEMS

The limited discussions of Sections 7-1 through 7-3 cannot give a complete picture of available op amps.   The engineer should become cognizant of all available types. For example, FET and super-gain input stages provide improved input characteristics for certain applications. Monolithic op amps with gains of greater than 10^6 are practical. Wider bandwidth devices are available. The Fairchild type μA741 is an example of a fully compensated circuit, with frequency compensation included in the chip. Units with temperature-regulated substrates, as discussed in Section 6-10, are available, as well as radiation-hardened units that can function well in severe environments.

Several applications of op amps were given in Chapter 6. It was shown that mathematical operations can be implemented, and that the op amp can be used for filtering and as an equalizer (compensator) in systems because it lends itself to the synthesis of a great variety of transfer functions.

In low-level measurement systems, the op amp is widely applied. It can function as an error detector in bridge circuits. It can perform the gain function in oscillators; and in control systems, the op amp can provide required low-frequency electrical amplification. In the communications field, it can be used to amplify the signal from a phonograph pickup or from an image orthicon tube in a TV camera system.

Applications of op amps will be further noted in later chapters. In Section 8-8, it is used as the major component in voltage regulator systems. The sense amplifiers and comparators treated in Section 10-6 are basically op amps. In Chapter 12, we find op amps in digital-to-analog and analog-to-digital converters and in sample-and-hold circuits.

We now turn our attention to other types of linear integrated circuits. Chapter 8 provides a sampling of such circuits, and also discusses characteristics of analog communications systems.

Problems

7-1   Confirm the entries in Table 7-2.

7-2   Discuss why the coefficient 2 appears on the Cs term in Eq. (7-2).

7-3   Develop a table similar to Table 7-2 for currents and voltages in the CA3033A op amp of Example III.

7-4   Derive Eq. (7-9).

7-5   The type 702 op amp is shown in the figure. Consider that the Gnd terminal is grounded and the power supplies are connected as shown. One means of frequency compensation is accomplished with the external 50 pF capacitance.

    (a) Analyze dc conditions in this amplifier and make a chart listing all direct collector currents.

    (b) List collector-to-emitter voltages for all transistors.

    (c) Determine the nominal standby power delivered by each power supply.

    (d) Estimate the voltage gain provided by this amplifier. Consider $h_{fe} = 100$.

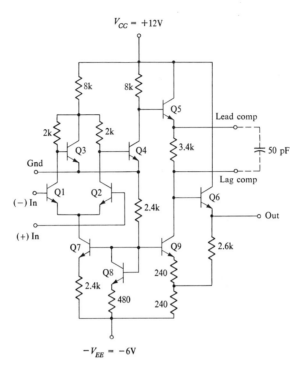

$V_{CC} = +12V$

8k   8k

Q5

Lead comp

2k   2k

Q3   Q4   3.4k   50 pF

Gnd

Lag comp

Q1   Q2   Q6

(−) In

2.4k   Out

(+) In

Q7   Q9   2.6k

Q8   240

2.4k

480   240

$-V_{EE} = -6V$

Problem 7-5

7-6   The IC shown within the dashed lines is the MC1545 gate-controlled dual input wideband amplifier.   A 50 Ω source is connected to Terminal 4.   External connections shown at Terminals 2, 3, and 5 have been selected so that the IC will operate as an amplifier whose gain is dependent upon the direct voltage applied to the gate, Terminal 1.   When the gate is grounded ($V_G = 0$) the amplifier is in an OFF state.   As V_G may be increased in value, transistor Q5 allows emitter current to reach diff amp Q1-Q2 and its gain will increase in proportion to the voltage V_G. Should V_G be made large enough to assure that D1 is OFF, the base-bias network for Q5 then allows sufficient current to Q1 so that the voltage gain of the entire amplifier will reach its maximum of 20 dB (10).   This gate control behavior is shown in (b) of the figure.

(a) Comment on the function Q8 and Q9.

(b) Comment on the function of Q10 and Q11.

(c) With this connection, is the Q3–Q4 diff amp providing any amplification?

(d) Assuming the output terminals 6 and 10 are at dc ground potential; determine the static emitter currents of Q8, Q9, Q10, and Q11.

(e) Q7 is a constant-current source.   Determine its collector current by assuming D2 and Q7 are identical structures.

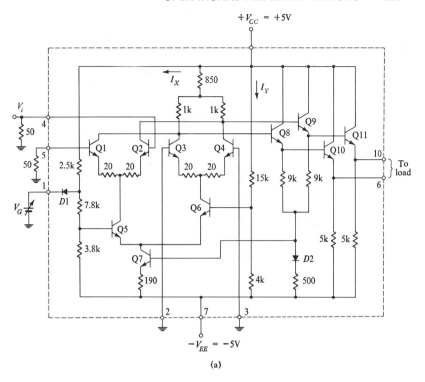

(a)

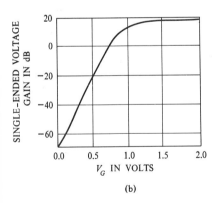

(b)

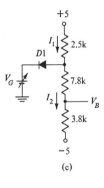

(c)

Problem 7-6

(f) Calculate the currents shown as I_X and I_Y. Consider D1 to be OFF. From this and the previously made calculations, determine the quiescent current drain of the entire amplifier. What is the standby power requirement?

(g) Virtually the entire gain of this connection is obtained from the Q1–Q2 diff amp. Express this voltage gain in terms of I_E, where I_E is the emitter current of either Q1 or Q2. Note the 20 Ω resistors, but omit $r_{bb'}$ and the source resistance.

(h) The action of the switch may be studied using the circuit shown in (c) of the figure. Calculate the voltage V_B at the base of Q5 when $V_G = 0$ and when the diode is fully OFF. These are the extremes.

(i) Assume that the emitter of Q5 is held at a constant level by the action of I_Y and a 0.7 V drop across the Q6 emitter. Use the answer to part (h) to determine the extremes in voltage that would appear between base and emitter of Q5.

7-7   Assume an instantaneous positive-going signal at an Invert input of the CA3033A amplifier. Show that the corresponding signal at output terminal 3 is negative-going by listing those transistors that provide phase inversion for the signal.

7-8   Design an op amp subject to the following constraints:
(a) Voltage gain: 1000 minimum
(b) Power supplies: ± 4 V
(c) Resistances limited to 20 kΩ
(d) Transistors: $h_{fe} = 100$ (nominal)
(e) Standby power: 50 mW maximum
(f) Differential input and output.

7-9   Design an op amp subject to the following constraints:
(a) Voltage gain: 500 minimum
(b) Power supplies: ± 8 V
(c) Resistances limited to 12 kΩ
(d) Transistors: $h_{fe} = 70$ (min), 150 (max)
(e) Standby power: 120 mW maximum
(f) Differential input, single output.

7-10   It is decided to replace the level-shifter–output circuit of the design example given in Fig. 7-13 by a system employing positive feedback having a voltage gain of 1.2. Design a circuit of the Fig. 5-16(b) type to provide that gain.

7-11   Design an enhancement-type MOSFET diff amp to be used at the input stage for an op amp. Consider that the static characteristics are given by $I_D = g_m V_{GS}$, with $g_m = 0.001$ mho. Bias for operation below $I_D = 10$ mA. What is the differential gain of your design?

8
LINEAR IC'S FOR COMMUNI- CATIONS

The IC op amp discussed in the preceding chapters is of importance to a great many electronic systems. In this chapter, we consider other forms of the linear IC. The range of special circuits that can be integrated is great; a limited number of examples of monolithic circuits useful in analog communications systems are discussed.

The development of the linear IC progressed slowly. Historically, the first step was to form two transistors on the same chip and to bring out six leads that correspond to the emitter, collector, and base terminals of each transistor. Such an arrangement is valuable to the circuit designer of discrete diff amps, for matched transistors are thereby possible and thermal problems are minimized by that arrangement.

This chapter begins with a discussion of the cascode amplifier. The emitter-coupled amplifier, an IC one step beyond the transistor pair, is next described. More complex linear circuits are then considered, beginning with a video amplifier. All aspects of an analog communications receiver are covered, including the use of linear IC's for nonamplifying service. An example of a chroma demodulator circuit useful in color TV reception is given. The final section discusses integrated voltage regulators.

8-1 CASCODE AMPLIFIER

The *cascode amplifier* is a simple yet widely accepted circuit that is avail-

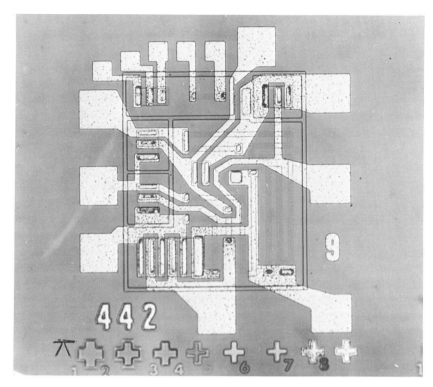

SA20 video amplifier. Note indexing symbols, bonding pads, and long resistance elements. This circuit is discussed in Section 8-3. (*Photo courtesy Sylvania Electric Products, Inc.*)

able itself in integrated form or may be incorporated into a more complex IC.

The cascode amplifier is simply a common-emitter stage feeding a common-base–connected transistor. Generally, the entire collector current of the common-emitter stage becomes emitter current for the common base, as can be seen in Fig. 8-1(a). External capacitances C are for bypass purposes; the load is a tuned circuit located at the second collector.

Let us determine the performance of a typical cascode amplifier. Fig. 8-1(b) depicts the amplifier active devices in terms of small-signal y-parameters. The parameters of the first or common-emitter stage are denoted y_{ie} and so on, and parameters of the common-base are y_{ib} and so on. For the pair, treated as a two-port network, we derive the following from the figure:

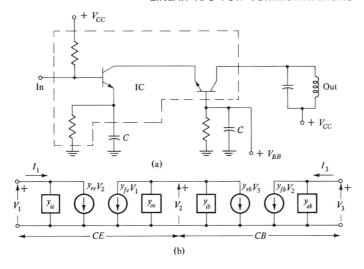

Fig. 8-1   Cascode   connection: (a)   typical   circuit;   (b)   y-parameter equivalent.

$$y_i = \frac{y_{ie}(y_{oe} + y_{ib}) - y_{fe}y_{re}}{y_{oe} + y_{ib}} \cong y_{ie}$$

$$y_r = -\frac{y_{re}y_{rb}}{y_{oe} + y_{ib}}$$

$$y_f = -\frac{y_{fe}y_{fb}}{y_{oe} + y_{ib}} \cong -y_{fe}$$ (8-1)

$$y_o = \frac{y_{ob}(y_{oe} + y_{ib}) - y_{rb}y_{fb}}{y_{oe} + y_{ib}} \cong y_{ob}.$$

From this set of equations we conclude that the input admittance is essentially that of a common-emitter stage. *A significant reduction in reverse transfer admittance has taken place, for y_r is dependent upon the product of two very small values.* The forward transfer admittance y_f is approximately equal to the first stage value, and output admittance is about equal to that of a common-base stage.

The cascode connection can provide a high level of voltage gain. Parameters for the RCA CA3028A are given at 10.7 MHz, $V_{CC} = 9$ V, $I_C = 4.5$ mA as

$$y_i = 0.6 + j1.6 \text{ mmho}$$

$$y_r = 0.0003 \text{ mmho}$$

$$y_f = 99 - j18 \text{ mmho}$$

$$y_o = j0.08 \text{ mmho}.$$

From Appendix II, we obtain the expression for gain:

$$A_v = \frac{-y_f Z_L}{1 + y_o Z_L}. \tag{8-2}$$

With the numbers given and a 3000-ohm load, the voltage gain $= 300$.

Discussion of the cascode circuit as an RF-IF amplifier is given in Section 8-5.

8-2   EMITTER-COUPLED AMPLIFIER

The emitter-coupled IC shown in Fig. 8-2(a) consists of only four elements: two transistors, a capacitance C, and a resistance R_e.   The silicon dioxide capacitance is connected between the collector of Q1 and the base of Q2.   It has a value in the Motorola MC1110 of 30 pF, and its purpose is to provide an ac short-circuit between those two terminals.

The emitter-coupled amplifier operates as a common-collector–common-base (CC-CB) pair.   It therefore has the following low-frequency characteristics: high output impedance; voltage gain of a CB stage; current gain of a CC stage.

A CC stage normally has a high input impedance except when its load is a low impedance.   Because the load on the first stage is the input of a CB, the CC stage, and therefore the entire amplifier, does not have a very high input impedance.   An ac functional representation of this IC is given in Fig. 8-2(b).   The capacitance is assumed to be a short circuit in that figure.

DC Analysis

In application, the positive supply voltage V_{CC} is fed to terminals 2 and 3, the transistor collectors.   A negative voltage V_{EE} is connected to terminal 5. When terminals 1 and 4 are grounded through moderate or low resistance values, the total direct emitter current is given by

$$I_{E(total)} = \frac{-V_{EE} - V_{BE}}{R_e} \tag{8-3}$$

For $R_e \cong 1$ kΩ, $V_{EE} = -5$ V, $V_{BE} = 0.7$ V, the total current is about 4 mA; each stage is operating at the 2 mA level.   If the value chosen for V_{CC} is $+5$ V, $V_{CE} = 5.7$ V for each transistor.

AC Analysis

The input signal is applied at terminal 1, the base of Q1, and the load connected at terminal 3.   Terminals 4 and 5 are grounded.   A y parameter

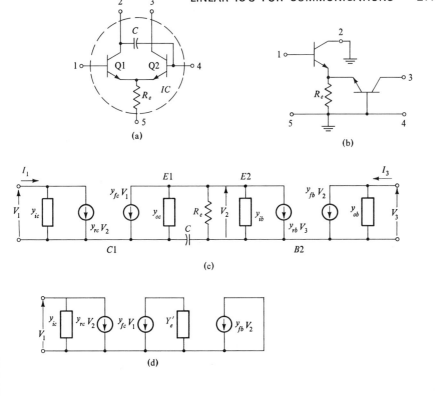

Fig. 8-2 Emitter-coupled IC amplifier: (a) circuit; (b) ac representation; (c) complete y-parameter equivalent circuit; (d)(e) circuits for derivations.

representation for the entire IC is given in Fig. 8-2(c).   Second subscript c is used for parameters of the CC stage, and b for parameters of the CB stage. Element R_e appears in parallel with y_{oc} and y_{ib}.   To simplify symbols, we use

$$Y_{e'} = y_{oc} + y_{ib} + 1/R_e. \qquad (8\text{-}4)$$

These three elements are in parallel when C is considered to be a short circuit.

In order to determine the composite y parameters for the pair, the circuit of Fig. 8-2(d) can be used for the derivation of y_i and y_f, and the circuit of Fig. 8-2(e) for determination of y_r and y_o. From those figures, we conclude that, *for the entire IC,*

$$y_i = y_{ic} - y_{rc}y_{fc}/Y_{e'}$$
$$y_r = -y_{rc}y_{rb}/Y_{e'}$$
$$y_f = -y_{fc}y_{fb}/Y_{e'}$$
$$y_o = y_{ob} - y_{rb}y_{fb}/Y_{e'}.$$

(8-5)

At low frequencies, these parameters are real and typical values are:

$$y_i = 0.2 \times 10^{-3} \qquad y_f = -38 \times 10^{-3}$$
$$y_r = 0.02 \times 10^{-3} \qquad y_o = 0.02 \times 10^{-3}$$

For the values given here, this IC feeding a 1000 Ω load from a 1000 Ω source would provide:

$$Z_i = 1000 \ \Omega \qquad A_v = 38$$
$$Z_o = 17000 \ \Omega \qquad A_i = -40$$
$$G = 1520$$

Fig. 8-3 shows the behavior of parameters with frequency.

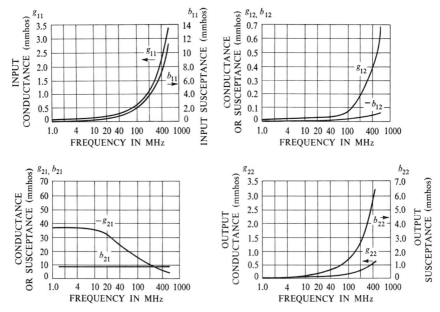

Fig. 8-3 Frequency behavior of the parameters of Motorola MC1110 emitter-coupled amplifier. Conditions are $V_{CB} = 5V$, $I_{E \text{(total)}} = 4\text{mA}$.

The MC1110 can be operated to 300 MHz.   It has a low noise figure. Fig. 8-4 shows the IC used as a 100 MHz radio frequency (RF) amplifier. At the input, L_1, C_1, and C_2 are tuned to the signal frequency. L_2, C_3, and C_4 form the output tuned circuit, with L_2 also acting as an RF choke to keep the output signal from the $+5$ V line.

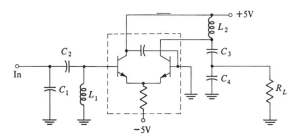

Fig. 8-4   100 MHz RF amplifier.

8-3   VIDEO AMPLIFIER

Video amplifiers are characterized by wide bandwidth and a large output signal-handling capability.   In analog communication systems, a wide bandwidth must be available when a large amount of information is to be amplified. In digital systems, a wide bandwidth is necessary to minimize distortion of pulse waveforms.   The sharp edges of a rapidly rising pulse can be transmitted to the load only if the bandwidth of the system is sufficiently broad.

Bandwidth extension is possible by the use of frequency-compensating networks that trade gain for bandwidth or by the use of negative feedback. Feedback also results in lowered gain in favor of wider bandwidth.

Let us consider the circuit shown in Fig. 8-5.   Biasing resistances have

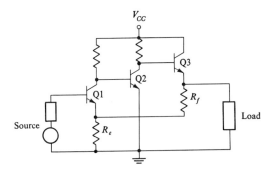

Fig. 8-5   Simplified schematic of video amplifier.

been omitted from the figure for clarity. Transistors Q1 and Q2 are connected common-emitter, and Q3 is an emitter-follower. A portion of output
stage emitter signal current flows through R_e along with the emitter current
of Q1.

Transistors Q1 and Q2 are each responsible for 180° of phase shift in the
signal; the emitter-follower Q3 does not cause phase inversion. Thus the
signal voltage at the load is in phase with the source voltage. When this
load signal is fed back to the input circuit, the fraction that appears across R_e
will buck or subtract from the source voltage. The resulting signal available
at the emitter-base junction of Q1 is therefore much smaller than the source
voltage. This condition is characteristic of negative feedback.

The feedback factor B is defined as the ratio of the voltage fed back to
the voltage at the load. For this circuit,

$$B = -\frac{R_e}{R_e + R_f}. \tag{8-6}$$

The negative sign is the result of the reference convention discussed in
Section 4-1. Equation (4-3) applies to this type of feedback:

$$A_f = \frac{A}{1 - BA}. \tag{8-7}$$

In order for the amplifier to be insensitive to changes in the forward path, the
open-loop gain A must be made as large as possible.

As an example of an IC designed primarily for video amplifier service,
the Sylvania SA20 Wideband Video Amplifier is shown in Fig. 8-6. Component values are given, and it can be observed that $R_e = R_4$ and $R_f = R_6$.

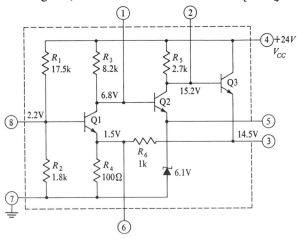

**Fig. 8-6   Complete circuit diagram of SA20 amplifier with direct voltages
at each terminal given.**

DC Analysis

Nominal direct voltages at measurable locations are given in the figure for $V_{CC} = 24$ V. These conditions will be used to determine the direct currents in the circuit. It will be assumed that all terminals are unconnected except for number 4 to V_{CC} and number 7 to ground.

Note that the stages are direct-coupled. A 6.1 V Zener diode is used in the emitter of Q2. Allowing 0.7 V for the Q2 base-emitter drop results in 6.8 V at the collector of Q1. Across R_3 is 17.2 V, so that the current through that resistance is 2.1 mA. This is essentially the collector current of Q1. V_{CE} for that transistor is $6.8 - 1.5 = 5.3$ V.

The collector current of Q2 is about equal to the current through R_5: $I_{C2} = 8.8/2.7$ k $= 3.25$ mA and $V_{CE} = 9$ V. Transistor Q3 has $V_{CE} = 9.5$ V. From the drop across R_6 of 13 V we conclude that I_E for Q3 is 13 mA.

The current through R_e is equal to $I_{E(Q1)} + I_{E(Q3)} \cong 15$ mA. This current, when multiplied by 100 Ω gives the emitter voltage of Q1, 1.5 V. The total power supply drain, neglecting that which goes to the R_1-R_2 network, is 18.5 mA, and the standby power consumption is about 0.44 W.

Bias levels can be changed at the will of the user. If an external resistance is connected between terminals 4 and 8, or between terminals 8 and 7, it would parallel either R_1 or R_2, and the result would be a different operating point not only for Q1 but for the other transistors as well.

Assume that the base of Q1 draws very little current from the R_1-R_2 voltage divider. The direct voltage at terminal 8 is

$$V_{(8)} \cong V_{CC} R_2/(R_1 + R_2) \qquad (8\text{-}8)$$

The quiescent voltage at the output terminal 3 is essentially equal to the voltage at the emitter terminal of Q1 times the gain of the amplifier:

$$V_{(3)} = \frac{R_e + R_f}{R_e} [V_{(8)} - V_{BE(Q1)}] \qquad (8\text{-}9)$$

Therefore $V_{(3)}$ can be changed by external shunting of any of the four elements, R_1, R_2, R_e, and R_f.

AC Operation

True open-loop performance would require that feedback elements R_e and R_f be in the circuit, but that each be connected to ground rather than joined at the emitter of Q1. While this connection is not possible, one can eliminate both the effect of the overall feedback and also the local feedback by connecting a 1 μF capacitor between terminal 6 and ground. The resulting voltage gain is greater than 1000 (60 dB). The normal closed-loop gain is 21 dB or 11, determined by internal R_f and R_e. By connecting an external resistance between terminals 3 and 6, in parallel with R_f, the feedback gain can be reduced to any value desired by the user.   A capacitance in series with the external resistance will assure that dc conditions are not modified.

Because of the base biasing network used for Q1, the input impedance at terminal 8 is primarily dictated by R_2.   That element is much lower in value than the other parallel paths, R_1 and the Z_i of the amplifier including feedback.   The real part of the input impedance is of the order of 1.6 kΩ; parallel input capacitance is about 2.5 pF.   For signals above 15 MHz, the input capacitance is most important.

An external compensating capacitance is normally to be connected between terminals 1 and 2 in order to shape the frequency response of the overall amplifier.   The effect of that capacitance upon performance is shown clearly in Fig. 8-7(a).   The circuit may oscillate if a capacitance is not used.

The output impedance looking back into terminals 3 and 7 is only one or two ohms at frequencies in the audio range.   With higher frequency signals, the magnitude of Z_{of} will increase to perhaps 100 ohms, as seen in Fig. 8-7(b).

The amplifier will operate successfully over the temperature range from $-55°C$ to $+125°C$.   It is capable of supplying an output signal of 14 V peak-to-peak.   Since the dc levels at input and output terminals are considerably above ground, the amplifier is usually capacitively coupled to the source and the load.

8-4   AN ANALOG COMMUNICATIONS SYSTEM

As an example of an analog communications system, consider the familiar broadcast-band AM radio receiver.   The various signal-processing functions of this receiver are available in IC form, but before discussing the specific circuits we shall describe the overall system in some detail.

Consider the block diagram shown in Fig. 8-8.   The electromagnetic wave radiated from the transmitting antenna and picked up by the receiving antenna is an *amplitude-modulated carrier* wave.   The amplitude of this radio frequency (RF) wave varies in accord with the audio intelligence to be conveyed.   The carrier is a high frequency, between 550 and 1600 kHz.   The intelligence is a low frequency, less than 5000 Hz, and represents sound generated in the broadcast studio.

An equation to mathematically represent the AM wave would be complicated if more than one audio frequency is present.   Therefore, for simplicity, consider the carrier frequency to be $f_c = 10^6$ Hz, and the audio modulation to be a single frequency, $f_m = 1000$ Hz.   Then the AM wave can be written as

$$v = V_{cm}(1 + m_a \cos \omega_m t) \cos \omega_c t, \qquad (8\text{-}10)$$

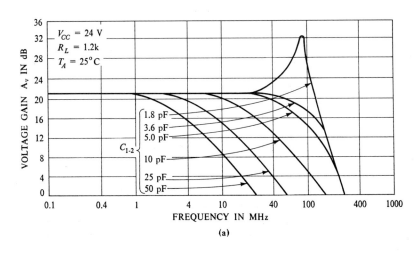

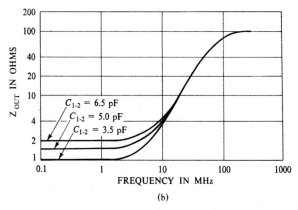

Fig. 8-7 Performance of SA20: (a) frequency response for various compensation capacitances; (b) output impedance vs. frequency.

where

$$v_c = V_{cm} \cos \omega_c t, \text{ the carrier wave,}$$

and

$$v_m = V_{mm} \cos \omega_m t, \text{ the audio modulating wave,}$$

and m_a is referred to as the *modulation index* and is a number between 0 and 1 that is proportional to V_{mm}/V_{cm}. When $m_a = 1$, we say that we have 100 percent modulation, for the Eq. (8-10) waveform would then vary from an

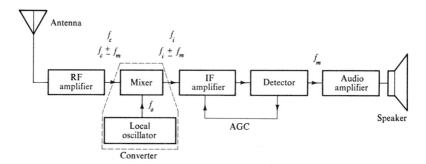

Fig. 8-8   Block diagram of superheterodyne receiver for audio communications.

instantaneous amplitude of zero to $2V_{cm}$. The AM wave shown in Fig. 8-9 represents $m_a \cong 0.5$. The line shown dashed in the figure is referred to as the *envelope*. To recover the intelligence (audio) from this waveform, it is necessary only to extract the envelope from the composite waveform.

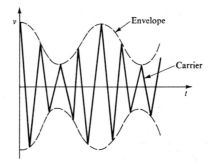

Fig. 8-9   AM waveform with $m_a = 0.5$.

A trigonometric expansion of Eq. (8-10) yields

$$v = V_{cm} \cos \omega_c t + \frac{m_a}{2} V_{cm} \cos(\omega_c + \omega_m)t + \frac{m_a}{2} V_{cm} \cos(\omega_c - \omega_m)t. \quad (8\text{-}11)$$

This equation represents the sum of three waveforms, one at the carrier frequency, one of a higher frequency than the carrier, and another lower. The higher frequency is part of the *upper sideband*; the lower frequency lies in the *lower sideband*. Should the modulating signal contain many frequencies, the sidebands too would be made up of many different frequencies.

The RF wave, carrier plus sidebands, is amplified by the RF amplifier which is tuned to accept only the narrow band of frequencies selected by the listener. The RF wave and the continuous wave output from the local oscillator (LO) at frequency f_o are *mixed* to form the modulated intermediate frequency or IF. The oscillator is adjustable in frequency by the listener. Since it is necessary for LO − RF = IF, a constant often equal to 455 kHz, LO and RF adjustments are "ganged." This function is usually accomplished by the station selector or tuning control.

The IF amplifying stages have "tuned" loads, usually L-C resonant tank circuits. The gain of this block is modified by the Automatic Gain Control (AGC) circuitry, a feedback system that uses dc proportional to the signal strength at the detector stage to control the gain of preceding stages.

The detector removes the carrier and reestablishes the audio frequency intelligence. In addition, it is the source of the AGC voltage. Audio amplifier stages raise the signal to the desired level for conversion to sound in a loudspeaker.

The system described is referred to as a "superheterodyne" receiver, primarily because it uses fixed tuned circuits for IF amplification. Adjustments are limited to RF and LO sections. The superheterodyne principle is also employed in frequency modulation (FM) reception and in TV reception where the picture information is AM and the audio is FM.

We now turn our attention to the contents of the blocks in Fig. 8-8.

Conversion

Conversion refers to the frequency shift that occurs in the mixer stage. The RF is transformed to an IF, thanks to the LO that acts in a way similar to a chemical catalyst.

If RF and LO waves are added and then applied to a linear device such as a small-signal amplifier, the mixing would be unsuccessful, for no IF would be generated. A nonlinear form of operation is required. A square-law characteristic between output and input such as

$$i_c = K v_b^2 \qquad (8\text{-}12)$$

satisfies the requirements. The symbol K represents a constant of proportionality. Voltage v_b is the sum of the two waveforms,

$$v_b = v_1 + v_2, \qquad (8\text{-}13)$$

with v_1 the local oscillator voltage,

$$v_1 = V_{om} \cos \omega_o t \qquad (8\text{-}14)$$

and v_2 the amplitude-modulated voltage, as given in Eq. (8-11),

$$v_2 = V_{cm} \cos \omega_c t + \frac{m_a V_{cm}}{2} [\cos(\omega_c + \omega_m)t + \cos(\omega_c - \omega_m)t]. \qquad (8\text{-}15)$$

Substitution of these voltages into Eq. (8-12) results in a host of terms. Those involving ω_c, ω_o, $\omega_c + \omega_o$, and $\omega_c + \omega_o \pm \omega_m$ and their harmonics will be discarded because it is generally desired that ω_i, the IF, be lower than the RF carrier, and because a tuned load at the output of the mixer will pass only those frequencies in the neighborhood of the IF. The angular IF is

$$\omega_i = \omega_o - \omega_c, \tag{8-16}$$

and in terms of ω_i, the output quantity of the mixer is

$$i_c = KV_{om}V_{cm}\left[\cos\omega_i t + \frac{m_a}{2}\cos(\omega_i + \omega_m)t + \frac{m_a}{2}\cos(\omega_i - \omega_m)t\right]. \tag{8-17}$$

Note that the audio modulation is now associated with ω_i.

If the local oscillator and mixer functions are included in a single stage, that stage is referred to as a converter. Fig. 8-10 shows separate IC mixer

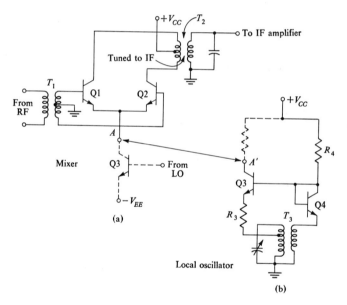

Fig. 8-10   Practical IC mixer and local oscillator. When joined to form a converter at points A and A', dashed elements are eliminated.

and local oscillator. The mixer gain varies in accord with the LO voltage. The oscillator shown works as a simple feedback system with emitter signal fedback in phase to the base of Q3. Transformer T_3 is tuned to the desired frequency of oscillation, so that the loop gain is greater than unity at that frequency. Diode Q4 and resistance R_4 establish the operating point for amplifier Q3. As noted in the figure, the oscillator could be joined with the mixer of Fig. 8-10(a) at points A and A'.

Further discussion of mixing is given in Section 8-6.   RF and IF amplifiers, along with AGC, are described in detail in Section 8-5.

Detection

The purpose of the *detector* stage in a communications receiver is to separate
the modulation from the IF, discard the IF, and pass the intelligence on to
stages that will provide further gain, such as audio amplifiers for radio reception or video amplifiers for television reception.   An example of detection
for broadcast-band radio receivers is given here.   A chroma demodulator
for TV reception is discussed later in the chapter.

In discrete circuits, detection of AM signals is almost always accomplished using a transformer, series diode, and R-C load, as discussed in the
literature.[3]   This arrangement can be pictured as performing rectification to
slice off one-half of the wave and filtering to remove the high-frequency IF.
The remainder is half of the low-frequency modulation envelope, and contains all the intelligence.

Detection, as conventionally performed, may be integrated, but the coil
would of necessity be separate.   An IC transistor detector can also work
with a transformerless IF output.   This circuit is shown in Fig. 8-11.   Transistor Q1 is diode-connected to provide stable bias for Q2.   The static

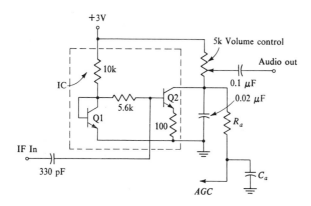

Fig. 8-11   IC detector for radio reception.

collector current of Q2 is less than 0.25 mA, to enable that transistor to
operate in a region favorable for detection.   The 5.6 kΩ resistance prevents
Q1 from loading down the IF amplifier output.   Coupling capacitors are
shown in the figure, for usually dc conditions in single-sided circuits are not
of value to adjacent stages.

The 0.02 μF capacitor forms part of the detector and removes the IF.

Because of rectification resulting from Q2 nonlinearity, the waveform at the collector of that transistor will be audio riding on top of dc. For the audio amplifier input, the 0.1 μF element blocks that direct component. The dc is used for AGC purposes and the audio removed by low pass filter composed of R_a and C_a.

Audio Amplifier

The purpose of the audio amplifier stage is to amplify the weak audio signal at the output of the detector to a level compatible with the loudspeaker and sound output requirement for the system.

Since an audio amplifier is concerned with a frequency span from perhaps 20 Hz to less than 20 kHz, any of the op amps discussed in Chapter 7 would suffice. If the power requirement is greater than 1/2 watt, a monolithic preamplifier could be used, followed by a discrete output stage, perhaps Class B. Another approach would be to use a film hybrid IC capable of the expected power dissipation.

The circuit shown in Fig. 8-12 is a monolithic IC preamp with a discrete output stage transformer-coupled to a speaker load. Balanced biasing is

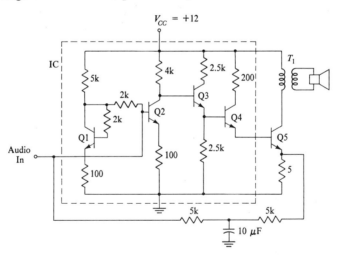

Fig. 8-12   IC audio amplifier with discrete output transistor Q5.

used, with Q1 and Q2 operating at $I_c = 2.2$ mA. The major portion of the gain is obtained from Q2. The 2-kΩ resistors help to reduce signal flow through Q1. Transistor Q3 is a buffer that provides high input impedance and low output impedance to emitter-follow Q4 and the discrete output transistor Q5. Transformer T_1 provides an impedance match between Q5 and the speaker. For compensation, negative feedback can be externally connected, as shown in the figure.

Another example of an audio amplifier is given in Problem 8-14.

8-5   RF-IF AMPLIFIER

The IC shown in Fig. 8-13 is extremely versatile, and can be utilized for many types of high-frequency service.   The circuit is the Motorola MC1550 RF-IF amplifier.   The entire circuit is constructed on a 30 × 32 mil die.

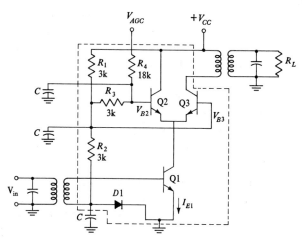

Fig. 8-13   Motorola MC1550 RF-IF amplifier with tuned terminations at source and load. IC is inside dashed square.

The terminations shown in the figure are applicable for a tuned amplifier— they are needed in order to visualize the IC in an environment suitable for service.   Capacitances symbolized by C are for ac grounding purposes.

At first glance, the circuit is very similar to diff amps considered earlier. We notice, however, that Q1 is performing an amplification function in addition to its usual role as dc constant-current source for the differential amplifier pair, Q2 and Q3.   The ac load is located between collectors of that pair.

Further observation indicates that Q2 is *not* operating as an amplifier. Both the base and the collector of that transistor are grounded—base through C, and collector through V_{CC}.

Transistor Q3 is a grounded-base stage between the common-emitter stage Q1 and the load, so that the Q1-Q3 pair functions as a cascode amplifier.

The function of Q2 is gain control.   By direct voltage feedback to the V_{AGC} terminal from another part of the system, the emitter current of Q2 can be varied. This entire function is referred to as *automatic gain control* (AGC). Because Q1 provides a constant direct current, the amount of that current available to gain stage Q3 will be dependent upon the effect of V_{AGC} upon Q2. Gain control of Q3 by varying its emitter current is used here.

DC Analysis

Unless other information is available, it is usually valid to consider the dc resistance of transformers and coils as nearly zero. Supply V_{CC} provides biasing voltage for Q1 through the two 3-kΩ resistances, and D1 gives the needed temperature compensation by the biasing technique given in Chapter 5. The value of collector current for Q1 is generally equal to the diode current that is established by V_{CC} and resistances R_1 and R_2. Transistors Q2 and Q3 share the Q1 collector current in accord with the magnitude of V_{AGC}.

With $V_{CC} = +6$ V, the diode current will be about 0.88 mA. This, then, is the approximate level of I_{C1}. Voltage division results in the voltage at the base of Q3 of about 3.3 V, so that the collector-to-ground potential for Q1 must be $3.3 - 0.7$ or 2.6 V. V_{CE} for Q3 is about 3.4 V.

AGC Operation

The percentage of the Q1 current that Q2 takes is highly dependent upon V_{AGC}. The AGC line also slightly affects the value of I_{E1}, an undesired result. Analysis of the circuit has shown that this sensitivity of I_{E1} to V_{AGC} has a value in the neighborhood of 22 μA/V. Since this corresponds to a change in I_{E1} for full AGC action of only 2 percent, it will be neglected from further consideration.

The base bias voltage required to change a transistor from full ON to full OFF may be referred to as the *transition width*. In the diagram of Fig. 8-14, the transition width is $\Delta V_{B(Q2)}$; it is defined as the voltage change

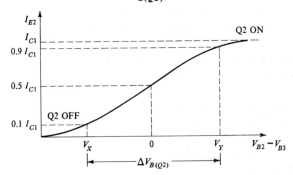

Fig. 8-14   Relation between $V_{B2} - V_{B3}$ and emitter current showing transition width $\Delta V_{B(Q2)}$.

required to cause a change in quiescent current of from 0.1 to 0.9 of the constant-current source value, I_{C1}.

We may use an equation previously derived for diff amps to relate the Q2 current to I_{C1} and the difference in base voltages. Eq. (5-26), repeated here, is:

$$I_{E2} = \frac{I_{C1}}{1 + \epsilon^{\Lambda(V_{B2} - V_{B3})}}. \tag{8-18}$$

The values for $V_{B2} - V_{B3}$ where $I_{E2} = 0.1 I_{C1}$ and $0.9 I_{C1}$ will be designated as V_X and V_Y. It follows that:

$$0.1 I_{C1}(1 + \epsilon^{\Delta V x}) = I_{C1} \quad \text{or} \quad \epsilon^{\Delta V x} = 9$$

$$0.9 I_{C1}(1 + \epsilon^{\Delta V y}) = I_{C1} \quad \text{or} \quad \epsilon^{\Delta V y} = 0.11. \tag{8-19}$$

Since we consider that V_{B2} is subject to variation,

$$V_Y - V_X = \Delta V_{B(Q2)}. \tag{8-20}$$

The solution of Eqs. (8-19), for room-temperature operation, yields

$$\Delta V_{B(Q2)} = 114 \text{ mV}.$$

This is the change required, or the transition width.

It has been shown in Chapter 3 that the transconductance parameter, g_m, for a transistor is directly proportional to the level of quiescent collector current. Thus the AGC condition that results in Q3 being fully ON but not saturated corresponds to high-gain operation, and the lowest-gain condition results when Q2 takes most of the current and Q3 is starved.

To study operation more thoroughly, we solve node equations written for currents at the base terminals of Q2 and Q3. If base currents are neglected, we obtain

$$\Delta V \equiv V_{B2} - V_{B3} = -\frac{V_{CC} + \Gamma_B V_D - V_{AGC}(1 + \Gamma_B)}{(1 + \Gamma_C)(1 + \Gamma_A + \Gamma_B) - \Gamma_A \Gamma_C}. \tag{8-21}$$

The diode forward drop is V_D and $\Gamma_A = R_1/R_3$, $\Gamma_B = R_1/R_2$, $\Gamma_C = R_4/R_3$. These ratios are dimensionless. The derivative of ΔV with respect to V_{AGC} yields the sensitivity:

$$\frac{d \Delta V}{d V_{AGC}} = \frac{(1 + \Gamma_B)}{(1 + \Gamma_C)(1 + \Gamma_A + \Gamma_B) - \Gamma_A \Gamma_C} \tag{8-22}$$

For values given in the figure, we obtain

$$\frac{d \Delta V}{d V_{AGC}} = 0.133 \text{ V/V}. \tag{8-23}$$

As mentioned previously,   114 mV change in ΔV will turn OFF an ON transistor.   Eq. (8-23) predicts that the required ΔV_{AGC} to accomplish this is 0.86 V.   It may be added that the purpose for resistances R_3 and R_4 is to widen the AGC voltage range from 0.114 to 0.86 V so that the circuit is less susceptible to external noise.

Tuned Amplifier

The frequency bahavior of the composite small-signal y parameters for the MC1550 IC is shown in Fig. 8-15. Real and imaginary parts of each para-meter are given: all data apply for $V_{CC} = 6$, $V_{AGC} = 0$ except the bottom two

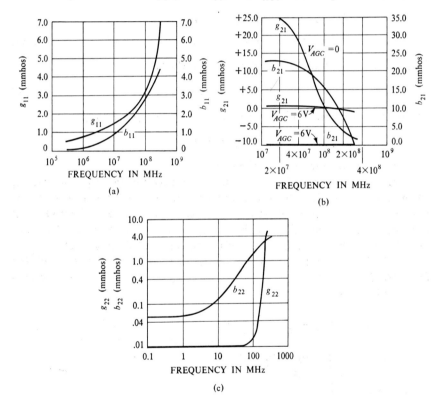

Fig. 8-15   Frequency behavior of the parameters of Motorola MC1550 amplifier. Conditions are $V_{CC} = 6$ V, $V_{AGC} = 0$ V except where noted.

curves in the (b) figure, which are valid for $V_{AGC} = 6$.   Curves showing the frequency variations in the y_{12} parameter are not given, because that ex-tremely low-valued parameter is difficult to measure except at high fre-quencies.

It is convenient to speak in terms of power gain for high frequency applications. Using Fig. 8-16, we now derive some useful expressions for power
levels. The power delivered to a load is simply

$$P_o = |V_2|^2[Re(Y_L)]. \tag{8-24}$$

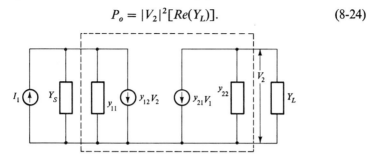

Fig. 8-16   Model for derivation of gain equations.

The absolute magnitude of output voltage V_2 in Fig. 8-16 is

$$|V_2| = \frac{|y_{21}| V_1}{|y_{22} + Y_L|}. \tag{8-25}$$

Therefore the output power is

$$P_o = \frac{|y_{21}|^2 [Re(Y_L)]}{|y_{22} + Y_L|^2} |V_1|^2. \tag{8-26}$$

Power delivered to the IC by the signal source is

$$P_i = |V_1|^2[Re(Y_{in})], \tag{8-27}$$

or, from Appendix II,

$$P_i = |V_1|^2 \left[Re \left(y_{11} - \frac{y_{12} y_{21}}{y_{22} + Y_L} \right) \right]. \tag{8-28}$$

The power gain becomes

$$G = \frac{P_o}{P_i} = \frac{|y_{21}|^2 [Re(Y_L)]}{|y_{22} + Y_L|^2 \left[Re \left(y_{11} - \frac{y_{12} y_{21}}{y_{22} + Y_L} \right) \right]}. \tag{8-29}$$

Note possible instability from the denominator of this equation. A stable
amplifier must exhibit

$$Re \left(y_{11} - \frac{y_{12} y_{21}}{y_{22} + Y_L} \right) > 0. \tag{8-30}$$

By employing the cascode convection or by neutralization, y_{12} is reduced to nearly zero.   Then Eq. (8-29) is

$$G = \frac{|y_{21}|^2 [Re(Y_L)]}{|y_{22} + Y_L|^2 [Re(y_{11})]}. \tag{8-31}$$

The *maximum available gain* (MAG) is obtained by conjugate matching, that is, $g_{11} + jb_{11} = g_S - jb_S$ and $g_{22} + jb_{22} = g_L - jb_L$.   For this matched case, we obtain

$$\text{MAG} = \frac{|y_{21}|^2}{4g_{11}g_{22}}. \tag{8-32}$$

Let us consider a numerical example.   A 50 MHz tuned amplifier is to be designed.   From Fig. 8-15, $g_{21} = 13$, $b_{21} = 21$, $g_{11} = 2.3$, and $g_{22} = 0.01$; all must be multiplied by 10^{-3}.   From these values, $|y_{21}|^2 = (g_{21}{}^2 + b_{21}{}^2) = 610 \times 10^{-6}$, and Eq. (8-32) gives MAG = 6600.   This is the theoretical limit on power gain achievable using the device in a circuit with matched load and source.

Video Amplifier

To use the circuit as a video or wideband amplifier, terminations are shown in Fig. 8-17(a).   The assumed load is resistance R_L in parallel with C_L. At the input, the signal source is represented by resistance R_G.   Since no AGC will be used, the AGC line is grounded.

Performance equations can be derived for the circuit using the model shown in Fig. 8-17(b).   Because of the large number of elements, the resulting equations are rather complex.   For an exhaustive analysis involving a multiplicity of calculations, digital computer assistance would be desirable.

In the model, Q1 is represented by a modified hybrid-π model.   Because the load on Q1 is the input impedance of two common-base stages in parallel, Q1 will provide a low level of voltage gain and therefore $C_{b'c}$ and r_{ce} can be omitted without serious affect upon accuracy.   The input resistances of CB stages Q2 and Q3 are symbolized by r_{e2} and r_{e3}.   Element $r_{e2} = 1/\Lambda I_{E2}$, and $r_{e3} = 1/\Lambda I_{E3}$; each may equal about 50 Ω.   For Q1 and Q3 transistors, C_S represents collector-to-substrate capacitance of about 5 pF.

It is convenient to simplify the analysis by using the following relations:

$$R_S = R_G + r_{bb'} \qquad C_{b'e} = g_m/\omega_T$$
$$r_{b'e} = h_{feo}/g_m \qquad C_o = C_S + C_L$$

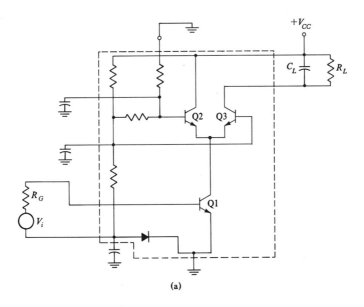

(a)

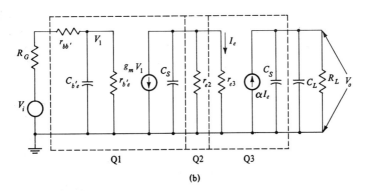

(b)

Fig. 8-17 Integrated video amplifier: (a) schematic; (b) approximate equivalent circuit.

Solution of equations written to determine the gain of the Fig. 8-17 circuit yields

$$A_v(s) =$$

$$-\frac{\alpha \omega_T / R_S r_{e3} C_O C_S}{(s + 1/C_O R_L)[s + (r_{e2} + r_{e3})/C_S r_{e2} r_{e3}][s + \omega_T(1/g_m R_S + 1/h_{feo})]}.$$

$$(8\text{-}33)$$

From examination of this equation, we conclude that the low-frequency voltage gain (also called the dc gain), obtained by setting $s = j\omega = 0$, is

$$A_v(0) = -\frac{\alpha g_m h_{feo} r_{e2} R_L}{(r_{e2} + r_{e3})(h_{feo} + g_m R_S)}. \tag{8-34}$$

Let us use some typical values to evaluate Eq. (8-34):

$\alpha \cong 1$ $r_{e2} \to \infty$

$g_m = 0.04$ mho $r_{e3} = 25\ \Omega$

$h_{feo} = 50$ $R_S = R_G + r_{bb'} = 50 + 50 = 100\ \Omega$

$C_S = 5$ pF $R_L = 1000\ \Omega$

$\omega_T = 2\pi(900 \times 10^6)$ Hz $C_L = 5$ pF

When transistor Q2 is drawing zero emitter current because $V_{AGC} = 0$, $r_{e2} \to \infty$. Then $A_v(0) = -37$.

To predict the frequency response of this circuit, the denominator of Eq. (8-33) is studied. So-called break frequencies exist when real and imaginary parts of a term are equal in magnitude. For the first term, the break frequency occurs when $\omega = 1/C_O R_L$, or $f = 10^8/2\pi = 16$ MHz. From the second term, the break is at $\omega = 1/C_S r_{e3}$, or $f = 1280$ MHz. The third term yields a break frequency of 243 MHz.

Since the break frequencies are so widely separated, they are acting independently. The bandwidth of the amplifier is clearly dependent upon the first term, which was calculated to be 16 MHz.

8-6   SIGNAL MIXING

It was shown in Section 8-4 that a square-law input-output characteristic can be useful to perform the signal mixing function. Many devices are linear for small signals, and a square-law relation is apparent only under large-signal conditions. On the other hand, an ideal FET, as noted in Chapter 3, may behave according to

$$I_D = I_{DSS}\left[1 - \left(\frac{V_{GS}}{V_P}\right)\right]^2. \tag{8-35}$$

This device is a useful mixer. The gain function, in this case transconductance, then varies with V_{GS} according to

$$g_m = \frac{2I_{DSS}}{V_P}\left(1 - \frac{V_{GS}}{V_P}\right). \tag{8-36}$$

In order to use this nonlinearity for mixing, the RF and LO are added, and their sum v_i is applied to the gate terminal, in addition to the bias voltage, V_{GS}. The output voltage, approximately equal to $g_m Z_L v_i$, contains a term of the form $(v_c + v_o)(v_c)$.   The product of the LO sinusoid v_o and modulated signal v_c terms gives rise to the difference frequency or IF term.

For a linear IC using bipolar transistors to function as a mixer, we seek a circuit wherein the gain can be varied in accord with a voltage such as the LO. The RF-IF amplifier shown in Fig. 8-18 can give this kind of operation if the

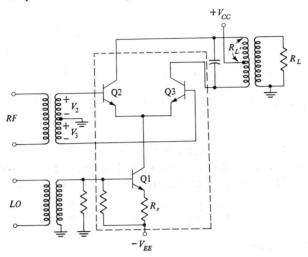

Fig. 8-18   RF amplifier used as a mixer.

local oscillator signal is fed to the base of Q1.   Often this is done using a tuned circuit, and the load may also be tuned, but at the IF.   Within the dashes, the circuit shown is the RCA C3005 RF-IF IC.

The mixer circuit works as follows.   Local oscillator signals cause time variations in the collector current of Q1.   This collector current is the emitter current of Q2 and Q3.   Since the gain of that differential pair is proportional to emitter current, a time-varying gain, acting upon the RF signal, generates the IF.

The goal of this mathematical development is to evaluate the mixing potentiality of the circuit of Fig. 8-18.   For simplicity, we consider the behavior of Q1 as linear.   The input resistance of the transistor alone is $R_{i(Q1)}$.   The instantaneous Q1 base current is then

$$i_{b1} \cong \frac{V_{om} \sin \omega_o t}{R_{i(Q1)} + \beta R_e}.$$ (8-37)

V_{om} is the peak value of the LO signal from base-to-ground, and βR_e represents the effective additional input resistance caused by R_e.   Since $i_c = \beta i_b$ and $i_C = i_c + I_C$, it follows that

$$i_{C1} = \frac{V_{om} \sin \omega_o t}{R_{i(Q1)}/\beta + R_e} + I_{C1}$$ (8-38)

This is the emitter-current variation in Q2 and Q3. The total emitter current for a differential pair is, from Eq. (5-25),

$$i_{C1} = i_{E2} + i_{E3} = I_S \epsilon^{\Lambda v_{BE3}}(1 + \epsilon^{\Lambda(v_{BE2} - v_{BE3})}).$$ (8-39)

This equation may be rearranged to yield

$$i_{C2} = \frac{\alpha_{dc} i_{C1}}{1 + \epsilon^{\Lambda(v_{BE3} - v_{BE2})}} \quad \text{and} \quad i_{C3} = \frac{\alpha_{dc} i_{C1}}{1 + \epsilon^{\Lambda(v_{BE2} - v_{BE3})}}.$$ (8-40)

It is assumed that Q2 and Q3 are identical, so their α's are equal. Voltages v_{BE2} and v_{BE3} are the total base-emitter potentials for the corresponding transistors.

We can extract the time-varying part of Eq. (8-40) and note that

$$v_o = (i_{c2} - i_{c3})R_L',$$ (8-41)

where R_L' is the load resistance reflected to one-half of the primary of the output transformer. If we further assume that biasing is identical, $V_{BE2} = V_{BE3}$; Eq. (8-41) becomes, after use of trigonometric identities,

$$v_o = \frac{\alpha_{dc} R_L' i_{C1} \sinh\Lambda(v_{be2} - v_{be3})}{1 + \cosh\Lambda(v_{be2} - v_{be3})}.$$ (8-42)

Now it is to be noted that

$$v_{be2} - v_{be3} = V_{cm} \sin \omega_c t.$$ (8-43)

If Eq. (8-42) is simplified by dividing denominator into numerator, one obtains a series of terms in the sinusoid $v_{be2} - v_{be3}$. Since i_{C1} also contains a sinusoid of frequency ω_o, the product is the mixing—the IF. Thus

$$v_o(\text{IF}) = \frac{\alpha_{dc} R_L' V_{om} \Lambda V_{cm}}{4(R_{i(Q1)}/\beta + R_e)} [\sin(\omega_c + \omega_o)t + \sin(\omega_c - \omega_o)t].$$ (8-44)

The conversion voltage gain, the ratio of the maximum value of IF to RF, is

$$A_{CG} = \frac{\alpha_{dc} R_L' \Lambda V_{om}}{4(R_{i(Q1)}/\beta + R_e)}.$$ (8-45)

Let us evaluate Eq. (8-45). For $\alpha_{dc} \cong 1$, $R_L' = 1000$, $\Lambda = 40$, $V_{om} = 1$, $R_{i(Q1)} = 5000$, $\beta = 50$, and $R_e = 1000$, we obtain $A_{CG} = 9$. The major reason for the rather low conversion gain is element R_e. If $R_e = 0$, the numbers cited would yield $A_{CG} = 200$.

8-7 IF AMPLIFIER DESIGN EXAMPLE

A simple design example is considered in order to indicate to the reader the nature of certain of the decisions required in the design process. All design constraints cannot be included in such an example because the design is somewhat dependent upon the manufacturing methods of a particular plant or operation.

The solution to the problem given here is not unique; alternate solutions may be more appealing to the reader.

Specifications for Design Example

Sound IF Amplifier for TV Service
1. Voltage gain: 1200 minimum.
2. Center Frequency: 4.5 MHz.
3. Bandwidth: ± 25 kHz.
4. Power supply: $+10$ V.
5. Single ended input and output.
6. Transistors: $h_{fe} = 100$ (nominal).
7. Resistors limited to 10 kΩ, capacitors to 20 pF.
8. Input impedance: 10 kΩ minimum.
9. Selectivity to be accomplished by external tuned circuits at source and at load.
10. Load impedance: 10 kΩ, resistive.

Solution

We shall use an emitter-follower input stage to provide the required input impedance. Two emitter-coupled pairs could provide the necessary gain, according to the discussion in Section 8-2. The output impedance of the second emitter-coupled pair will be large because the final transistor operates in the CB configuration, and thus this connection will be used to meet the impedance specification implied in no. 9.

Emitter-Follower

The single power supply specification implies that the base terminal of emitter-follower Q1 in Fig. 8-19 will be at a direct voltage above ground. The base is fed from constant voltage source V_R. Because the base-bias resistor R_2 is connected to the junction of T_1 and C_1, and C_1 provides an ac ground, that resistor will not degrade the input impedance. The static emitter current of Q1 is determined by the following:

$$V_R = I_B R_2 + I_E R_1 + V_{BE(Q1)}.$$

V_R is selected to be 2 V, $V_{BE(Q1)} = 0.7$ V, $I_E = \beta_{dc} I_B$. We select $I_E = 0.2$ mA. It follows from the equation that $R_1 + R_2/\beta_{dc} = 6500$ Ω. A practical size

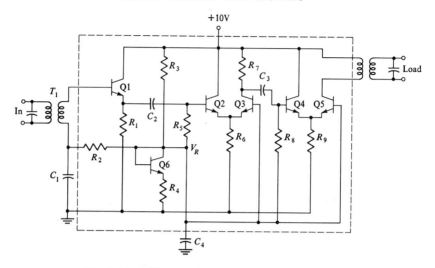

Fig. 8-19   TV sound IF amplifier design example.

for R_2 is 10 kΩ.   Then $R_1 = 6.4$ kΩ.   The operating point coordinates for Q1 are $I_C = 0.2$ mA, $V_{CE} = 8.7$ V.

Emitter-Coupled Pairs

Capacitance coupling between stages is feasible, for we are not concerned with low frequency response. C_2 and C_3 can be determined after impedance levels are calculated.

Since the base of Q3 is at $+2$ V, the drop across R_6 must equal 1.3 V. Let each collector current be 0.25 mA; it follows that $R_6 = 2.6$ kΩ.   A reasonable value for R_5 to assure an acceptable impedance level is 5 kΩ. The input impedance of Q2 is about 10 kΩ at the selected operating point. These two effects taken in parallel yield $R_{i'(2)} = 3.3$ kΩ.

The output impedance of Q1, with its effective R_G of 5 kΩ, is perhaps 170 Ω.   Capacitor C_2 can be calculated from Eq. (3-45).   Thus for $f_L = 4$ MHz,

$$C_2 = \frac{1}{2\pi f_L(R_o + R_{i'(2)})} = 11 \text{ pF}.$$

To assure a large gain, R_7 is selected to be 10 kΩ; $V_{CE(Q3)}$ is therefore 6.2 V.

The Q4-Q5 emitter-coupled pair can be designed as before with $R_9 = 2.6$ kΩ, $R_8 = 5$ kΩ.   To determine C_3, we again use Eq. (3-45):

$$C_3 = \frac{1}{2\pi f_L(R_o + R_{i'(4)})}.$$

Let $f_L = 4$ MHz, $R_o = R_7 \| R_{o(Q3)} = 5$ kΩ, and $R_{i'(4)} = 3.3$ kΩ. Then $C_3 = 5$ pF, a feasible size.

Gains

The load reflected into the collector circuit of Q5 is given as 10 kΩ. For a CB stage operating at 0.25 mA, $h_{ib} \cong 1/g_m = 1/0.01 = 100$. $A_v \cong -h_{fb}R_L/h_{ib} = 100$, for $h_{fb} \cong -1$. The voltage gains of Q4, Q2, and Q1 are about unity. Transistor Q3 sees a load of $10\,\text{k} \| R_{i'(4)} = 2.5$ kΩ; its gain is therefore about 25. Overall voltage gain will be near 2500, under the assumptions made.

Because f_L was selected to be 4 MHz for the calculations to assure that the capacitances would be small, a gain loss is apparent at the center frequency, 4.5 MHz. The circuit will most likely provide a gain of perhaps 1500 when this attenuation and when Z_o of the transistors are considered.

Reference Voltage

The reference voltage V_R was selected to be 2 V. If the current through the R_3-R_4-Q6 branch is to be 1 mA, and we allow 0.7 for the base-emitter drop, $R_4 = 1.3$ kΩ and $R_3 = 8$ kΩ.

Capacitor Sizes

Using 0.3 pF/mil^2, we find that C_2 could be 6 × 6 mils, and C_3 about 4 × 4 mils. If slightly larger capacitances are tolerable, f_L could be chosen to be lower.

We have completed a "first design." Based upon the calculations and upon test data on discrete "breadboard" units, other more refined designs will follow until the circuit is completely satisfactory.

8-8 CHROMA DEMODULATOR

The reception of color television signals makes use of the superheterodyne principle discussed in a preceding section. The receiving system is complicated by the fact that it must process not only frequency-modulated sound information but also amplitude-modulated picture information and synchronization signals to keep the receiver in time agreement with transmitter operation. Following RF amplification, the composite signal is mixed with a local oscillator voltage to form the IF. After IF amplification, the audio is separated from picture information. The video detector that follows the IF section serves to remove picture intelligence from the IF, for the IF has completed its mission as a vehicle for the low level amplification of the signal.

In the color TV receiver, the *chroma (chrominance or color)* signal must be separated from the *luminance* or brightness signal. The chroma is processed after video detection to remove its subcarrier and to reestablish one color that was suppressed at the transmitter, and then it is fed to the picture tube. As shown in Fig. 8-20, the luminance may be applied to the cathodes of that tube and the chroma supplied to three grids.

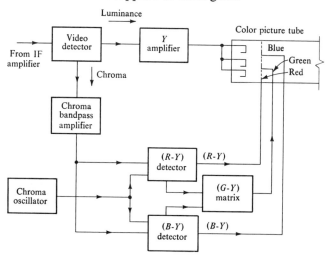

Fig. 8-20   Block diagram of portion of color TV receiver.

Chroma Signal

The chroma signal is contained in a band 1 MHz wide centered about 3.58 MHz. This frequency is referred to as the *chroma subcarrier.* Thus while the main function of the chroma bandpass amplifier is comparable to that of a video amplifier, it is really a tuned circuit that must cut off signals below 3.08 MHz. and above 4.08 MHz. Luminance is not amplified by this amplifier, because it contains frequencies below this pass band. On the other hand, the chroma signal is not passed by the Y amplifier, because its purpose is amplification of the relatively low frequency luminance signal.

At the video detector output, the composite signal contains *synchronization* (sync) and *color burst* signals in addition to chroma and luminance. Sync pulses are used to tie together horizontal and vertical sweep oscillators in transmitter and receiver. The purpose of the color burst is to synchronize the local *chroma oscillator* to the transmitter. The chroma oscillator provides a 3.58 MHz reference in the receiver with which the phase of the chroma signal can be compared.

To describe a color, or to differentiate between colors, there are three characteristics to be considered, *hue, saturation*, and *brightness*. These may be thought of as being analogous to certain characteristics of a radio wave to assist in understanding.   Hue, which defines the wave length of a color, is synonymous with frequency.   Saturation, the purity of the color and a measure of its dilution with white light, is similar to signal-to-noise ratio. Brightness, a measure of light energy, is similar to the amplitude of a radio wave.

The additive process of color mixing in the picture tube requires three primary colors, red, green, and blue.   Most other colors can be generated from combinations of these three.   The picture reproduction is accomplished by using a screen with a pattern of phosphor dots that change the energy of the three electron beams into three light emissions.   Because the phosphor dots are close together, the human eye provides the function of addition that gives rise to colors other than the three primaries.

Three colors are picked up at the color camera, but a luminance signal is not specifically viewed.   Therefore luminance must be assembled from, and expressed in terms of, these colors.   The relation between luminance voltage V_Y, and red, green, and blue voltages has been determined to be

$$V_Y = 0.30V_R + 0.59V_G + 0.11V_B. \tag{8-46}$$

The chrominance signal is most easily discussed in terms of *color differences*. Thus from Eq. (8-46), we can easily obtain

$$V_R - V_Y = 0.70V_R - 0.59V_G - 0.11V_B$$
$$V_G - V_Y = -0.30V_R + 0.41V_G - 0.11V_B \tag{8-47}$$
$$V_B - V_Y = -0.30V_R - 0.59V_G + 0.89V_B.$$

It can be shown (Problem 8-16) that it would be redundant to transmit these three voltages.   The green color difference is therefore not transmitted; it can be made up from the other two:

$$V_G - V_Y = -0.51(V_R - V_Y) - 0.19(V_B - V_Y). \tag{8-48}$$

The green color difference signal is always developed in the receiver according to Eq. (8-48) and the characteristics of the picture tube.

The simultaneous transmission of two color difference signals is accomplished using a process called *divided carrier modulation* whereby these signals modulate the same chroma subcarrier, at 3.58 MHz.   The amplitude of the chroma signal represents color saturation; its phase represents hue.

Demodulator

Demodulation of the chroma signal can be done efficiently using the Motorola XC-1325P IC shown in Fig. 8-21.   In addition to the chroma signal, reference

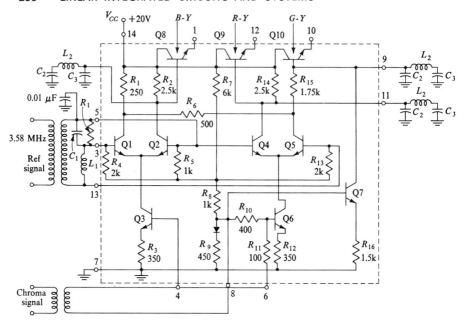

Fig. 8-21 XC-1325P chroma demodulator with external circuitry for inputs and filters. Element values: $L_1 = 30$ uH, $C_1 = 82$ pF, $R_1 = 680$, $C_2 = 39$ pF, $C_3 = 100$ pF, $L_2 = 47$ uH.

B-Y and *R-Y* sine waves are supplied to the demodulator. These references are obtained from the local 3.58 MHz chroma oscillator that is synchronized with the transmitter by the color burst. The oscillator generates a single sine wave, but elements L_1, C_1, and R_1 are chosen so that the *R-Y* voltage between terminals 13 and 5 differs in phase from the *B-Y* voltage between terminals 3 and 5 by 105.8° (discussed later).

The Q1-Q2 diff amp is base-driven by a sufficiently large voltage so that Q1 or Q2 each will be fully ON for one half cycle. During the alternate half cycle, the corresponding transistor will be fully OFF.

Consider the chroma signal being applied between terminals 4 and 8. If it is in phase with the reference voltage that turns Q2 ON, a half cycle of collector current will flow through R_2. The filter composed of L_2, C_2, and C_3 presents a low impedance to 3.58 MHz, and removes the time-varying part of that current; the dc level remains. If the phase of the chroma signal is 90° different from the Q2 base drive, the R_2 current will consist of equal positive and negative excursions, and its average value will be zero. For signals 180° out of phase with the reference, the result is a negative-going direct current through R_2.

The demodulator as discussed provides an output voltage change at the base of Q8 that depends: (a) upon the amplitude of the chroma signal for large signals result in large currents; (b) upon the phase of that signal, for, as we have seen, the dc level is directly related to the phase.   The B-Y voltage has been detected and appears at the base of Q8.

During half cycles when Q1 is ON, B-Y detection will take place at the collector of Q1.   This voltage will be used in generating the G-Y signal.

Transistors Q4 and Q5 operate upon the chroma signal in the same manner as the B-Y channel.   The reference voltage obtained from terminals 13 and 5 has the phase necessary to detect R-Y in the chroma.   The R-Y voltage, after filtering, is fed to the base of Q9.

A voltage divider, $R_{10}/(R_{10} + R_{11}) = 0.8$, is used to reduce the $(R$-$Y)/$ $(B$-$Y)$ ratio.   If further reduction is desired, a small external resistance can be added in series with terminal 6.   The amount of reduction used is determined by the picture tube phosphor selected.

Quiescent current levels for this circuit may be directly determined.   With $V_{CC} = 20$ V, the current through the R_7 branch is about 2.5 mA.   Current sources Q3 and Q6 are therefore generating about 3.2 mA each, and the collector current of Q7 is 0.75 mA.

Matrixing

The G-Y information, as previously mentioned, is dependent upon the chroma B-Y and R-Y levels.   It is the purpose of the resistance matrix composed of R_1, R_6, and R_{15} to reestablish G-Y.   Transistor Q7 draws collector current through R_{15} to provide a base bias for Q10.

To analyze the performance of the matrix, we consider that Q1 and Q5 are constant-current signal sources with currents $I(B$-$Y)$ and $I(R$-$Y)$, as shown in Fig. 8-22.   In parallel with $I(B$-$Y)$ is 250 Ω and $Z_{o(Q1)}$; $Z_{o(Q5)}$ parallels

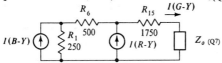

Fig. 8-22   Equivalent circuit for G-Y signal.

the Q5 generator.   If we consider that these output impedances are large compared to other resistances in the circuit, we arrive at the diagram of the figure.   We consider that the output impedance of Q7 will be very small. Solving the network for the current through R_{15} yields

$$I(G\text{-}Y) = -0.1I(B\text{-}Y) - 0.3I(R\text{-}Y). \qquad (8\text{-}49)$$

The output transistors Q8, Q9, and Q10 are each fed from sources of the same internal impedance.   The bases of Q8 and Q9 look back into 2500 Ω, the sum of $R_1 + R_6 + R_{15}$.

The requirements for the "9300°K" color tube are summarized in Fig. 8-23. The relative gains and phase angles for the three colors are noted. One may calculate $|(R\text{-}Y)|/|(B\text{-}Y)| = 0.75$. This is accomplished approximately by the R_{11}-R_{10} network in the demodulator. $\underline{/(R\text{-}Y)} - \underline{/(B\text{-}Y)} = 105.8°$. This is accomplished by the L_1-R_1-C_1 network. Further, the $G\text{-}Y$ requirement shown is obtained by the resistive matrix as given by Eq. (8-49). (A vector addition of $-(R\text{-}Y)$ and $-(B\text{-}Y)$ in Fig. 8-23, multiplied by the 0.3 and 0.1 weighting factors, will yield that equation.)

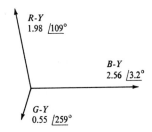

Fig. 8-23   Requirement for "9300°K" picture tube.

Output of Demodulator

The three output emitter-followers are shown feeding three high voltage common-emitter stages in Fig. 8-24. The collectors of these video output

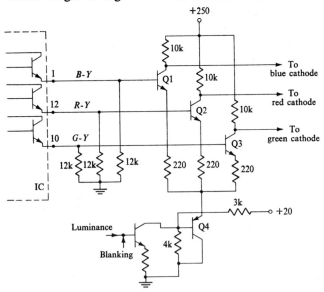

Fig. 8-24   Demodulator output circuitry.

stages may then directly feed the three cathodes of the picture tube with the color signals and also the luminance information that is coupled into the output stages from Q4.   In this system, color and luminance are joined before reaching the cathode ray tube (CRT).

Transistor Q4 is a *pnp* operating as an emitter-follower.   Its output impedance is therefore very small, and it can be considered as a voltage source that adds its output voltage to the video amplifier network.   Since transistors Q8, Q9, and Q10 of the IC are also emitter-followers, a linear addition of chroma and luminance is accomplished at the input of each video amplifier.

8-9   VOLTAGE REGULATORS

All IC's that have been discussed require at least one supply of direct voltage. Often it is necessary for that supply to be well regulated.   *Regulation* refers to the ability of the supply to maintain a constant direct voltage even though the current demand on the supply may vary over a wide range, or the line voltage supply may be subject to fluctuations.

Fig. 8-25 depicts in block form the rudiments of a voltage regulator.   A sample of the load voltage to be held constant is compared to the output of a

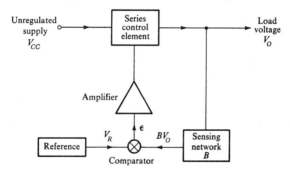

Fig. 8-25   Block diagram of voltage regulator system.

reference element; the difference between these levels is amplified and fed to the series control element.   That element or circuit must act in a manner similar to a valve in a water line; its resistance and therefore the series voltage drop must correspond to the amplifier output command.   Hence, when $V_R = BV_O$, the control element must not change; when $V_R > BV_O$, it must reduce the series voltage drop; and when $V_R < BV_O$, the series voltage drop must be increased.

A circuit to realize the functions of the block diagram is given in Fig. 8-26(a).   The sensing network is simply a potentiometer R_P.   Adjustment of the tap permits the setting of the desired voltage level.   The Zener diode

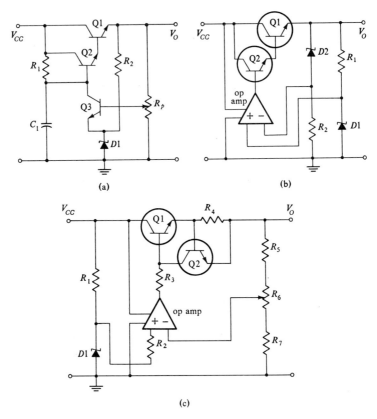

Fig. 8-26 (a) Simple three-transistor regulator; (b) regulator using op amp for comparator and amplifier; (c) system with current limit network.

D1 provides the necessary reference voltage. Reverse current for the diode flows through R_2.

Transistor Q3 performs the comparator function. At its base is a sample of V_O, and its emitter voltage is the reference V_R. Should V_O vary for any reason, current conduction through R_1 and its collector-emitter path is affected. Transistors Q1 and Q2 are connected as a Darlington pair. When Q3 conducts more heavily to increase the current through R_1, the drop across that resistance tends to reduce the voltage available to the base-emitter circuit of the pair, tending to reduce the conduction of Q1. The corresponding increase in series resistance lowers V_O. Capacitor C_1 is used to stabilize the gain of the control loop to eliminate possible oscillations.

The circuit as shown in Fig. 8-26(a) has been integrated; C_1 and R_P are discrete elements. An effective form of regulator is possible using an op amp to perform the comparator-amplifier functions. Such a system is shown in Fig. 8-26(b). The diode networks D1-R_1 and D2-R_2 can be identical; they form a bridge. Should $V_O = V_{D1} + V_{D2}$, nearly equal voltages are available at the $(+)$ and $(-)$ inputs to the op amp, and the output voltage of the op amp is just sufficient to maintain Q1 in the required ON stage. An increase in V_O results in a rise in the voltage across R_2. This causes the inverting terminal voltage to exceed the voltage at the noninverting terminal. The op amp output level then decreases in the direction necessary to cause the drop across Q1 to increase, thus lowering V_O.

While no adjustment of the regulator level is used in the system of Fig. 8-26(b), the choice of Zener diodes determines that level, since $V_O = V_{D1} + V_{D2}$.

An adjustable form of regulator is shown in Fig. 8-26(c). A Zener diode voltage is available at one input to the op amp and a fraction of V_O at the other. Elements R_3 and Q2 along with R_4 provide current limiting. A large load current causes the drop across R_4 to turn Q2 ON. The collector current for that transistor flows through R_3, therefore prohibiting Q1 from taking any additional base current.

Switching Regulators

For some applications, the linear voltage regulator is highly inefficient. For example, if the line voltage is nominally 28 V and a 1 ampere, 10 volt supply is needed, the series pass element of a conventional regulator would have to dissipate 18 W, more than the load power. In such circumstances, it is sometimes desirable to use a regulator system that uses a switching (ON-OFF) mode of operation.

In the circuit of Fig. 8-27, Q1 is the series control element, and discrete elements L and C form a filter. Since Q1 will be saturated part of the time and cutoff at other times, the current waveshape requires filtering. Diode

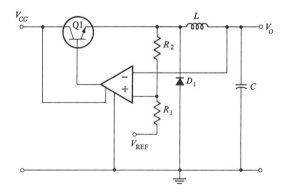

Fig. 8-27   Switching regulator using op amp.

D1 forms a path for the direct load current when Q1 is OFF.   With Q1 ON, D1 is reverse-biased.

The regulator shown in self-oscillating.   With Q1 ON, the voltage at the noninverting input is approximately $V_{CC}(R_1/R_2) + V_{REF}$ for $R_2 \gg R_1$.   When the output V_O exceeds the voltage level at the (+) terminal, the high gain amplifier turns Q1 OFF.   The inductor current now decreases and V_O declines.   When V_O falls to a level slightly below V_{REF}, the error amplifier again turns Q1 ON.

A linear regulator can be used to provide the amplifier and reference functions for the switching regulator.

Our discussion of the linear IC is now complete.   We turn our attention to digital circuits and systems.   The reader will find that certain of the topics covered in Part III have foundations in the linear circuits that have been covered.   This is as expected because most natural sources of electrical waveforms are analog (for example, biological voltages, the locations and velocities of physical devices, and so on).

Problems

8-1   Derive Eqs. (8-1).

8-2   Use the y-parameter values given in Section 8-1 and the equations available in Appendix II to calculate Z_i and Z_o for the cascode amplifier feeding a 3000 Ω load from a 300 Ω source.

8-3   Derive Eqs. (8-5).

8-4   Use 100 MHz parameters for the emitter-coupled amplifier, $R_G = 50\ \Omega$, $R_L = 50\ \Omega$, to determine A_i, A_v, Z_i, and Z_o at that frequency.

8-5   Consider that the SA20 video amplifier is to be modified by connecting an external resistance between terminals 7 and 8 so that the direct voltage at the base of Q1 drops to 1.5 V.   All transistors are still ON, and their V_{BE} drops are to be considered as 0.7 V.   Determine the new set of currents I_{c1}, I_{c2}, and I_{c3}.

8-6   On a sketch of the layout of the SA20 video amplifier as given in the photograph at the beginning of this chapter, locate the 3 transistors, the diode, and the 6 resistances.

8-7   Derive Eq. (8-21).

8-8   Derive Eq. (8-33).

8-9   For $s = j\omega = 0$ in Eq. (8-33), verify Eq. (8-34).

8-10   Using the information in Section 8-5, design a video amplifier with a 25 MHz bandwidth.   What must be the maximum value of the load resistance if $C_O = 10$ pF?   What is the dc gain of your circuit?

8-11   For the video amplifier example given in Section 8-5, consider that we wish to redesign it so that the gain falls off at twice the normal rate at the first break frequency, 16 MHz.   The value of source resistance R_G will be modified to accomplish this.   What change in value of R_G is necessary?

8-12   Draw a composite ac equivalent circuit for the proposed circuit shown in the figure.   Consider that the transistors are identical and each can be described by two parameters, h_{ie} and h_{fe}.

(a) Prove that the short-circuit input admittance is given by

$$\frac{1}{h_i} = \frac{1}{h_{ie}} + \frac{1}{2R} + \frac{h_{fe}}{2h_{ie}(1 + h_{fe})}.$$

(b) Show that the short-circuit current gain is

$$h_f = \frac{2R(1 + h_{fe})}{h_{ie} + R\left(\dfrac{2 + 3h_{fe}}{1 + h_{fe}}\right)} - 1.$$

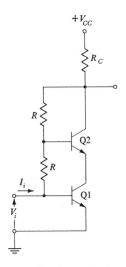

Problem 8-12

8-13   The IC shown within the dashed line is for amplifier service.   Answer the following questions after studying the circuit:

(a) Find $I_{C(Q1)}$ and $I_{C(Q2)}$.   Consider all transistors to be identical.
(b) What is the purpose of R_6 and R_7?

(c) What is the probable purpose of C_1, C_2, and C_3?

(d) What is the voltage gain of this circuit?   The output V_o is single-ended; the input V_i is differential.

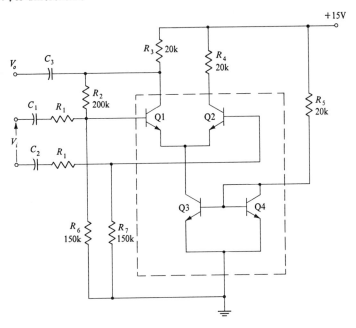

Problem 8-13

8-14   The circuit of the figure is similar to the type $\mu A745$ ac preamplifier.   It is useful as a low-level audio amplifier.   Consider that $V_{CC} = +6$ V, and each transistor has $h_{fe} = 100 = h_{FE}$, $h_{ie} = 4000\ \Omega$.   Answer the following concerning this circuit:

(a) In which configuration is each transistor operating?

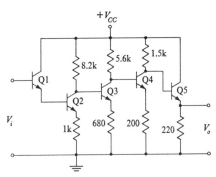

Problem 8-14

(b) If the dc level at the emitter of Q5 is $+0.9$ V, calculate and list all direct voltages in the circuit.

(c) Assume that no external load is connected at V_o. Calculate the small-signal voltage gain from base of Q4 to emitter of Q5.

(d) Determine Z_i at base of Q4.

(e) Using (d), calculate the voltage gain from Q1 base to Q3 collector.

(f) How does the product of (c) and (e) compare with the published nominal gain of 46 dB for this circuit?

8-15   A synchronous demodulator is often analyzed mathematically by considering that the ouput voltage is proportional to the product of signal and reference voltages.   Thus if the signal is

$$v = V(t) \sin \omega t$$

and the reference is

$$v' = K \sin(\omega t + \theta) ,$$

show that the product of these waves contains

$$v_o = KV(t) \cos \theta$$

if a low pass filter rejects the $2\omega t$ term in the product.

8-16   Show that the green color difference voltage given in Eqs. (8-48) can be derived from the other color differences as given in Eq. (8-47).

8-17   Show that the vectors shown in Fig. 8-23 satisfy Eq. (8-49).

8-18   The type 751 differential video amplifier is shown.   This circuit has a bandwidth of 30 MHz, a nominal voltage gain of 600, and requires no external frequency compensation.   The circuit is useful as a READ amplifier in high-speed thin-film memories, and as a general purpose pulse amplifier.

For analysis, consider In Gnd and Out Gnd terminals to be grounded and the external load is 1.2 kΩ between each output terminal and ground.

(a) Estimate the low-frequency input resistance between input terminals.

(b) Determine direct collector current for each of the 11 transistors. Note symmetry.

(c) Determine currents supplied by V_{CC} and V_{EE}.

(d) Calculate voltage gains of CE stages Q1 and Q5.

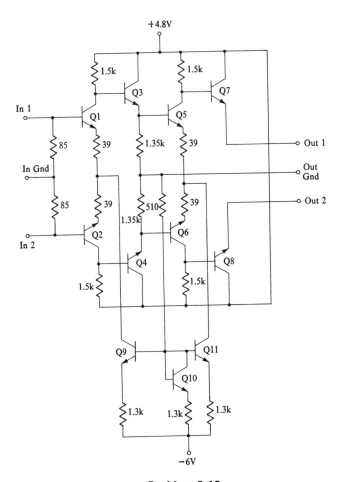

Problem 8-18

8-19 A linear *4-quadrant analog multiplier*, such as Motorola's MC1595, is depicted in the figure.   The system output represents the product of voltage input X and Y.   If X is fed to each input, the output is proportional to X^2.

Consider the working of the system. $i_x = K_1 X, v_y = K_2 Y$.   The transistors Q1 and so on may be considered to be signal current sources, with $i_{c(Q1)} = K_3 XY = i_{c(Q3)}$ and $i_{c(Q2)} = -K_3 XY = i_{c(Q4)}$.

(a) Discuss the polarity of the output voltage to show that the system will respond correctly to both polarities of X and Y.   This is the reason the system is called a 4-quadrant multiplier.

(b) The constant K_3 is dependent upon the value of the direct constant-current source used in the Y diff amp.   Explain why this is so.

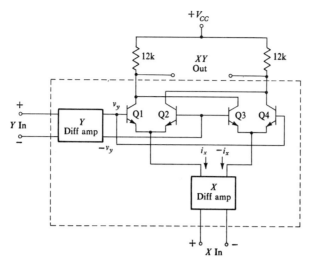

Problem 8-19

8-20   The P. R. Mallory type MIC 0102 preamplifier is shown within the dashed lines in the figure.   The circuit is particularly applicable to phonograph applications. Answer the following questions about this circuit:

(a) Is the main function of Q3 amplification, constant-current bias supply or isolation?

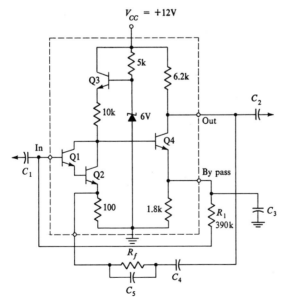

Problem 8-20

(b) What connection or configuration is each transistor operating in?

(c) The direct voltage at the BYPASS terminal is known to be about 1.4 V. From this information calculate I_C and V_{CE} for each transistor.

(d) What function does R_1 and C_3 serve?

(e) What is the purpose of C_4, R_f and C_5?

(f) List those elements that contribute to low-frequency gain roll off.

(g) List those elements that affect high frequency operation.

(h) R_1 is in a dc feedback loop, and R_f is in an ac loop. Do these loops provide positive or negative feedback? Why?

(i) The open-loop voltage gain of this amplifier is said to be 64 dB. Assume $h_{fe} = 50$ for each transistor, h_r and h_o are negligible for Q1 and Q2, but h_o equals 140×10^{-6} mho for Q4. Calculate the gain, and compare with the published figure.

PART THREE

DIGITAL INTEGRATED CIRCUITS AND SYSTEMS

9
LOGIC
GATES

Part III, "Digital Integrated Circuits and Systems," is concerned with large-signal transistor operation and with those IC's applicable to digital systems.

In digital computers, digital control systems, and digital communications systems, the electrical waveforms are pulses. In the generation, switching, storage, and shaping of pulse waveforms, the transistor must act as a *binary device*, a device having two states, ON and OFF. The ON state is most often associated with the transistor saturation region, that is, high collector current and low collector voltage. The OFF state is closely allied to cutoff, the state of minimum I_C and maximum V_{CE}.

In the common-emitter connection, because of the amplification afforded by the transistor, little power is required in the base circuit to switch between the two states. The states differ greatly in their current and voltage levels, so that it is not possible to analyze switching circuits using only the small-signal models discussed in earlier chapters. Therefore, our discussion commences with a study of transistor switching characteristics. The *logic gate*, the most fundamental digital building block, is considered in its four most widely used forms. FET gates are also treated, as they are gaining a place of importance in digital circuitry. Assemblages of gates to perform mathematical functions, storage, generation, and shaping are treated in later chapters.

9-1 TRANSISTOR SWITCHES

The common-emitter collector characteristics for a low-power transistor are illustrated in Fig. 9-1.   A 2500 Ω load line has been drawn linking $V_{CC} = 15$ V with $V_{CC}/R_L = 6$ mA.   Point A represents the upper limit of the OFF region,

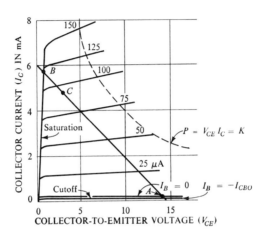

Fig. 9-1   Load line on collector characteristics with points of interest noted.

and point B represents saturation or full ON.   Switching could occur between these extremes, and should base current assume values of $125\mu A$ or greater to drive collector current to point B, the transistor would be called a *saturating switch*.   A *nonsaturating switch* is one in which base current variations cause operation between the region of point A and a point such as C. C represents the ON condition, and is located in the active region of the characteristics.

The dashed line shown in Fig. 9-1 is referred to as the *maximum dissipation contour*; it is the locus of points where the product of collector voltage and current is a particular constant, in this case 40 mW.   The contour represents a limit on the ability of a transistor to remove the heat generated at the collector junction.   Steady-state operation above the contour is not possible, for damage will result to the transistor.

A saturating switch requires more base current "drive" and longer switching time, but results in a much lower ON resistance than does the nonsaturating type.   With respect to collector dissipation, the saturating type of switch has advantages.   Both types exhibit low dissipation in the standby or OFF condition, but when ON, the saturating switch has a smaller V_{CE}-I_C

product.   Should the load line as shown in Fig. 9-1 cut across the maximum-dissipation contour, it is possible that the maximum allowable junction temperature of the transistor will not be exceeded, provided that the switching time is fast.   Therefore, because their average dissipation is low, switching transistors can handle large voltages and currents without exceeding the rated maximum dissipation.   A nonsaturating switch with a load line intersecting the maximum-dissipation contour must be carefully designed in order not to exceed the junction temperature limit.

There are several possible OFF conditions for the common-emitter–connected transistor, three of which are shown in Fig. 9-2.   In (a) of the

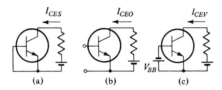

(a)   (b)   (c)

Fig. 9-2   OFF conditions for common-emitter–connected transistor: (a) zero input voltage; (b) zero input current; (c) restraining input voltage.

figure, zero base-to-emitter voltage is available because of the obvious short circuit across the input terminal pair; the resulting collector current is often given the symbol I_{CES}.   When the base terminal is open, as in Fig. 9-2(b), no base current can exist, and collector current is symbolized by the familiar I_{CEO}.   Collector current I_{CEO} is two or more times greater than I_{CES}.   A third method of turning OFF the transistor applies a restraining input voltage to reverse-bias the base-emitter diode.   Under a reverse bias, the collector current can be most effectively turned OFF; I_{CBO}, the inevitable leakage, is then the remaining collector current.

To summarize switching states, it may be concluded that in the OFF condition input and output resistances are high, small leakage currents are apparent, and collector voltage approaches the value of the supply voltage V_{CC}.   The saturated ON condition results in $I_C \cong V_{CC}/R_L$, $V_{CE} = V_{CE(sat)} < 1$ V and a fairly low input resistance.   Several hundred millivolts must be supplied the input terminal pair to provide the base current necessary for the saturation of a low-power transistor.

9-2   SWITCHING SPEED

A universal test for transistors used as switches is their response to a *step* input function.   A step function is a voltage or current that at $t = 0$ changes immediately and instantaneously from a value of zero to a value A, and remains at that level afterwards.

Naturally, the ideal step function can only be approached in practice, for a train of ideal pulses, if described by a Fourier series, contains terms of frequencies to infinity. A device that would pass an ideal pulse must have an infinite bandwidth. The transistor does not follow an ideal rectangular pulse because of transit time, charging time, and minority-carrier storage in its base.

The base current produced by a given base-to-emitter voltage will vary because of production tolerances. To eliminate this problem, a constant current source may be used to provide a controlled value of I_B in saturating switching circuits.

Consider an ideal square wave of voltage, shown in Fig. 9-3(a), applied between base and emitter of an *npn* transistor. The transistor was in the

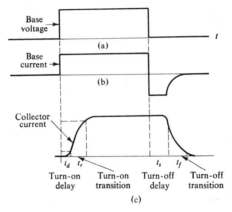

Fig. 9-3 Transistor step-function response: (a) step of base voltage; (b) resulting base current; (c) resulting collector current.

OFF condition, and the pulse is turning it ON. Base current is immediately evident, as in (b) of the figure, but collector current does not instantaneously respond. Electrons from the emitter must travel across the *p*-type base, and a finite time is required for their travel. This time is shown in (c) of the diagram as "turn-on delay." Electrons arriving at the collector travel by various paths; some paths are longer than others, and consequently a finite time is again required before normal operation is achieved. This effect, wherein faster electrons traveling shorter paths arrive at their destination ahead of the less-energetic and dispersed carriers, is referred to in the diagram as "turn-on transition." Delay time is generally measured from the beginning of the input pulse to the 10 percent point on the output waveform. Transition or rise time has, for its limits, the 10 percent and 90 percent marks.

When the transistor is driven into saturation, the collector junction is forward-biased, and the collector then emits electrons into the base (for *npn*). This causes an excess of minority carriers in the base region. A saturated transistor cannot be effectively turned OFF until this "stored base charge" is reduced. Therefore the high current level in the collector is supported by this charge during the time interval immediately after the OFF command is given to the input terminals. *Minority-carrier storage effect* is the name given to the "turn-off delay" encountered. The "turn-off transition" results from different velocities and path lengths that affect the arrival time of the last electrons to reach the collector. Note that the base current suffers a reversal in direction when the base-driving voltage is removed; stored charge accounts for this base current.

To measure stored base charge, a voltage pulse is applied to a circuit such as shown in Fig. 9-4. If the pulse cuts off the transistor, and the output

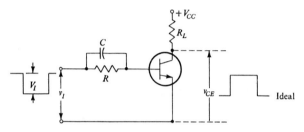

Fig. 9-4   Turn-off of a transistor with a voltage pulse.   A speed-up capacitor is included, and ideal v_{CE} is shown.

(v_{CE}) waveform is observed for various test values of C, a "clean" or best turn-off will be observed with a particular value of C. In this instance "clean" means that v_{CE} will be straightsided. The amount of stored base charge is then given by

$$Q_s = CV_I.$$

The capacitor, C, is often referred to as a *speed-up* capacitor. In circuit design, knowledge of Q_s from given data or from a test such as just described can be combined with information concerning the voltage amplitude of the pulse and base current required for saturation (I_B''). Then R can be determined from

$$R = V_I/I_B''. \tag{9-1}$$

It is of course desirable to minimize the total time required for a switching operation to be performed. Turn-on delay may be reduced by driving the base with a high current, for this provides a greater number of available carriers in that region. The rise time t_r is dependent in part upon the frequency-response characteristics of the device, but can be reduced by *over-driving*, supplying the base with a current pulse of sufficient amplitude to drive the transistor deep into saturation, rather than just to V_{CC}/R_L.

Figure 9-5 shows the collector-current response to three input pulses. The A curve rises to a value of V_{CC}/R_L, or just to saturation. Curves B and C would rise (dashed lines) to twice and thrice the saturation collector

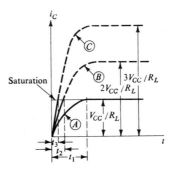

Fig. 9-5   Collector-current response to input pulses.

current, respectively, were it not for saturation, which prevents their complete rise. It is obvious from examination of the figure that curve C has the shortest transient time (t_3), and curve A has the longest (t_1).

When turning OFF a transistor that has been in the ON state, the base region must be swept clean of minority carriers before collector current can cease. The turn-off period (t_f) is characterized by the same parameters that dictated turn-on, except that an initial collector current is apparent. Just as overdriving can speed up the turn-on operation, so it can also be used, in the opposite sense, to turn off a transistor quickly.

Overdriving, when used to saturate or cut off a transistor, may tend to cause a higher input-junction dissipation than is apparent for normal operation, and consideration must often be given to this additional power conversion. An ideal base-current waveform for fast turn-on and turn-off is shown in Fig. 9-6. This waveform results in a speed-up of the drive into saturation, levels off at a value just necessary for saturation so that storage

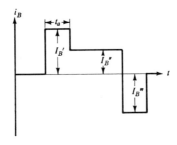

Fig. 9-6   Ideal base-current drive for a saturating switch.

effects will be minimized, and overdrives into the cutoff region when shutoff is required.

A waveform approaching this ideal can be achieved when the speed-up capacitor previously discussed is incorporated into the input circuit of a switching transistor. If the saturated base-current level is again designated as I_B'' and the entire pulse height as I_B', then it is desired that the overdrive, represented by $I_B' - I_B''$, be supplied by the capacitive branch during the transient. The duration of time, designated as t_a, during which an overdrive is required need be only as long as is necessary for the collector current to reach saturation.

Consider the circuit of Fig. 9-7. What appears to be base bias, namely, the V_{BB}—R_2 branch, also seems at first glance to be backward, for V_{BB} is a

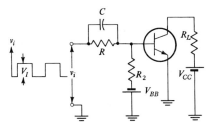

Fig. 9-7   Switching circuit.

negative potential and the circuit uses an *npn* transistor. The function of V_{BB} and R_2 is to provide a base bias at cutoff; that is, to reverse-bias the input junction. This reverse bias on the emitter-base diode will allow only I_{CBO} as the collector current.

Now, turning our attention to the *C-R* branch, we recall that the instantaneous current through a capacitor is given by

$$i = C \frac{dv}{dt}.$$

For the present problem the increment of voltage is V_I; the signal is going to change, from 0 to $+V_I$ volts. The desired capacitive current, as previously stated, amounts to $I_B' - I_B''$, and if the switching-time increment necessary and desired is designated as t_a, then the capacitance may be determined from

$$C = \frac{(I_B' - I_B'')t_a}{V_I}. \tag{9-2}$$

A nonsaturated switch with its ON level in the active region of the collector characteristics can provide faster operation because storage time t_s

will be extremely short.   Circuits have been proposed to "clamp" collector voltage at an unsaturated value.

9-3   SWITCHING TIME CALCULATIONS

In order to estimate the time delays present in a transistor switch, several different models are used and the switching time broken into segments that correspond to the available models.   Here, a brief summary of the most important aspects of switching time calculations is given; the interested reader is referred to the literature for a more extensive treatment of the subject.[15]

Delay Time

Consider a positive voltage pulse applied at $t = 0$ to the base of an *npn* transistor.   The initial base voltage available is V_1 (it can be zero or negative); the pulse rises to V_2.   Refer to Table 9-1.   The first portion of the delay time is the time required to charge junction capacitances, t_{d1}.   The model for this situation is the result of initial operation at $I_C = 0$; thus $g_m \to 0$, $r_{b'e} \to \infty$ and diffusion capacitance $\to 0$.   Remaining are elements representing space charge capacitances C_{ib} and C_{ob}.   Since collector voltage does not change in t_{d1}, that terminal is shorted to the emitter in the model.   Delay time is calculated to be the time required for $V_{b'e}$ to rise to the cutin level, $V_\gamma \simeq 0.5$ V for silicon devices.

The second component of delay time, t_{d2}, represents transit time, the time required for minority carriers to cross the base and reach the collector junction.   A good estimate for t_{d2} is $1/3\omega_T$.

Rise Time

As previously mentioned, rise time is somewhat dependent upon overdrive. We define an overdrive factor $N_1 = h_{FE} V_2 / I_{CS} R_S$.   The ratio $h_{FE}(V_2/R_S)$ represents the theoretical level that collector current would reach if it were not for saturation.   Division by the actual saturated value of collector current, $I_{CS} = V_{CC}/R_C$, causes N_1 to equal or be greater than unity.   Using the model shown in Table 9-1, Eq. (9-5) may be derived for the 0 to 10 percent rise time, referred to as t_{d3}, and Eq. (9-6) derived for the 10 percent to 90 percent rise time t_r.

Storage Time

When the transistor is turned OFF, minority carrier storage effects result in a delay.   A charge control model can be used for derivation of t_{s1}.   The derivation will not be given here; it is available in the literature.[15]   Symbols

TABLE 9-1   SWITCHING CALCULATIONS

Time	Model	Equation
t_{d1}	R_S, b' C_{ob}, R_G, $r_{bb'}$, C_{ib}, V_2, V_1, e	$t_{d1} \cong R_S(C_{ib} + C_{ob})\ln \dfrac{V_2 - V_1}{V_2}$ (9-3)
t_{d2}	none	$t_{d2} = \dfrac{1}{3\omega_T}$ (9-4)
t_{d3}	R_S, $V_{b'e}$ $C_{b'c}(1+g_m R_C)$, $r_{b'e}$, $C_{b'e}$, $g_m V_{b'e}$, R_C	$t_{d3} = \tau_r \ln\left[\dfrac{1}{1 - 0.1/N_1}\right]$ (9-5)
		$\tau_r = h_{FE}(1/\omega_T + C_{b'c} R_C)$
t_r	same as t_{d3}	$t_r = \tau_r \ln\left[\dfrac{1 - 0.1/N_1}{1 - 0.9/N_1}\right]$ (9-6)
t_{s1}	Charge control	$t_{s1} = \tau_s \ln\left[\dfrac{I_{B2} - I_{B1}}{I_{CS}/h_{FE} - I_{B1}}\right]$ (9-7)
		$\tau_s = \dfrac{\omega_N + \omega_I}{\omega_N \omega_I(1 - \alpha_N \alpha_I)}$
t_{s2}	same as t_{d3}	$t_{s2} = \tau_r \ln\left[\dfrac{1 + 1/N_2}{1 + 0.9/N_2}\right]$ (9-8)
t_f	same as t_{d3}	$t_f = \tau_r \ln\left[\dfrac{1 + 0.9/N_2}{1 + 0.1/N_2}\right]$ (9-9)

used in Eq. (9-7) and not previously defined are α_N, the normal transistor α, and α_I, the inverted alpha. When the functions of emitter and collector are interchanged, α_I is the short-circuit current gain. The symbol $\omega_N = \omega_{hfb}$, and ω_I is the cutoff frequency when emitter and collector are interchanged.

Symbol $I_{B1} = V_1/R_S$, and $I_{B2} = V_2/R_S$; they will have different signs if V_1 and V_2 differ in sign.

Fall Time

Calculations for the time required for the collector current to change from 100 percent to 90 percent of its final value, t_{s2}, and 90 percent to 10 percent,

t_f, are similar to rise time calculations, and use an over-drive factor N_2 defined as $-h_{FE} I_{B1}/I_{CS}$. N_2 will be positive, for I_{B1} is normally negative. Equations are given in the table.

EXAMPLE. The switching circuit has the following characteristics:

$V_1 = 0$ V	$R_C = 50\ \Omega$	$C_{ib} = 2$ pF	$\omega_N = 2\pi(120)$ MHz
$V_2 = 2$ V	$r_{bb}' = 100\ \Omega$	$C_{ob} = C_{b'c} = 4$ pF	$\omega_I = 2\pi(10)$ MHz
$V_{CC} = 4$ V	$h_{FE} = 100$	$f_T = 100$ MHz	$\alpha_N = 0.99$
$R_G = 50\ \Omega$			$\alpha_I = 0.6$

Initial calculations give

$$I_{CS} = 80 \text{ mA}, \qquad I_{B2} = 13.3 \text{ mA}, \qquad N_1 = 16.7, \qquad N_2 = 0$$
$$\tau_r = 0.18\ \mu\text{s}, \qquad \tau_s = 0.04\ \mu\text{s}$$

Therefore, the rise switching times are

$$t_{d1} = 0 \qquad\qquad t_{d3} = 1 \text{ ns}$$
$$t_{d2} = 0.5 \text{ ns} \qquad t_r = 9 \text{ ns}$$

The storage and fall times are

$$t_{s1} = 120 \text{ ns} \qquad t_f = 394 \text{ ns}$$
$$t_{s2} = 19 \text{ ns}$$

Thus poor turn-off behavior occurs because the transistor is deep in saturation, and because $N_2 = 0$.

9-4 BINARY ARITHMETIC AND BOOLEAN ALGEBRA

To perform the mathematical manipulations necessary in high-speed electronic computing systems, the transistor switch is used with great effectiveness. In order to understand the application of this device to processes such as counting and addition, we begin by discussing the type of algebra applicable to switching networks.

Binary Addition

Because most types of realizable switching devices can be in only one of two *states*, ON or OFF, computation using such devices is generally limited to use of the *binary* number system. The binary system consists of two symbols, typically 0 and 1, and makes use of these digits to symbolize any decimal number. A *binary digit* is often referred to as a *bit*.

A binary number is made up of bits that represent powers of 2. Thus the number 1010 in binary represents the decimal number *ten*, for reading 1010 from right to left, we find $0 \times 2^0 + 1 \times 2^1 + 0 \times 2^2 + 1 \times 2^3$, and their decimal sum is ten.

In Table 9-2, we note the equivalence of some decimal and binary numbers.

TABLE 9-2

Decimal	Binary	Decimal	Binary
0	0	6	110
1	1	7	111
2	10	8	1000
3	11	9	1001
4	100	10	1010
5	101	11	1011

It is immediately clear from the table that rules for the addition of two binary numbers are very simple. Thus to add the bits 1 and 1, we obtain 0 and carry 1 to the next left-hand column. Some examples of binary addition follow:

101	5	11	3	1010	10
101	+5	01	+1	1111	+15
1010	10	100	4	11001	25

Boolean Algebra

The branch of mathematics called Boolean algebra is especially applicable to electronic computing, for it is based upon only two values or states, often represented by 0 and 1. In switching terms, 0 state could represent an open contact or circuit, while 1 could represent a closed connection. When a Boolean variable is discussed, it exists in either of the two possible states.

Because Boolean algebra differs in many ways from conventional mathematics, the symbolism differs too. Three basic operations are necessary in the use of Boolean algebra for switching circuits; these are AND, OR, and NOT. For the operation referred to as OR, we use the $(+)$ symbol. Thus $A + B$ means "variable A OR variable B." The AND operation is symbolized by juxtaposition; thus AB means the Boolean product of variable A and variable B, and is usually referred to as A AND B. Where needed for clarity, a product dot may be found useful. *Complementation*, a third basic operation, is referred to as NOT, and symbolized by a bar above the variable. Thus $\bar{A}$ is NOT A. For two states represented by 0 and 1, $\bar{0} = 1$ and $\bar{1} = 0$.

To interpret the three Boolean algebra operations noted in the preceding paragraph, the following discussion may be of assistance. In terms of switches, complementation of a closed switch represented by 1 is $\bar{1} = 0$, and 0 is an open switch. The OR operation for two variables, $A + B$, can represent two *parallel* switches. If either A or B close, continuity will result. Thus $1 + 1 = 1$, $0 + 1 = 1$, $1 + 0 = 1$. AND can be envisioned as *series*-connected switches. All must close before continuity exists. Thus $1 \cdot 0 = 0$, $0 \cdot 1 = 0$, $1 \cdot 1 = 1$.

In addition to the three operations, AND, OR, and NOT, certain laws for the manipulation of Boolean algebraic expressions are valuable. The applicable relations are given here:

Commutative Laws	$A + B = B + A$	$AB = BA$
Associative Laws	$A + (B + C) = (A + B) + C$	$A(BC) = (AB)C$
Distributive Laws	$A(B + C) = AB + AC$	$A + BC = (A + B)(A + C)$
Idempotent Laws	$A + A = A$	$AA = A$
Unit and Zero Laws	$0 + A = A$	$1 \cdot A = A$
	$0 \cdot A = 0$	$1 + A = 1$
Complementarity Laws	$A\bar{A} = 0$	$A + \bar{A} = 1$
Involution Law	$\overline{(\bar{A})} = A$	
Dualization Laws (deMorgan's Theorems)	$\overline{AB} = \bar{A} + \bar{B}$	$\overline{A + B} = \bar{A}\bar{B}$

Truth Table

To determine the Boolean function that describes the behavior of several variables, consider the following example. Three variables are present, A, B, and C. We desire the function that describes the case when any two are in the *high* or 1 state. A truth table can be filled out, and the various combinations of variables assigned.

Variables

A	B	C	F
0	0	0	0
0	0	1	0
0	1	0	0
0	1	1	1
1	0	0	0
1	0	1	1
1	1	0	1
1	1	1	1

Ones are noted in the F, or function, column when two variables are in the 1 state. Thus, from the table,

$$F = \bar{A}BC + A\bar{B}C + AB\bar{C} + ABC. \tag{9-10}$$

Using the distributive law, this may be somewhat simplified to

$$F = \bar{A}BC + A\bar{B}C + AB.$$

To implement this function, electronic gates may be employed.

9-5 IMPLEMENTATION OF LOGIC

When a logic function is performed by an electrical network, an electrical quantity, such as current or voltage, must be assigned to represent binary variables. It is most convenient to assign the binary values 1 and 0 to two different *voltage* levels. For example, we can assign the more positive voltage level to correspond to logical 1, and the least positive level to logical 0. This convention is generally referred to as *positive voltage logic*. Negative logic as well as current logic are also possibilities, but will not be used here. The *logic swing* is the magnitude of the voltage difference between the logic states.

The levels referred to in the preceding paragraph are accomplished in a simple common-emitter transistor stage by driving the base terminal from an appropriate source of voltage (see Fig. 9-8). Zero base voltage (v_B) will result in zero collector current, and the corresponding output level at the collector, v_C, will equal V_{CC}. This is the logical 1 or high state at the collector. Recall that for a silicon unit about $+0.65$ volt is required to turn ON the input junction. To drive the collector current to saturation requires about $+0.8$ V. This input saturation level is referred to as $V_{BE(sat)}$, and the collector voltage at saturation is $V_{CE(sat)} \cong 0$ V, or logical 0.

Now, from the preceding discussion, we conclude that the circuit discussed,

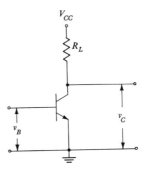

Fig. 9-8   Simple common-emitter inverter stage.

a simple common-emitter stage, is an *inverter* or NOT circuit, for it performs the complementation function.   The output of an inverter takes on the 1 state if—and only if—the input to the inverter does not take on the 1 state. A NOR gate with single input, as discussed in Section 9-6, can function as an inverter.   The amplification afforded by an active inverter is usually welcomed in digital system design.

A *buffer* is sometimes needed to solve loading problems.   It would, ideally, have a high input impedance and low output impedance.   Inversion is not required from a buffer, and therefore an emitter-follower stage or two cascaded common-emitter stages are used.   The buffer is often contained within an IC logic gate package.

9-6   GATING

To realize binary operations involving two or more variables, electronic *gates* are employed.   AND, OR, NAND (not AND), and NOR (not OR) gates are available, and form the basis for digital computations.   We continue to be concerned with voltage-mode logic, wherein the state of a device is represented by the voltage level at its output; generally, a high voltage corresponds to the 1 or ON state, low voltage to the 0 or OFF state.   Table 9-3 is a truth table for these logic functions.   Fig. 9-9 gives generally accepted circuit symbols for these functions.   A more complete treatment of symbols is given in Section 11-2.

Although diode-resistor circuits can be used to perform logical operations, amplification provided by transistor logic circuits has made the former obsolete.

To introduce logic gates, we begin with *direct-coupled transistor logic*, commonly referred to as DCTL.   An example of DCTL circuitry is shown in Fig. 9-10.   Consider that the input lines may be connected to the collectors of transistors that are initially saturated, and that the saturation

TABLE 9-3   LOGIC FUNCTIONS

Variables	AND	OR	NAND	NOR
A B C	$F = ABC$	$F = A + B + C$	$F = \overline{ABC}$	$F = \overline{A + B + C}$
0 0 0	0	0	1	1
0 0 1	0	1	1	0
0 1 0	0	1	1	0
0 1 1	0	1	1	0
1 0 0	0	1	1	0
1 0 1	0	1	1	0
1 1 0	0	1	1	0
1 1 1	1	1	0	0

AND   NAND   OR   NOR

INVERT

Fig. 9-9   Symbols for mathematical operations.

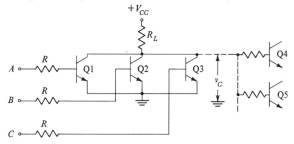

Fig. 9-10   Three-input DCTL circuit.

voltage level is insufficient to turn ON any of the transistors (Q1, Q2, and Q3) in the circuit of the figure. Therefore, the level at v_C will be high, nearly equal to V_{CC}, and this condition, in turn, will be sufficient to cause the load transistors Q4 and Q5 to be in the ON state. Collector voltage v_C is not exactly equal to V_{CC} because of the voltage drop across R_L caused by base current flowing through that resistance to Q4 and Q5.

A positive signal on any input line, A, B, or C, will turn ON the corresponding transistor, drop the voltage level at v_C to $V_{CE(sat)}$, and turn OFF

Q4 and Q5.   At the common collector, the NOR function is being generated:

$$F = \overline{A + B + C}. \tag{9-11}$$

This is clearly the case because a logical 1 input yields a logical 0 output, and vice versa.

Resistances R shown in the figure are used to make the base current drawn by an ON transistor less dependent upon the input diode characteristic.

The term *fan-out*, symbolized by N, is used to indicate the number of identical gates that may be successfully driven (switched) by a single gate. *Fan-in*, M, is the number of identical gates connected at the input of a gate under consideration.   The circuit conditions represented by N and M strongly affect gate behavior.

Propagation Delay

Transistor transient performance was reviewed in Section 9-2, and the usual definitions of delay and transition times were given.   Other terms have become accepted for gate behavior.   They are the *propagation delay* $\overline{t_d}$, (also t_{pd}), and *delay time* t_d.

Refer to Fig. 9-11, which shows a non-ideal voltage pulse waveform and the resulting output voltage waveform.   Phase inversion is obvious.   We are

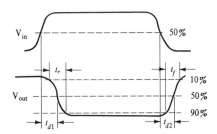

Fig. 9-11   Definitions of delay time t_d, rise time t_r, and fall time t_f.

concerned with the levels at which 50 percent of the ultimate change has taken place.   The time delay between v_{IN} reaching 50 percent and v_{OUT} reaching 50 percent is t_{d1} at the rising end of the pulse and t_{d2} for the falling edge.   Propagation delay is defined as the average of these two delays. That is,

$$\overline{t_d} = \frac{t_{d1} + t_{d2}}{2}. \tag{9-12}$$

This quantity depends upon internal capacitance and charge storage, loading including fan-in and fan-out, and stray capacitances.

Rise time t_r and fall time t_f are conventionally defined as the transition time between 10 percent and 90 percent of the ultimate values, as discussed earlier.

Interesting behavior is shown in Fig. 9-12, where switching time is plotted against operating temperature for the Raytheon type RC-103 NOR gate.

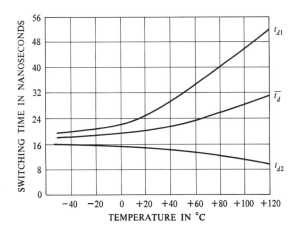

Fig. 9-12   Propogation delay $\bar{t}_d$ and delay times t_{d1} and t_{d2} for RC-103 DCTL gate.

Noise

Noise, undesired disturbances, may be sufficiently strong to cause a logic gate or an inverter to switch its state and therefore falsely indicate signal conditions.   Noise may have several sources: nearby power equipment may induce noise voltages into the digital system; or coupling may be present from other lines in the same equipment, and, under certain conditions, a quiescent line may temporarily be switched because of such coupling.

In the circuit of Fig. 9-13(a), gate Q1 feeds Q2.   When Q1 output is in the 1 or high state, Q2 is fully ON, and because of the load that transistor presents to the collector of Q1, the voltage v_{C1} will not be V_{CC}, but instead will be $V_{BE(sat)}$.   When, on the other hand, Q1 is ON, its collector voltage is $V_{CE(sat)}$ and v_{B2} is also low.   For such an iterative circuit, the operating points on transfer characteristic axes are equal voltages, that is, $A = A'$ in Fig. 9-13(b).

Distance $0A$ represents $V_{BE(sat)}$ and since this quantity is easily determined, $0A'$ is made equal.   This locates the ON operating point.   For the OFF condition, distance C is measured on the characteristic and marked off as C' $(C = C')$.   Thus both points are specified.

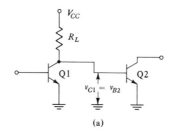

(a)

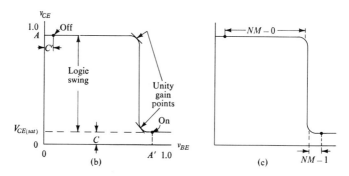

Fig. 9-13   Definitions: (a) relevant circuit; (b) forward transfer characteristic; (c) noise margins.

Unity gain points are also noted in the figure.   They are indicative of a differential gain of one ($\Delta v_C / \Delta v_B = 1$).

The noise margin (NM) is the v_{BE} voltage difference between the operating point and the closest unity gain point.   This definition allows two noise margins: designated *NM*-0 and *NM*-1 in Fig. 9-13(c).

9-7   DIRECT-COUPLED TRANSISTOR LOGIC (DCTL)

A single DCTL gate is shown in Fig. 9-14(a), with typical values given for resistive elements.   As shown, using resistance R_B, *this circuit is also referred to as RTL* (resistor-transistor logic).   With a speed-up capacitor across R_B, this circuit is referred to as RCTL (resistance-capacitance transistor logic). The dashed lines in the figure are connections to other gates within the IC; they share a common collector and R_C is common.

Parasitic elements are associated with the resistances, and also exist between collector and ground or substrate and between base and ground. Normal biasing effectively transforms active parasitics into passive parasitics: the result is shown in Fig. 9-14(b).   The typical capacitance values shown are for parasitics and do not include normal transistor interelectrode capacitances.

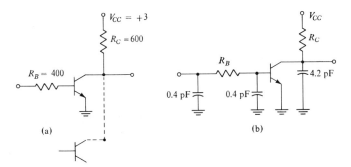

Fig. 9-14   DCTL gate with typical values: (a) physical components; (b) parasitic capacitances included.

Worst-case Considerations

Transistor circuit characteristics vary from their nominal values because of manufacturing tolerances, temperature, aging, power-supply variations, and are affected by circuit loading. The influence of one of these effects, temperature, is graphically portrayed in Fig. 9-15.

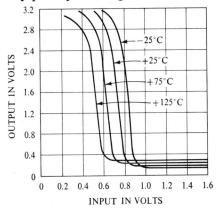

Fig. 9-15   Effect of ambient temperature upon transfer characteristics of DCTL gate.

The design of logic circuitry usually proceeds under a "worst-case" philosophy, in which the worst or extreme conditions to be expected in application are included in the mathematical design procedure. The worst-case design example included here is concerned with dc circuit requirements. It can be extended to transient response and to other types of circuits.

In order for a transistor to be saturated, it is necessary for

$$h_{FE} I_B \geq I_{C(\text{max})}. \tag{9-13}$$

Symbols will be *underscored* to represent the minimum value of a quantity, and *overscoring* will be used to represent a maximum value. Therefore, a worst-case equation for Eq. (9-13) is

$$\underline{h_{FE}} \underline{I_B} \geq \overline{I_C}. \qquad (9\text{-}14)$$

This relation must hold over the desired operating range of the circuit, and must take into account manufacturer's tolerances and aging. The circuit designer using DCTL must be aware of the worst-case gates he will use in a given application. Worst-case information is often available in the form of transfer characteristic variations.

For IC logic gates, it is convenient to transform the worst-case performance into calculation of the maximum fan-out possible with the circuit under consideration. This will provide a design limit of considerable use.

Consider the circuit shown in Fig. 9-16. Q1 is feeding N stages, one of

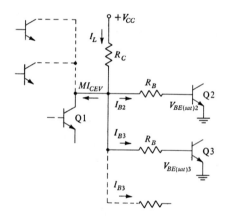

Fig. 9-16   DCTL for worst-case example.

which is considered a starved transistor because its base-emitter drop is larger than the others, that is, $V_{BE(sat)2} > V_{BE(sat)3}$ because of production variations.

The driving transistor Q1 is OFF, and the total OFF current taken by the M transistors of the gate is MI_{CEV}. The N load transistors are all in the ON state. Thus the current through R_C is

$$I_L = I_{B2} + (N - 1)I_{B3} + MI_{CEV}. \qquad (9\text{-}15)$$

A summation of voltages yields

$$R_C I_L = V_{CC} - (I_{B2} R_B + V_{BE(sat)2}). \qquad (9\text{-}16)$$

For simplicity, we define $\Delta V_B = V_{BE(\text{sat})2} - V_{BE(\text{sat})3}$. Then

$$I_{B3} R_B = I_{B2} R_B + \Delta V_B. \tag{9-17}$$

From the foregoing equations, we may obtain an expression for N:

$$N = \frac{V_{CC} - I_{B2} R_B - V_{BE(\text{sat})2} - MI_{CEV} R_C + \Delta V_B R_C/R_B}{(R_C/R_B)(I_{B2} R_B + \Delta V_B)}. \tag{9-18}$$

Now the maximum number of load gates, the fan-out, can be determined from Eq. (9-18) by using the appropriate values for quantities on the right-hand side of that equation. We use the minimum numerator value (a worst case) and maximum denominator value (another) to obtain the maximum N. Thus

$$\overline{N} \le \frac{\underline{V_{CC}} - \overline{I_{B2}}\,\overline{R_B} - \overline{MI_{CEV}}\,\overline{R_C} - \overline{V_{BE(\text{sat})2}} + \underline{\Delta V_B}\,\underline{R_C/R_B}}{(\overline{R_C/R_B})(\overline{I_{B2}}\,\overline{R_B} + \overline{\Delta V_B})}. \tag{9-19}$$

To evaluate this equation, one can proceed as follows. The maximum desired level of I_C is selected. Collector resistance R_C may be determined from

$$\underline{R_C} \ge \frac{\overline{V_{CC}} - \overline{V_{CE(\text{sat})}}}{I_C}, \tag{9-20}$$

and $\overline{I_{B2}}$ evaluated from

$$\overline{I_{B2}} = \frac{\overline{V_{CC}} - \overline{V_{CE(\text{sat})}}}{\underline{h_{EF2}}\,\underline{R_C}}. \tag{9-21}$$

A numerical example can be of assistance. We wish to calculate the maximum permissible fan-out of an IC DCTL gate of the Raytheon RC103 type. The element values and parameters are:

$$\overline{R_B} = 440\ \Omega \qquad \overline{V_{CC}} = \underline{V_{CC}} = 3\ \text{V}$$

$$\underline{R_B} = 360\ \Omega \qquad MI_{CEV} = 0$$

$$\overline{R_C} = 650\ \Omega \qquad h_{FE2} = 50 \qquad\qquad \underline{R_C} = 600\ \Omega$$

$$\overline{V_{BE(\text{sat})2}} = 0.8\ \text{V} \qquad V_{CE(\text{sat})} = 0.2\ \text{V}$$

Assume $\overline{\Delta V_B} = \underline{\Delta V_B} = 0.05\ \text{V}$. Calculation with Eq. (9-21) yields $\overline{I_{B2}} \cong 1$ mA. Then, from Eq. (9-19) we obtain

$$N \le 13.7,\ \text{so}\ N = 13\ \text{gates}.$$

For the case of $\Delta V = 0$, calculations using Eq. (9-19) and the parameters given yield the allowable fan-out to be about 30. The equation is very sensitive to ΔV.

Other calculations are of course possible. It is interesting to note that the allowable fan-out may be maximized if it is possible to select R_B, for that maximum usually occurs at values of R_B over 1000 ohms. Unfortunately, a large value of R_B degrades transient performance, and in most instances a design compromise is called for.

9-8 DIODE-TRANSISTOR LOGIC (DTL)

In the DTL gate shown in Fig. 9-17, diodes D1 perform the logic operation, while D2 and D3 provide an offset voltage. The single transistor Q1 not only amplifies but also inverts the signal, whether that function is desired or not.

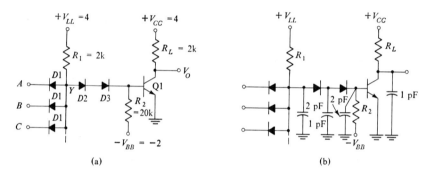

Fig. 9-17 Basic form of DTL gate: (a) typical element values given; (b) circuit of (a) with parasitic capacitances included.

The circuit operates as follows. With all inputs at a low or 0 voltage level, diodes D1 are ON, and the voltage at point Y is slightly positive ($+ V_F$ volts, where V_F is the forward voltage across an ON diode). Diodes D2 and D3 are also ON because of the forward biasing provided by $- V_{BB}$. The transistor Q1 is OFF because its base-emitter voltage is essentially $- V_F$, the drop across D3. With Q1 OFF, its collector voltage is about equal to V_{CC}.

In order to change the conditions at Y, all D1 diodes must be turned OFF simultaneously. The gate, therefore, yields the AND function, plus inversion. DTL basically is NAND logic:

$$F = \overline{ABC}. \tag{9-22}$$

With the D1 diodes OFF, the V_{LL}-D2-D3-R_2-V_{BB} circuit current must be designed so that V_{BE} is sufficiently positive to provide the necessary 700 or 800 mV to change the state of Q1 to full ON. With Q1 ON, V_O drops to $V_{CE(sat)}$, which corresponds to the low or 0 level. To assure successful operation, it is therefore necessary that $V_{CC} > 3V_D$ (one V_D representing the full ON Q1 input junction voltage), and $V_{LL} > 3V_D$, again to insure that Q1 is turned ON.

The basic DTL gate with parasitic capacitances included is shown in Fig. 9-17(b). Raytheon supplies DTL NAND gates with R_2 internally connected to the emitter terminals of Q1 in its RM241 type. Element R_L is nominally 4 kΩ, and values for R_1 and R_2 are 3.6 kΩ and 6 kΩ, respectively. At 25°C, the typical acceptable fan-out is 16 at $V_{CC} = 6$ V, and the average delay time is 40 ns.

Modifications to the basic DTL gate of Fig. 9-17(a) are shown in the circuit of Fig. 9-18. Transistor Q2 eliminates diode D2, resistance R_3 has

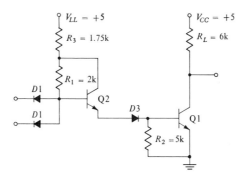

Fig. 9-18   Modified DTL.

been added, and R_L increased from its former value of 2 kΩ to about 6 kΩ. R_2 has been reduced in value and internally connected to the emitter of Q1.

The modifications mentioned in the preceding paragraph have the effect of increasing the maximum permissible fan-out. The total resistance in series with the input diodes has been almost doubled to 3.75 kΩ, thus reducing their ON current. Base current for Q1 is virtually unaffected, for it now flows through 1.75 kΩ rather than 2 kΩ, as previously. Discharge of stored base charge is now dependent upon 5 kΩ rather than 20 kΩ.

Worst-Case Design

In the worst-case design of a DTL gate of the Fig. 9-17(a) type, we must be assured that the transistor type and parasitic capacitances allow the desired speed capability. The maximum value of collector current must be determined consistent with switching speed requirements.

The amount of reverse bias (V_R) on the input diodes when all input lines are high is determined by the required noise immunity. Solution for the reverse input diode voltage from Fig. 9-19(a) yields

$$V_R = V_{CC} - (\overline{V_{BE(\text{sat})}} + 2\overline{V_F}).\qquad(9\text{-}23)$$

where $2V_F$ is the total voltage drop across the two level-shifting diodes. It is assumed that the driving transistors are OFF and their collector voltages are equal to V_{CC}.

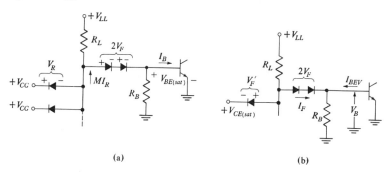

(a)

(b)

Fig. 9-19   Circuits for worst-case equation derivations, DTL.

Still assuming the input lines to be high, we can determine I_B from the appropriate equations:

$$I_B = \frac{V_{LL} - 2V_F - V_{BE(\text{sat})}}{R_L} + MI_R - \frac{V_{BE(\text{sat})}}{R_B}.\qquad(9\text{-}24)$$

Since, as has been mentioned,

$$\underline{I_B} \geq \frac{\overline{I_C}}{\underline{h_{FE}}},\qquad(9\text{-}25)$$

then it must be true that

$$\frac{\underline{V_{LL}} + \underline{MI_R}\,\underline{R_L} - 2\overline{V_F} - \overline{V_{BE(\text{sat})}}}{R_L} - \frac{\overline{V_{BE(\text{sat})}}}{R_B} \geq \frac{\overline{I_C}}{\underline{h_{FE}}}.\qquad(9\text{-}26)$$

This worst case occurs at the lowest operating temperature.

Assume that one of the inputs is in its low state. Refer to the circuit of Fig. 9-19(b). The input level is now $V_{CE(\text{sat})}$, the diodes are forward-biased with drops $V_F{'}$ and $2V_F$, and a small reverse current I_{BEV} flows in the base lead. Let us solve this circuit for the minimum base-to-ground voltage, $V_{B(\text{OFF})}$, required to keep that transistor OFF. In terms of the I_F, the maximum level-shift-diode current at a forward voltage equal to $\overline{V_{CE(\text{sat})}}$ + $\overline{V_F} - V_{B(\text{OFF})}$, this voltage is

$$V_{B(\text{OFF})} \geq (\overline{I_F} + \overline{I_{BEV}})\overline{R_B}.\qquad(9\text{-}27)$$

The worst case usually occurs at the highest operating temperature.

Refer to Fig. 9-20 in order to determine the maximum collector current when the driving transistor is ON and saturated. This current is composed

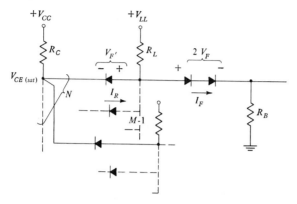

Fig. 9-20 DTL circuit for derivation of Eq. (9-29).

of two components, the current from V_{CC} through R_C and the fan-out (N) current to the load gates. The maximum collector current is

$$\overline{I_C} = \frac{\overline{V_{CC}} - \overline{V_{CE(sat)}}}{R_C} + \overline{N}\left[\frac{\overline{V_{LL}} - \overline{V_{CE(sat)}} - V_F'}{R_L} + (\overline{M} - 1)\overline{I_R} - \underline{I_F}\right]. \quad (9\text{-}28)$$

The final term, $\underline{I_F}$ is the minimum level-shift diode current at a forward voltage approximately equal to $V_{CE(sat)} + V_F'$. Solving this equation for the maximum permissible fan-out yields

$$\overline{N} \leq \frac{\overline{I_C} - [\overline{V_{CC}} - V_{CE(sat)}]/R_C}{[\overline{V_{LL}} - V_{CE(sat)} - V_F]/R_L + (\overline{M} - 1)\overline{I_R} - I_F}. \quad (9\text{-}29)$$

In a circuit-design problem, the characteristics of the transistors used are usually available from test data.

AND-OR-NOT Function

Let us directly join the collectors of two DTL gates of the Fig. 9-17(a) type. By doing this, the common collector voltage will be high when inputs A, B, C to gate 1 are all OFF, and when the inputs D, E, F to gate 2 are also all OFF. The function available at the common collector is

$$F = \overline{ABC} \cdot \overline{DEF}. \quad (9\text{-}30a)$$

By deMorgan's theorem given in Section 9-4, this function can be rewritten as

$$F = \overline{ABC} + \overline{DEF}. \tag{9-30b}$$

Eq. (9-30b) has been referred to as the AND-OR-NOT function, with those three functions (two levels of logic) being accomplished in one gate-switching time. The symbolism is shown in Fig. 9-21.

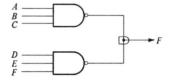

Fig. 9-21 AND-OR-NOT function generation.

9-9 TRANSISTOR-TRANSISTOR LOGIC (T²L)

Transistor-Transistor Logic, referred to as T^2L or TTL, is similar in many respects to DTL. In Fig. 9-22, we note the input diodes of DTL have been

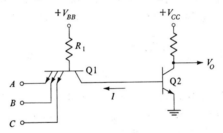

Fig. 9-22 Basic T²L gate.

replaced by a multiple-emitter transistor; and the offset diodes have been eliminated, their function being performed by the collector-base diode of Q1.

The T^2L gate works as follows. Low or logical 0 input is assumed on lines A, B, and C. The corresponding emitter-base diodes of Q1 will be forward-biased because of the polarity of V_{BB}. It is not possible for the reverse base current of Q2 to support full collector-emitter currents in Q1. Since current I shown in the figure cannot be larger than a few microamperes, Q1 is in a saturated state, and the collector-base diode of Q1 provides the offset. This offset voltage is not positive enough to turn Q2 ON.

If all inputs to Q1, A, B, and C, rise to the 1 level, the corresponding emitters are turned OFF, and the collector junction of Q1 is forward-biased. Now the base current of Q1 flows through that junction and on to Q2 in the direction opposite to I in the figure. Transistor Q2 turns ON. Thus the function of the gate is NAND, and

$$F = \overline{ABC}. \qquad (9\text{-}31)$$

A current-hogging problem can exist with both DCTL and T²L because the driving gate must supply input current to an ON T²L gate.

Design modifications have been made to the basic T²L gate of Fig. 9-22. An important circuit is the one shown in Fig. 9-23(a). Element values shown pertain to type SN54HOO marketed by Texas Instruments, Inc. The corresponding voltage-transfer characteristic is sketched in Fig. 9-23(b). This circuit has the ability to drive heavy loads in either state, and it can function well even if the loads are capacitive.

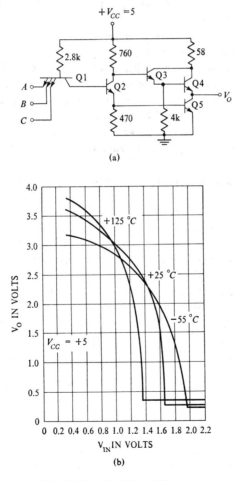

(a)

(b)

Fig. 9-23 Modified T²L gate.

The transfer characteristic shows the logical 1 level at 3.6 volts. With $V_{CC} = 5$ V, and Q2 OFF, $V_0 \cong V_{CC} - V_{BE(Q3)} - V_{BE(Q4)} \cong 3.6$ V, since each base-emitter drop can be considered to be about 0.7 V. Both Q3 and Q4 are functioning as emitter-followers: therefore, in the 1 state, Q4 is able to supply the necessary load current. (The 58 Ω resistance is used to limit the collector current of Q4.)

Consider that the input is low; Q2 and Q5 are OFF, Q3 and Q4 are ON. With the input voltage increasing to about 0.7 V, Q2 will come ON, but Q5 remains OFF. The gain of Q2 is not sufficiently large to turn Q5 ON until the input has increased to two diode drops, or about 1.4 V. When Q5 comes ON, its input resistance lowers, and since its base circuit is in the emitter leg of Q2, this lowered resistance increases the gain of Q2. This is the region in Fig. 9-23(b) with the steep slope.

Transistor Q5 will eventually saturate. As the input voltage further increases, the voltage at the collector of Q2 is sufficiently reduced to begin to turn Q3 and Q4 OFF. The collector current of Q5 is therefore reduced, and this effect is fed back to Q2 because that transistor's emitter voltage has been reduced. Q2 is turned ON harder by the regenerative feedback, and eventually saturates. Q3 and Q4 are then OFF.

The switching behavior described here is not accomplished perfectly, however. Large power-supply transients may result when Q4 switches from ON to OFF.

Typical propagation delays have been reported to be in the neighborhood of 5 ns.

The voltage-transfer characteristic for a T^2L gate of the Fig. 9-23(a) type is objectionable for some applications because of the "extra" breakpoint or bend that exists in the neighborhood of $V_{IN} = 1.25$ V. Actually, the circuit starts its turn-ON at a value of V_{IN} considered to be too low for some applications. This results from premature turn-ON of transistor Q2.

Modifications have been made to reshape the transfer characteristic. The bypass network shown in Fig. 9-24(a) may be added between the emitter of Q2 and ground. A low voltage on that emitter will keep Q6 OFF, and thus the network will restrict Q2 from conducting, even though its emitter junction may be forward-biased. Therefore the collector voltage of Q2 remains at a high value.

The bypass network effectively turns ON when the emitter voltage of Q2 has reached about 0.7 V. At this voltage, Q5 also turns ON. Since the network requires one $V_{BE(sat)}$ level for operation, and Q2 requires a like amount of voltage, we now see that the break in the transfer characteristic has been essentially doubled in value by this modification.

Another modification to the basic T^2L is the addition of diodes between input lines and ground, as shown in Fig. 9-24(b). These elements have no

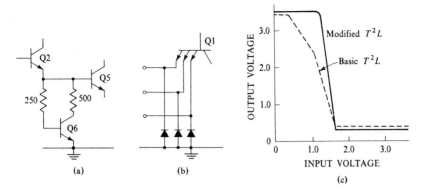

(a)   (b)

(c)

Fig. 9-24   (a) and (b): Further modifications to T²L gate; (c) transfer characteristic.

effect upon dc conditions, but serve to control undesired noise transients. For example, any noise spikes that appear on the input lines greater than negative 0.7 V will be conducted to ground by the diodes.

While T²L logic provides the NAND function, a single stage, acting as an inverter may be added during manufacture between Q1 and Q2. The result is an AND gate.

9-10   EMITTER-COUPLED LOGIC (ECL)

A simplified schematic diagram of an emitter-coupled logic (ECL) gate is shown in Fig. 9-25. A constant current source is shown, with a value of I_0. Distribution of I_0 to the various transistors is dependent upon the state of inputs A, B, and C.

With all input levels low, the transistors Q1, Q2, and Q3 will be OFF, and V_{01} is high, at V_{CC} volts. I_0 will therefore flow through Q7, and $V_{02} = V_{CC} - \alpha I_0 R_C$.

The logical 1 input level is considered to be greater than V_{BB}. If any

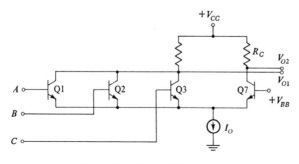

Fig. 9-25   Basic ECL gate.

input line goes high, the corresponding transistor turns ON, and I_O flows through the emitter-collector path of that transistor: Q7 gets no current and is turned OFF. The V_{O1} level drops and the V_{O2} level is raised. It follows that the outputs are

$$V_{O1} = \overline{A + B + C} \qquad (9\text{-}32a)$$

and

$$V_{O2} = A + B + C. \qquad (9\text{-}32b)$$

As discussed previously, the outputs are not referenced to zero volts, but, instead, are relative to V_{CC}. Thus it has become standard to connect emitter-followers to the Q3 collector and to the Q7 collector. The outputs of these transistors are derived from their emitter terminals, and will be zero volts when the base is low and about V_{CC} volts when the base is high.

Another simple modification to the idealized circuit of Fig. 9-25 is the substitution of a resistance for I_O. This element will serve the function of a constant-current source if its value is sufficiently large.

The values shown in Fig. 9-26 are representative of the Motorola type MC1001. In that gate, the V_{CC} line is grounded, V_{EE} should be -5.2 V.

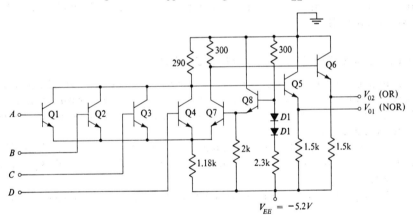

Fig. 9-26   Practical ECL gate.

The internal bias V_{BB} is designed for -1.175 V, and is accomplished by using the drop across forward-biased diodes D1 to provide the base bias for emitter-follower Q8. The emitter current of that transistor flows through a 2 kΩ resistance—the drop across that resistance is V_{BB}.

With V_{BB} equal to -1.175, and $V_{BE} = 0.75$ V for Q7, it follows that the static voltage at the emitters of Q1 and so on equals -1.925 V. This provides for $I_E = 2.77$ mA through the 1.18 kΩ resistance. The OR output is at -1.58 V and this level represents logic 0. The base of the NOR output emitter-follower is essentially at zero volts, so that the output is about −0.75 V, and this is the logic 1 level.

Assume that one or more of the inputs are switched to the high level (> −1.025 V). Allowing 0.75 V for the transistor base-emitter voltage, this leaves 3.4 V across the emitter resistance so that the current through that element has become about 3 mA. The NOR output is $0 − 0.75 − (0.003)$ (290), or at about −1.62 V. As seen in the transfer characteristic of Fig. 9-27, that output continues to decrease with increased base drive on an input transistor until that transistor saturates.

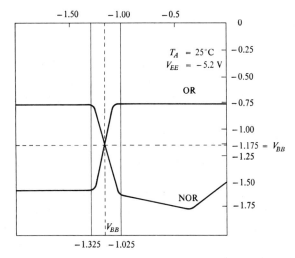

Fig. 9-27 ECL transfer characteristic.

The OR output under the same conditions is one base-emitter drop below ground, or approximately −0.75 V.

Since in normal operation the transistors in ECL gates do not saturate, switching delays associated with saturation effects are eliminated. Propagation delays of 4 ns per gate are reported. Fan-outs of 25 are obtainable.

9-11 FET GATES

The MOSFET is the most widely used FET type for IC gate applications. It is small and easily fabricated, and performs the function of load resistance as well as that of the switching device.

Consider the NOR circuit of Fig. 9-28(a). The devices used are p-channel enhancement type MOSFET's. The polarity of V_{DS} for normal operation is negative; to operate in the ON state, V_{GS} must typically take on values from −8 V to −12 V.

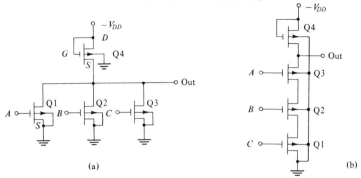

Fig. 9-28   NOR and NAND MOSFET gate circuits using p-channel devices.

A negative-going pulse of sufficient amplitude on line A, B, or C will turn ON the corresponding MOSFET. The output voltage will consequently rise in value from its large negative OFF value toward zero. Load resistance Q4 may represent 200 kΩ; it is physically different from the gate devices in order to achieve high noise immunity.

A NAND gate is given in Fig. 9-28(b). All input lines must be simultaneously fed with negative pulses in order for the output line to switch.

The value of load resistance provided by the Q4 MOSFET can be determined by considering that the transistor is operating in its common-drain configuration (source-follower). Low-frequency analysis makes use of only two parameters, according to the model shown in Fig. 3-10, current generator $g_m V_1$ and resistance r_{ds}. If we study the output impedance of the common-drain stage (looking into the source terminal) with gate shorted to drain, we obtain

$$Z_o = \frac{1}{g_m + g_{ds}} \cong \frac{1}{g_m}. \tag{9-33}$$

The approximation is valid because the operating point in the pentode region of the drain characteristics dictates that g_{ds} must be very small.

Eq. (9-33) cannot predict Z_o with high accuracy, because it is based upon a small-signal model while operation in a gate circuit is concerned with large pulses. Crawford states that the MOSFET load resistor is given by $R_L = 2/G_m$, where G_m is a dc transconductance.[16]

Complementary MOS Gates

Complementary MOS circuits are circuits in which both p- and n-channel FET's are used. Such circuits have been successfully integrated, and are available commercially as digital gates, inverters, flip-flops, counters, and memories. Complementary MOS gates have the advantage of requiring almost zero standby power when in either the ON or OFF state.

An enhancement-type p-channel device is OFF and an enhancement n-channel device is ON when a positive voltage is applied to the gate terminal. The drain-to-source voltage must naturally be negative for p-channel and positive for n-channel devices. With zero voltage at the gate, either type will be in the OFF state.

Let us commence our discussion of practical IC's by considering the inverter of Fig. 9-29(a). Note both p- and n-channel devices in the circuit.

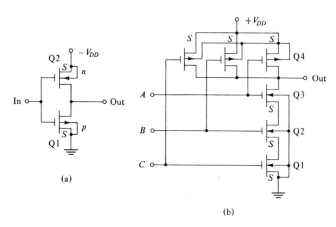

Fig. 9-29 (a) Complementary MOSFET inverter; (b) complementary MOSFET NAND gate.

The input logic levels will be either 0 or $-V_{DD}$ volts, and $-V_{DD}$ is also supplied to the source terminal of the n-channel device. Zero volts on the input line will assure that the p device is OFF, but the n device will be in the ON state because its source is negative, and zero at the gate represents an effective positive gate-to-source voltage. With Q2 ON, the voltage at the output terminal is $-V_{DD}$.

Now when the input is supplied with the other logic level, $-V_{DD}$ volts, Q1 will be turned ON. MOSFET Q2 will be OFF because its gate and source terminals are at the same voltage. The output now is 0 volts.

In either state, one transistor is OFF, prohibiting current flow from the $-V_{DD}$ supply.

The circuit shown in Fig. 9-29(b) performs the NAND operation. Inputs are available to the gates of the series-connected n-channel transistors Q1, Q2, and Q3, and the parallel connected p-channel elements Q4, Q5, Q6. Input levels are $+V_{DD}$ and 0. With $+V_{DD}$ on all input lines, Q1, Q2, and Q3 will be ON, all the p-channel units will be OFF, and the output is zero volts. If only one or two of the inputs are fed positively, at least one p channel will be ON and one n channel OFF and the output is at V_{DD}. Thus the NAND function is implemented. Gates of this type have fan-out capability of greater than 50, high noise immunity of about 50 percent of V_{DD}, nanowatt quiescent power dissipation, typical t_{pd} of 35 ns, high input impedance, and have the further advantage of operation from one supply voltage.

9-12 FUNCTION GENERATION

We have seen that the basic T^2L and DTL circuits perform the NAND function, ECL yields NOR and OR, and DCTL is basically a NOR gate. By added circuitry within the IC package, it is possible, for example, to construct T^2L AND and NOR gates when necessary.

The AND function may be generated simply by following a NAND gate with an inverter. This has the disadvantage of adding another stage of propagation delay.

Using only NAND gates, the NOR function may be generated as in Fig. 9-30. As shown, $\bar{A}$ and $\bar{B}$ supplied to a NAND gate gives an output

$$\overline{\bar{A}\bar{B}} = A + B.$$

The final inverter yields NOR, $\overline{A + B}$.

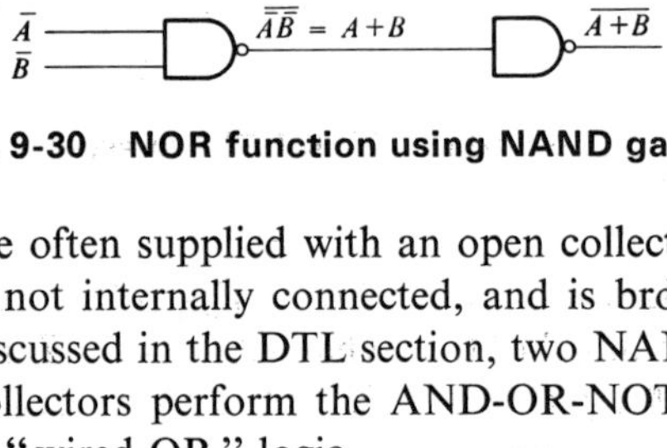

Fig. 9-30 NOR function using NAND gates.

NAND gates are often supplied with an open collector; the collector of the output stage is not internally connected, and is brought out to a connection pin. As discussed in the DTL section, two NAND gates connected together at their collectors perform the AND-OR-NOT function. This is often referred to as "wired-OR" logic.

The *EXCLUSIVE OR* function is defined by the following statement: *The output of a two-input EXCLUSIVE OR circuit assumes the* 1 *state if one and only one input assumes the* 1 *state.* Thus the corresponding truth table is

A	B	F
0	0	0
0	1	1
1	0	1
1	1	0

This operation is often symbolized

$$F = A \oplus B \tag{9-34}$$

From the table, we conclude

$$F = A\bar{B} + B\bar{A} \qquad (9\text{-}35a)$$

By using deMorgan's theorem, it can also be recognized that EXCLUSIVE-OR has other equivalents:

$$F = \overline{AB + \bar{A}\bar{B}} \qquad (9\text{-}35b)$$

and

$$F = (A + B)(\overline{AB}) \qquad (9\text{-}35c)$$

and

$$F = (A + B)(\bar{A} + \bar{B}). \qquad (9\text{-}35d)$$

This function is sometimes realized by an AND-OR-INVERT gate. Thus in the circuit of Fig. 9-31, the two logic stages perform EXCLUSIVE-OR according to Eq. (9-35b). An application is discussed in Chapter 11.

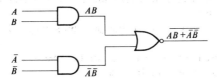

Fig. 9-31. Exclusive-OR logic implementation.

Problems

9-1   For a transient of the form

$$v = K(1 - \varepsilon^{-t/T}),$$

with T the time constant, and K the steady state value of v, make a listing of the percentage of total change occuring in times of $T/2$, T, $2T$, $3T$, $4T$, and $5T$.

9-2   Compare the time required for an exponential transient to build up from 10 percent of its final value to 90 percent of that value with the time required for 0 to 90 percent and 0 to 95 percent. Express your results in terms of the time constant T.

9-3   Mathematically determine the base current waveshape when a voltage step is applied to the parallel combination of speed-up capacitance C_1 and R_1, in series with the base input of a transistor. Represent the transistor by $C_t \| R_t$. Show that the current is made up of three components: an impulse, a decaying exponential, and a rising exponential.

9-4   Explain why capacitances in the t_{d1} model in Table 9-1 are given the symbols C_{ib} and C_{ob}. These elements are $C_{b'e}$ and $C_{b'c}$, but at what level of I_E?

9-5   Derive Eq. (9-3) for t_{d1}.

9-6   Derive Eq. (9-5) for t_{d3}. What assumptions are made in the derivation?

9-7   Boolean variables A and B are available. Use only NOR gates to implement the EXCLUSIVE OR function, $A\bar{B} + \bar{A}B$.

9-8   Simplify by algebraic means; use a truth table if helpful:
(a) $F = A\bar{B} + AB\bar{D} + AB\bar{C}D$
(b) $F = BC\bar{D} + \bar{B}EF + \bar{C}EF + DEF$
(c) $F = AC + BC + \bar{A}B$
(d) $F = \bar{A} + B + \bar{C} + A\bar{B}C$

9-9   Using positive logic, a gate performs the NAND function, $F = \overline{ABC}$. Show that with negative logic, the same gate will accomplish the NOR operation, $F = \overline{A + B + C}$.

9-10   Explain how a DTL of the type shown in Fig. 9-17(a) can be operated successfully if R_2 is attached to the emitter terminal of Q1 rather than to $-V_{BB}$.

9-11   Draw a discrete-component T²L gate without making use of multiple-emitter transistors.

9-12   A manufacturer of T²L gates of the Fig. 9-23(a) type recommends three possible procedures for connection of unused input leads so that the distributed capacitance associated with a floating-emitter will not degrade switching times and noise susceptibility:
(a) Connect unused inputs to an independent supply. How large should this supply be? Why?
(b) Connect unused inputs to a used input. How will this affect maximum permissible fan-out of the driving transistor?
(c) Connect unused inputs to V_{CC} through a series protecting resistance. What will the resistance protect against?

9-13   Implement, with NAND gates, the function $F = AB + BD + D\bar{C} + \bar{A}\bar{B}$. Consider A, B, C, D available. Use inverters where necessary.

9-14   Use NOR and AND gates to implement the following logic equations;

$$D = B(XY + \bar{X}\bar{Y}) + \bar{B}(X\bar{Y} + \bar{X}Y)$$
$$B' = \bar{X}Y + B(XY + \bar{X}\bar{Y})$$

Consider $\bar{X}$, Y, and B available. Try to accomplish this logic with seven gates, including one output OR gate. Your solution is a "full subtractor" performing $D = X - Y - B$, where B is "borrow."

9-15   Develop a truth table for the full subtractor equations given in Problem 9-14.

9-16   Generate the AND-OR-NOT function by conventional gate connections instead of the common collector given in Fig. 9-21. Show the logic diagram for inputs R, S, T and X, Y, Z.

9-17   Discuss reasons why a common op amp may not be suitable as an inverter even though the inverting input was used for the signal.

9-18   Sketch a layout for the positioning of elements of the IC DTL circuit shown in Fig. 9-18. Show interconnection pattern.

9-19   Derive Eq. (9-33) for Z_o of MOSFET load resistance.

9-20   The figure shows a complementary MOS IC. Numbers given refer to external terminals. List those connections necessary to utilize this array as a NAND gate.

9-21   What connections are necessary to make the circuit of the figure into a NOR gate?

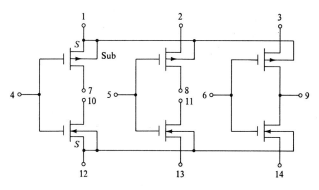

Problems 9-20, 9-21.

9-22   The Siliconix Type SI3001 *analog switch with driver* circuit and logic diagram is shown in the figure.   Normal voltage levels are $V_{CC} = +10$ V,

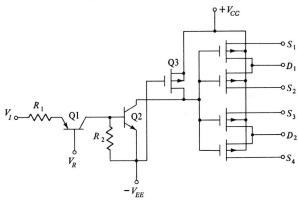

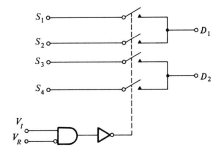

Problem 9-22.

$V_{EE} = -20$ V, $V_R = 0$ V.   The p-channel MOSFET's are normally OFF.   Answer the following questions regarding this circuit:

(a) What combination of voltage polarities on input and output terminals is needed to turn the MOSFET's ON?

(b) Discuss the function of R_1, R_2, and Q1.

(c) How do Q2 and Q3 function?

(d) The V_R line is sometimes referred to as the ENABLE line.   Comment upon this.

10
DIGITAL IC BUILDING BLOCKS

This chapter considers a set of logic functions and electronic circuits that are the building blocks for complex digital IC's and for digital systems. Attention is given to circuits directly related to the implementation of digital system logic as well as to circuits that are of a supporting nature in digital systems.

The previous chapter outlined the binary number system and introduced digital logic gates. These fundamental concepts serve as the basis for the design of the integrated flip-flops, Schmitt triggers, sense amplifiers, input-output drivers, logic translators, and gate expanders considered in this chapter.

10-1 THE FLIP-FLOP

Manipulation of binary numbers in digital systems requires that the numbers be stored, either temporarily or permanently. The *flip-flop*, also referred to as a multivibrator, a bistable, or an Eccles-Jordan trigger, is the digital logic component that provides the required storage. A binary number is stored in a series of flip-flops. One binary digit of a binary number, referred to as a bit, is stored in each flip-flop. In flip-flop circuits the term state refers to a binary digit of 1 as a 1-state and to a 0 as 0-state. Once the flip-flop circuit has stored either a 1 or 0 it changes *state* only if new information is presented at the inputs or if power is removed from the circuit.

A block diagram representing the flip-flop circuit is shown in Fig. 10-1.

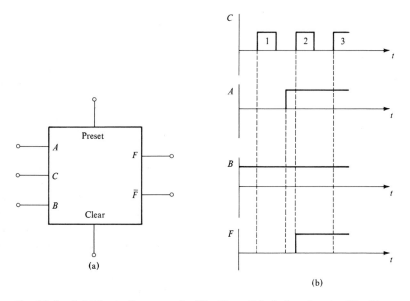

Fig. 10-1   (a) Block diagram of a flip-flop; (b) timing for the flip-flop.

The terminals marked *A*, *C*, *B*, PRESET, and CLEAR are input terminals, and the two marked *F* and $\bar{F}$ are outputs.   The two separate outputs available from the flip-flop are complements.   It is impossible for both outputs simultaneously to have the same value.

Digital systems using ICs are either *synchronous* or *asynchronous*.   A synchronous system is one that performs logic functions in accord with a train of *clock* pulses.   In asynchronous systems, the logic functions are completed sequentially.   Upon completion of one logic operation, an initializing pulse is generated that starts the next operation.   Generally, asynchronous systems are faster as the operations immediately follow each other, whereas, in clocked systems, the speed at which the binary information flows is determined by the rate of the clock.

Terminal *C* in Fig. 10-1(a) is the clock or synchronous input, and the PRESET and CLEAR terminals are the asynchronous inputs.   *A* and *B* are the binary digit inputs.   If the unit is being used in a clocked mode, then the output will change state to comply with binary data present at *A* and *B* only after the occurrence of a clock pulse.   Figure 10-1(b) shows timing for this operation.   The output does not reflect the state of *A* until clock pulse 2 occurs.

The asynchronous inputs, PRESET and CLEAR, override all other inputs.   A 1 on the PRESET line sets the *F* output to 1.   Conversely, the *F* output is set to a 0 by a 1 applied to the CLEAR input.

The term "trigger" is used in so many different ways in technical literature that its usage must be clearly defined. In flip-flop circuits, "to trigger" means to start the change of state of the device. A wide variety of methods are used for triggering on the clock and asynchronous pulses; leading and trailing edge triggering are the most common, but there are systems where multiple level trigger pulses are used. In Fig. 10-1(b), leading edge triggering was used. The output, F, changed state when the leading edge of clock pulse 2 occurred.

10-2   RS FLIP-FLOP

The simplest flip-flop is the *reset-set*, or *RS*, flip-flop. The block diagram of Fig. 10-1(a) can be converted to an RS flip-flop by letting A be the reset and B the set input. The logical function of the set line is to drive the F output to a 1 state and the $\bar{F}$ output to a 0. Conversely, a pulse on the reset line drives F to 0 and $\bar{F}$ to 1. The following truth table clearly sets forth the logical function of the unit. In this table, the subscript n denotes the bit time, and $n + 1$ one bit time later. The output of the flip-flop F occurs one bit time after the inputs R and S have been set to their respective values.

RS Truth Table

t_n		t_{n+1}
S	R	F
0	0	F^n
1	0	1
0	1	0
1	1	indeterminant

As the truth table indicates, the case of 0 on both inputs does not affect a change on the output lines. The symbol F^n is used at t_{n+1} to indicate that F maintains the value it had at time t_n. A 1 at the S input sets a 1 at the F output, and a 1 at the R input resets the output to 0. The case of a 1 at both inputs does not have a definite relationship to a given output, and so it is indeterminant. This particular input combination must be avoided in systems employing RS flip-flops. The JK flip-flop considered in Section 10-3 is designed to overcome this limitation.

The logic of an RS flip-flop can be understood by representing the flip-flop as a cross-connection of two gates. Whether the gates be NAND or NOR is unimportant as either type will provide the flip-flop function. A cross-coupled combination of gates using NOR logic is shown in Fig. 10-2.

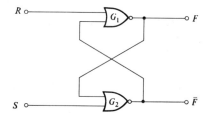

Fig. 10-2   Reset-set flip-flop represented by cross-coupled NOR gates

The truth table sets forth those logical relationships that this pair of gates can provide. Another way of expressing this relationship is to write function equations for the logic circuit. From Fig. 10-2, it is seen that the input to G_1 is $I_1 = R + \bar{F}$. The input for G_2 is expressed by $I_2 = S + F$. Using the NOR gate function, the output F is then given by

$$F = S + \bar{R}F. \tag{10-1}$$

For example, let S be a 1 and R a 0 at bit time n. For this pair of inputs, the output F, at bit time $n + 1$, is a 1, and $\bar{F}$ a 0. All the other states listed in the truth table can be derived from the function equation. The truth table is useful for this example since all the possible inputs and outputs are listed.

Flip-Flop Circuits

One of the simplest and most common RS flip-flop circuits uses resistor-transistor-logic (RTL). The basic gate circuits for this type of logic were discussed earlier in Section 9-7. Dual input RTL NOR gates, such as shown in Fig. 9-14(a), are cross-coupled, as suggested by the logical diagram of Fig. 10-2. The result of this combination is the RS flip-flop of Fig. 10-3.

As a starting point for analysis, assume that the initial state of the circuit, with 0 inputs applied to the reset and set terminals, has Q2 held at cutoff and Q3 in saturation. For this condition there will be a 1 at F and a 0 at $\bar{F}$. By definition the flip-flop is in a set state. It is reset by applying a 1 to the reset line and retaining a 0 on the set line.

A 1 on the reset line biases Q1 so that it starts to draw base current through R_B and the transistor enters the active region. As the base current increases, collector current begins to flow and the collector voltage for Q1 decreases because of the drop across R_C. This decreased collector voltage is coupled to the base of Q3 through the coupling resistor R_B.

The decrease in voltage on the base of Q3 drives it from saturation toward cutoff. Since transistor Q3 is now biased in a region where it has finite gain, its collector voltage begins to rise. This rise in the Q3 collector voltage causes increased drive at the base of Q2 toward saturation, aiding the original

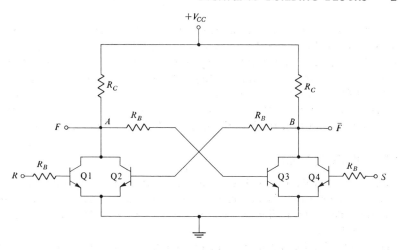

Fig. 10-3   RS flip-flop (RTL).

change in Q1.   The voltage changes just described occur around a loop from
point A to the base of Q3 to point B and back to the base of Q2.   For the
case of a loop gain greater than 1, the switching action will stop only when
Q2 is saturated and Q3 is cut off.   Transistor Q4 is held OFF by the 0 at S
and is not involved in the switching action.

Figure 10-4 shows an RS flip-flop circuit that uses ECL gates.   The basis
of this ECL circuit is the bistable section composed of transistors Q3, Q4,
Q5, and Q6.   Q3 and Q6 are connected in the classical differential configuration, and cross-coupling is through Q4 and Q5.   Circuit action may be appreciated by considering that a 1 is applied to either of the set terminals A

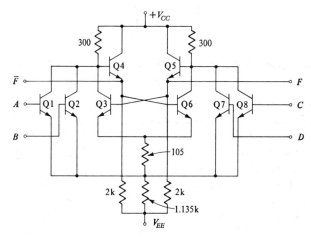

Fig. 10-4   Motorola MC352 ECL Reset-set flip-flop.

or B, and a 0 to the reset terminals C and D.   For this condition, there is base drive for Q5 because the collectors of Q6, Q7, and Q8 are high, and so Q5 conducts, placing a positive voltage at the base of Q3, insuring that Q3 is in saturation.   The low collector voltage of Q3 holds Q4 and Q6 OFF.   The emitter of Q5 is taken as the output F, which, because of the 1 applied to a set terminal, is in a 1 state.   Likewise, the $\bar{F}$ terminal is at 0.   The other logical relationships set forth in the truth table can be verified by the reader.

10-3   JK FLIP-FLOP

RS flip-flops provide a logical function that is adequate for many digital designs.   There are cases, however, where it is necessary to have a flip-flop that has outputs defined for all possible input combinations.   The *JK flip-flop* has this unique feature.

Logically, the JK flip-flop performs the same function as the RS unit, except that the indeterminate state is eliminated.   A method whereby the output state of the flip-flop is sampled and used to gate the inputs yields this desired feature.

A truth table relating the input and corresponding output states illustrates the logical action of the JK flip-flop.   There is a specified output for all input combinations.

JK Truth Table

t_n		t_{n+1}
J	K	F
0	0	F^n
1	0	1
0	1	0
1	1	$\bar{F}^n$

The function equation that mathematically illustrates the logic of the unit is

$$F = \bar{K}F + J\bar{F}. \tag{10-2}$$

Logic diagrams for the JK are given in Fig. 10-5.   The basic memory of the flip-flop is achieved by cross-coupling two logic gates, as in the RS circuit. The output of these gates is fed back to the input gates G_1 and G_2, where the flow of the input signal is controlled.   Addition of these input gates is what gives an output for every set of inputs.

From the truth table, it is seen that if J and K equal 0, the state of the flip-flop remains unchanged. F is set to a 1 by applying a 1 to J and a 0 to K. A 0 is set on the F line by interchanging inputs, $J = 0$, $K = 1$.   Note that when both inputs are 1, the flip-flop operates in a complementary mode. The output changes state each bit time.

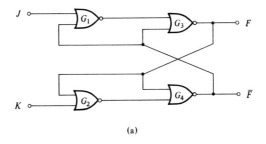

(a)

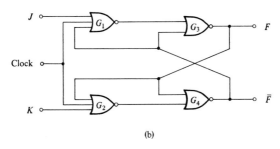

(b)

Fig. 10-5   Logic diagram of the JK flip-flop: (a) unclocked; (b) clocked.

A practical RTL JK flip-flop circuit is illustrated in Fig. 10-6.   The circuit is best discussed by starting with a 0 at F.   According to the truth table, the application of a 1 at J and a 0 at K will set F to a logical 1.   At the start of the cycle, assume that the clock pulse is such that Q13 is full ON.   Therefore, the collectors of Q5 and Q6 are at a low potential.   A 1 at J and a 0 at K holds Q7 and Q8 ON and OFF, respectively.   The collector-base junction of Q6 is now forward-biased, and current will flow from base to collector. Because of this current, charge is stored in the base region.   The clock pulse now goes low, turning Q13 OFF and raising the voltage on the collectors of Q5 and Q6.   Transistor Q6 draws base current which is supplied by the stored charge.   This current drives Q3 toward saturation, lowering its collector voltage, causing the regenerative action necessary for switching. F is set to 1 and $\bar{F}$ to 0.   After about one propagation time delay, transistor Q9 and Q12 turn ON; this removes the base charge from Q6 and holds the base of Q5 low.   This action not only removes the stored charge trigger but places the gate in an inhibit state as well.   An inhibit state does not allow the flip-flop to change state.

The memory function of a JK flip-flop can be performed using other types of logic circuitry.   The logic diagram of a typical T^2L JK flip-flop is shown in Fig. 10-7.   When the AND-OR combination of gates, G_3, G_4, G_7 and

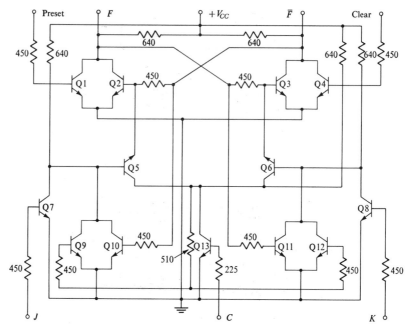

Fig. 10-6   Motorola MC916 RTL JK flip-flop.

G_5, G_6, G_8 illustrated in Fig. 10-7 are considered as one gate, then the logic diagrams of Fig. 10-7 and Fig. 10-5 are nearly identical.

The classical cross-coupling that is characteristic of logic gate flip-flops is seen to connect the output F at gate G_7 to the input of gate G_2. The other path of crossed feedback is from gate G_8 to G_1. Gates G_1 and G_2 are the

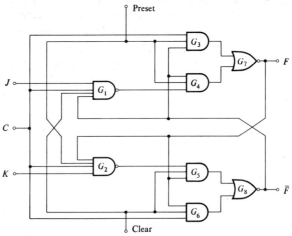

Fig. 10-7   Logic diagram of T²L JK flip-flop.

input gates that sample the output at bit time n and then, along with the applied J and K inputs, determine the output state.

Gates are triggered in different ways, depending on their design. The T^2L JK flip-flop presented here uses negative edge triggering. This means that the transition of data from the input gates to the bistable section occurs during the fall of the clock pulse. For this reason the data must be stored in the input gates for a time prior to clock pulse.

Three time periods are of major interest in clocked gates: the *setup time*, which is the time data must be present before the clocking edge; the *hold or release time*, which is the time data must be held after clocking edge; and *propagation delay*, the time required for data to be set into the bistable section.

Fig. 10-8 shows the timing relationships for logical memory units. For the MC3061 flip-flop, the following numerical values would apply:

$$t_c = 30 \text{ ns}$$
$$t_s = 15 \text{ ns}$$
$$t_h = 0 \text{ ns (can be negative)}$$
$$t_d = 18 \text{ ns}$$

The hold time, t_h, can be negative, since the input pulse can change before the clock pulse transition, and the gate will still give the proper output.

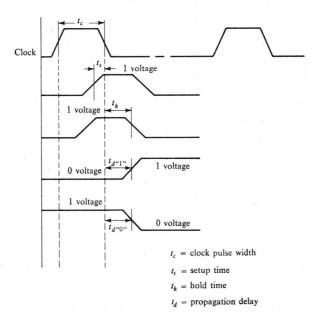

t_c = clock pulse width

t_s = setup time

t_h = hold time

t_d = propagation delay

Fig. 10-8   Timing for a T^2L JK flip-flop.

If an overriding input voltage is applied at the PRESET or CLEAR terminals, there will be a maximum propagation delay of 18 ns before the data appear at the output terminals.

A T^2L circuit which implements the logic of Fig. 10-7 is illustrated in Fig. 10-9. The gates specified in the logic diagram are set apart by the broken lines on the schematic diagram. The reader will recognize that for the most part the gates are similar to the typical circuits that have already been presented. A few of the gates look somewhat different only because of variations in

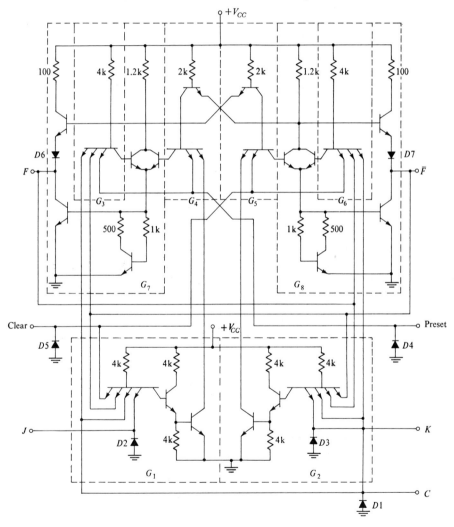

Fig. 10-9   Schematic diagram of T^2L JK flip-flop.

output circuitry.   Gates G_1 and G_2, the input gates, are classical T^2L NAND gates which have two transistor drivers in their output circuitry.   The transistors included in gates designated G_3, G_4, G_5, and G_6 are expanders which, along with the input transistors of G_7 and G_8, make up the AND-NOR function.   Finally, G_7 and G_8 are identified as NOR gates which have the modified form of Fig. 9-24.

By careful comparison of the logic wiring in Fig. 10-7 and the actual wiring between the integrated circuit gates in Fig. 10-9, the reader can identify the function of each gate.   This is required if the action of the circuit is to be understood.   Follow the JK input lines, the PRESET and CLEAR lines, and the clock lines, on both the schematic and the logic diagrams.

The cross-coupling is not so obvious.   The line leading from the F output can be traced to the inputs of gates G_2 and G_6, but what about G_5 ?   The coupling to G_5 is through transistor Q21   This is legitimate since the voltage at F is in phase with the voltage on the collectors of Q11 and Q12.   It is this collector voltage, rather than the output voltage at F, that is being fed back.

Diodes D6 and D7 serve to increase the threshold level of the output gates.   This is similar to the output section of the modified T^2L gate, Fig. 9-23, with only an increased level requirement at the base of Q4.   For silicon diodes, the direct voltage shift is 0.7 volt.

10-4   D FLIP-FLOP

The *delay* or D or shift flip-flop is a single input flip-flop that provides a one-bit storage or memory.   The output is equal to the input only it is delayed one bit-time.   The logic function for the D flip-flop relates the input D at bit-time n to the output F at bit-time $n + 1$.   The truth table for such a memory device is as follows.

D Truth Table

t_n	t_{n+1}
D	F
0	0
1	1

A logic diagram for a D flip-flop is presented in Fig. 10-10.   Notice that the input gate G_1 is set by the input signal.   At the proper interval during the clock period, the information is set into the cross-coupled bistable section of the flip-flop.

The only time when it is not permissible to apply or change the binary information at the input is between the setup and hold times.   Otherwise, when the clock pulse is low, the input information is positioned, by the steering

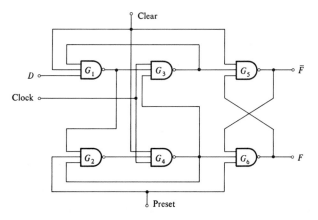

Fig. 10-10   Logic diagram of D flip-flop.

section, and then is set into the bistable section of the flip-flop between the set-up and hold times.   When the clock is high, the input is inhibited and binary data applied to the flip-flop are not retained.   Therefore, the unit will remain in a given state or store a specific binary bit until the inhibit condition is removed.

As with the other circuits that are designed to operate synchronously and asynchronously, there is a line available for directly PRESETTING and CLEARING the flip-flop.   These direct inputs override all other controls of the flip-flop regardless of the state of the clock.

The schematic diagram for a D flip-flop is shown in Fig. 10-11.   The interconnections between the different gates for the circuit can be easily traced.   For example, note how the CLEAR line leading to gates G_1, G_4, and G_5 can be identified on the schematic.   All the other lines can be followed equally well.   The cross-coupling from the output of G_5 to the input of G_6 and symmetrically from the output of G_6 to the input of G_5 is coupled through transistors Q17 and Q18.   Unlike the T^2L JK circuit of Fig. 10-9, these transistors are the only coupling necessary to make up the bistable section of the circuit.

Timing for this D flip-flop is identical to that already presented for the JK flip-flop.   Refer to Fig. 10-8.

Master-Slave Flip-Flop

The flip-flops that have been discussed in the above sections operate error-free only if the clock, PRESET, and CLEAR pulses are not distorted.   If there is rise-time distortion the units may fail to trigger properly.   The *master-slave* flip-flop eliminates this type of error.   It is triggered by voltage

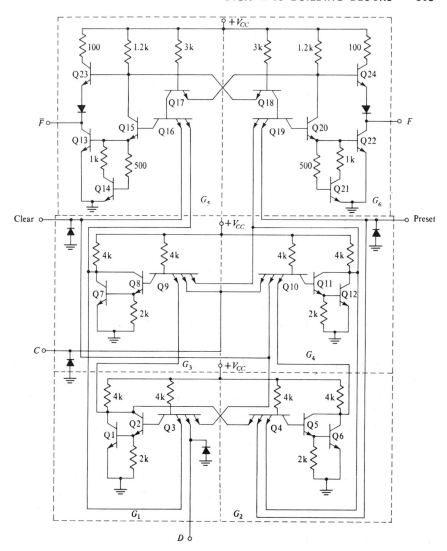

Fig. 10-11 Schematic diagram of D flip-flop. (Motorola MC3060).

levels, thereby removing sensitivity to rise-times. The triggering waveforms can be distorted, but the units will still operate correctly.

The logic diagram of Fig. 10-12 illustrates a master-slave flip-flop. For this representation, NOR gates are used; the reader must recognize that other logic gates would also be suitable. The unit depicted in the figure is wired to operate in a JK mode. If the couplings from F and $\bar{F}$ to K and J, respectively, were eliminated, then the unit would operate with RS logic.

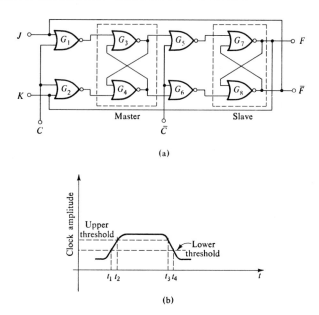

(a)

(b)

Fig. 10-12   (a) Logic diagram of JK Master-Slave flip-flop; (b) clock waveform.

A clocked mode is considered first.   As shown in the logic diagram, the clock pulse controls the data at the input and again at the gates between the two sections of the flip-flop.   The clock pulse waveform that regulates the unit is illustrated in Fig. 10-12(b).   As the voltage waveform changes from minimum to maximum and back to a minimum, there are two voltage thresholds, an upper, and a lower, that mark critical timing points for the flip-flop.   The first level at time t_1 serves to isolate the slave from the master. At time t_2, the binary data present in the input gates G_1 and G_2 are entered into the master section of the flip-flop.   As the clock pulse voltage falls below the upper threshold level, at time t_3, the input gates to the master are inhibited, stopping data flow from these gates to the master section.   Finally, at time t_4, the information is transferred from the master to the slave flip-flop, which completes the activity for one clock pulse duration.

If the master-slave is to be operated in a synchronous or clocked mode, the logic diagram as shown is complete; but for asynchronous operation, the PRESET and CLEAR inputs must be added.   A PRESET input applied to gates G_2, G_3, and G_7 and a CLEAR input applied to G_1, G_4, and G_8 would provide for asynchronous logic.   If these connections are made, then a 1 applied to the PRESET terminals would set F to a 1, and a zero applied to the CLEAR terminals would set F to a 0.   The input pulse for the PRESET and CLEAR operation is routed to both the master and slave sections; so the only requirement is that signals be of sufficient voltage to cause triggering of these bistables.

10-5 SCHMITT TRIGGER

The *Schmitt trigger* is a two-state regenerative circuit where the output state is determined by the level of the input.   A Schmitt trigger is used for squaring voltage waveforms and shifting voltage levels of waveforms.

A graphical relationship between the input and output voltages for a Schmitt trigger is illustrated in Fig. 10-13.   As a starting point for analysis,

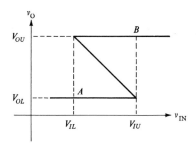

Fig. 10-13   Transfer curve for a Schmitt circuit.

assume that the input voltage is less than the value V_{IL}.   At this same time, the output will be in a low state, defined by V_{OL}.   For the range of inputs $v_{IN} < V_{IL}$, an output is uniquely defined.

For input voltages in the range between V_{IL} and V_{IU}, $V_{IL} \le v_{IN} \le V_{IU}$, uniqueness is lost.   For every input voltage in this range there are three possible output voltages.   However, since the operation is restricted to the transfer curve, the output remains at V_{OL} until $v_{IN} \ge V_{IU}$.   At this input voltage, the circuit enters its active region, which is unstable, and a quiescent operating point cannot be maintained.   Regenerative switching action occurs, and the output increases to the voltage V_{OU}, depicted by point B in Fig. 10-13.   The circuit will remain at this point until the input voltage is altered.   If the input voltage is increased beyond V_{IU}, the output remains at V_{OU}.

As one can see from the transfer curve, a decreasing input voltage eventually causes a transition in output voltage from V_{OU} to V_{OL}.   The output will remain at V_{OU} until the input voltage becomes less than the voltage V_{IL}, at which time the circuit switches back to its initial state $v_O = V_{OL}$.   If the input voltage remains at V_{IL}, the circuit will settle to point A on the curve. Further activity would now depend on the input voltage that is applied to the circuit.

The preceding description has shown how the Schmitt trigger operates as a voltage level detector.   A two-state binary output voltage is obtained from an analog input voltage.

The basic Schmitt trigger circuit is shown in Fig. 10-14.   This is recognized as an ECL circuit.   The operation of the circuit is given in the following discussion.

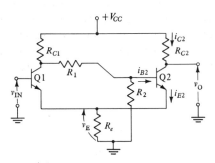

Fig. 10-14   Schmitt trigger circuit.

With $v_{IN} = 0$, the circuit will be in the initial state, where Q1 is OFF and its collector voltage forward-biases Q2.   Until v_{IN} is increased to V_{IU}, Q1 will remain OFF and Q2 conducting, with the resulting output at V_{OL}.   When $v_{IN} = V_{IL}$, the lower input trigger voltage, Q1 will begin to draw current, lowering its collector voltage.   This decrease in voltage is coupled to the base of Q2 through R_1.   Transistor Q2 now starts to turn OFF.   The voltage change of Q2 is fed back to Q1 via the common emitter resistor R_e.   This positive feedback causes the regenerative switching action of the circuit.   The input temporarily loses control because of the regenerative action, and the operating point moves to B on the transfer curve, as shown in Figure 10-13. The circuit is now switched, and the output is at a high value of V_{OU} ; Q2 is OFF, Q1 is ON, with the output voltage at V_{OU}.

If the input voltage is now decreased below the lower triggering level, V_{IL}, the circuit regeneratively switches back to the original state, Q1 OFF, Q2 ON, with an output of V_{OL}.

Expressions are now derived that relate the trigger points V_{IL} and V_{IU} to the circuit parameters.   An equation for V_{IU} will be derived first, followed by a derivation for V_{IL}.   Let v_E be the voltage across the emitter resistor R_e, and the cut-in voltage V_V be defined as that magnitude of voltage when applied to the base that causes appreciable collector current to flow.   A typical V_V is 0.5 V for silicon.   Since V_{IU} is the voltage where Q1 starts conduction, and V_V is the cut-in voltage of Q1, then $V_{IU} = V_V + v_E$.   When Q1 is OFF, the voltage across the common emitter resistor R_e is fixed by the emitter current flow of Q2.   For a transistor, the base, emitter, and collector currents are related by the following equation:

$$i_{E2} = i_{C2} + i_{B2} \tag{10-3}$$

and ignoring leakage current

$$i_{C2} = h_{FE} i_{B2} \tag{10-4}$$

for Q2 in the active region.   Then, by substitution, we obtain

$$i_{B2} = \frac{i_{E2}}{1 + h_{FE}}. \tag{10-5}$$

The base current of transistor Q2 is found by calculating Thevenin's equivalent circuit for the resistive network to the left of the base.   Thevenin's equivalent circuit for the base network is given by V_T and R_T in Fig. 10-15(a). Fig. 10-15(b) shows the circuit used for the calculation.   From this circuit, it is seen that

$$V_T = \frac{V_{CC} R_2}{R_{C1} + R_1 + R_2}, \tag{10-6}$$

and the Thevenin equivalent resistance is found to be

$$R_T = \frac{(R_{C1} + R_1)R_2}{R_{C1} + R_1 + R_2}. \tag{10-7}$$

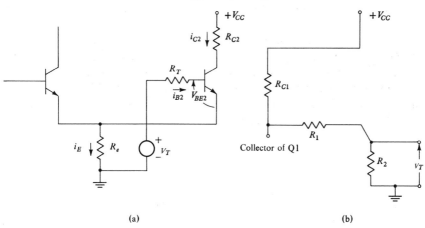

(a)   (b)

Fig. 10-15   (a) Circuit for calculating V_{IU}; (b) circuit used for finding Thevenin's equivalent.

A Kirchhoff voltage equation written around a loop involving the base of Q2, the Thevenin equivalent base source and emitter resistor, R_e, gives the following equation. V_{BE2} is the base-to-emitter saturation voltage, and the saturation resistance is ignored.

$$V_T - i_{B2} R_T - V_{BE2} - i_E R_e = 0. \tag{10-8}$$

Substituting Eq. (10-5) for i_{B2} and solving for i_E gives

$$i_E = \frac{(V_T - V_{BE2})(1 + h_{FE})}{R_e(1 + h_{FE}) + R_T} \tag{10-9}$$

and

$$v_E = i_E R_e = \frac{(V_T - V_{BE2})(1 + h_{FE})R_e}{R_e(1 + h_{FE}) + R_T}. \tag{10-10}$$

The voltage given by Eq. (10-10) plus the voltage drop base-to-emitter when Q1 starts conducting fixes V_{IU}. For the case where $R_e(1 + h_{FE}) \gg R_T$ and $V_{BE2} = 0$,

$$V_{IU} = v_E = V_T = \frac{V_{CC} R_2}{R_1 + R_2 + R_{C1}}. \tag{10-11}$$

Equation (10-11) indicates that *it is possible to design a circuit where the upper trigger voltage is fixed by components other than the active devices.* This is a very important feature to keep in mind to minimize the effects of variations in the transistors due to aging, temperature, manufacture, and the like.

When the Schmitt trigger is in a switched state, v_O high, or Q1 ON and Q2 OFF, the circuit can return to its quiescent condition only when v_{IN} becomes less than V_{IL}. An equation for this voltage is now derived.

Consider the circuit of Fig. 10-16 where Q1 is ON and Q2 is OFF. Recognize that Q2 will turn back ON when the base-to-emitter voltage is greater than the transistor cut-in voltage.

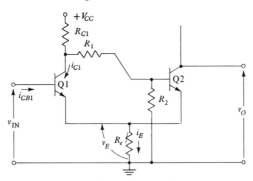

Fig. 10-16   Circuit used for calculating V_{IL}.

When Q2 is OFF, it does not load the collector of Q1, and so by using superposition where i_{C1} acts as a current generator the collector voltage of Q1 can be expressed as

$$v_{C1} = [V_{CC} - i_{C1}R_{C1}]\frac{R_1 + R_2}{R_{C1} + R_1 + R_2}. \tag{10-12}$$

For Q1 active,

$$i_{C1} = h_{FE}\,i_{B1}$$

and

$$i_E = i_{C1} + i_{B1}; \tag{10-13}$$

so

$$i_E = i_{C1}\,(1 + 1/h_{FE}).$$

A voltage equation is written relating the collector voltage of transistor Q1 and the cut-in voltage, V_{V2}, of Q2.   The voltage from base to emitter of Q2 must be increased to its cut-in value before Q2 will turn back ON.

$$v_{C1}\left(\frac{R_2}{R_1 + R_2}\right) - V_{V2} - i_E R_e = 0. \tag{10-14}$$

Equations (10-12) and (10-13) are substituted into (10-14) and then solved for i_{C1}, the collector current of Q1 when Q2 just starts turning ON.

$$i_{C1} = \frac{R_2 V_{CC} - V_{V2}(R_{C1} + R_1 + R_2)}{R_{C1}R_2 + (1 + 1/h_{FE})(R_{C1} + R_1 + R_2)R_e}. \tag{10-15}$$

The lower trigger point is

$$V_{IL} = V_{BE1} + v_E, \tag{10-16}$$

where

$$v_E = i_E R_e$$
$$= i_{C1}(1 + 1/h_{FE})R_e. \tag{10-17}$$

Equations (10-15), (10-16), and (10-17) are combined, giving

$$V_{IL} = V_{BE1} + \frac{[R_2 V_{CC} - V_{V2}(R_{C1} + R_1 + R_2)](1 + 1/h_{FE})R_e}{R_{C1}R_2 + (1 + 1/h_{FE})(R_{C1} + R_1 + R_2)R_e}. \tag{10-18}$$

This expression can be simplified somewhat if it is assumed that $h_{FE} \gg 1$, $V_{V2} = 0$, and $V_{BE1} = 0$.   Under these circumstances, the trigger voltage becomes

$$V_{IL} = \frac{V_{CC} R_2 R_e}{R_{C1}R_2 + (R_{C1} + R_1 + R_2)R_e} \tag{10-19}$$

Equations (10-11) and (10-19) are used as design equations where the circuit can be designed to give a particular transfer curve. The design can include a choice for the so-called voltage hysteresis, $V_{IU} - V_{IL}$, at the input and a suitable output voltage range, $V_{OU} - V_{OL}$. The hysteresis can be used to advantage in a comparator circuit, because once V_{IU} is exceeded, the output will remain high until the input falls below V_{IL}. The lower trigger voltage can be designed as the lowest acceptable limit of an input voltage waveform. When the input is less than V_{IL}, the change in state of the output can be used to activate an indicator.

Design Example

As a practical example, we shall use the circuit of Fig. 10-14 and design a Schmitt trigger circuit. Assume that a matched pair of silicon transistors are used and that the device parameters are defined as follows:

$$h_{FE} = 50$$
$$V_{BE} = 0 \text{ V}$$

The circuit must have trigger points of $V_{IU} = 6.0$ V and $V_{IL} = 5.0$ V, with a supply voltage of 20 V.

In order that regeneration starts when Q1 begins to conduct, Q2 must not be in saturation. This condition is met if $v_{C2} \geq v_{B2}$. For the circuit,

$$i_E = \frac{v_{B2} - V_{BE2}}{R_e}$$

But from Eqs. (10-3) and (10-4),

$$i_{C2} = \frac{v_{B2} - V_{BE2}}{R_e(1 + 1/h_{FE})}.$$

Now the collector voltage is the supply voltage minus the drop across R_{C2}:

$$v_{C2} = V_{CC} - i_{C2} R_{C2}$$
$$= V_{CC} - \frac{v_{B2} - V_{BE2}}{R_e(1 + 1/h_{FE})} R_{C2}.$$

This equation is solved for the resistive ratio:

$$\frac{R_{C2}}{R_e(1 + 1/h_{EF})} = \frac{V_{CC} - v_{C2}}{v_{B2} - V_{BE2}},$$

and for the boundary condition of

$$v_{C2} = v_{B2}$$

we obtain

$$\frac{R_{C2}}{R_e(1 + 1/h_{FE})} = \frac{V_{CC} - v_{B2}}{v_{B2} - V_{BE2}}.$$

When Q1 comes into conduction, the voltage $v_{B2} = V_T$ if $R_T \ll (1 + h_{FE})R_e$. V_T is related to V_{IU} by Eq. (10-11); so

$$\frac{R_{C2}}{R_e(1 + 1/h_{FE})} = \frac{20 - 6.0}{6.0} = 2.33,$$

or, for $h_{FE} = 50$,

$$R_{C2} = (2.33)(1.02)R_e$$
$$= 2.37\ R_e.$$

If R_e is chosen as 1 kΩ, then it follows that

$$R_{C2} = 2.37\text{ kΩ}.$$

Since

$$R_e(1 + h_{FE}) = 51\text{ kΩ},$$

then a Thevenin resistance of

$$R_T = 5.1\text{ kΩ}$$

is acceptable. From Eq. (10-6), with

$$V_T = V_{IU} = 6.0\text{ V},$$

the relationship between R_{C1}, R_1, and R_2 is

$$R_{C1} + R_1 = 2.33\ R_2.$$

A value for R_2 is now calculated from Eq. (10-7):

$$R_2 = 7.28\text{ kΩ}.$$

From the lower trigger level of 5.0 V and Eq. (10-19), the remaining resistive values are calculated:

$$R_{C1} = 0.666\text{ kΩ}$$
$$R_1 = 16.34\text{ kΩ},$$

which completes the design.

The ECL family of logic gates can be interconnected, as in Fig. 10-17, to give a Schmitt trigger circuit. Most of these circuits will have more than one input, thereby allowing triggering from more than one input function. The switching speed will be characteristic of that for the gate which is used.

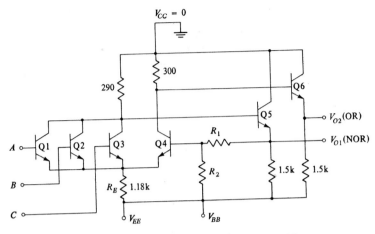

Fig. 10-17 Schmitt trigger formed from ECL gates.

10-6 SENSE AMPLIFIERS AND COMPARATORS

Digital systems are currently being designed with magnetic tape, magnetic disk, solid-core, and active memories. The IC *sense amplifier* considered in this section is used primarily with solid-core memories to amplify low-level binary data. The logic voltages that represent the binary numbers in those memories are generated by a changing magnetic field, and can be expected to have a magnitude of only a few tens or a few hundreds of millivolts.

The small signals obtained from the cores are categorized by the sense amplifiers as 1's or 0's, according to specified threshold levels. The digital output of the sense amplifier is compatible with the digital logic of the system.

Digital words are stored in memory as composites of several bits. A parallel readout of a word is effected by simultaneously reading those memory planes that contain the bits of the word. Each bit requires a sense amplifier. When the memory cores are interrogated, the sense lines must detect the presence of a 1 or 0. If 0's are in memory, the sense lines remain quiescent and the output of the sense amplifier remains unchanged. When 1's are present, the cores are " flipped " on interrogation and a voltage is impressed on the sense line, which sets a 1 in the sense amplifier output logic. There is a very real possibility that the process of interrogation or extraneous noise may produce small voltages on the sense line. For this reason, the sense amplifier must distinguish between legitimate 1 voltages and other, illegitimate, voltages.

Usually the memory cores are set in a matrix where coincidence of select currents produce a 1 voltage on a sense line. The sense amplifier must be capable of making a distinction between the 1 and 0 levels. Fig. 10-18

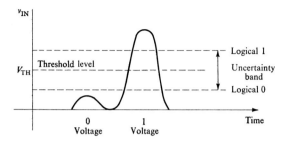

Fig. 10-18   **Core output and logic voltage levels.**

pictorially shows the relationship between the signal and the amplifier selection levels. The threshold voltage is defined as that voltage level which when exceeded will categorize the input voltage as a 1. Ideally, it would be desirable to have a threshold level V_{TH} above which the output would be a logical 1 and below a logical 0. However, because of nonideal circuit characteristics, there exists a finite uncertainty region. If the voltage on the sense line falls in this region, one cannot be sure which output will occur, a 1 or a 0. The typical threshold is 16 millivolts, with an uncertainty band of ± 2 mV.

A block diagram for a typical sense amplifier is illustrated in Fig. 10-19. Three blocks depict the functions required of the unit. One stage is a

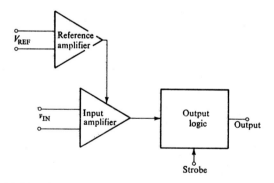

Fig. 10-19   **Functional block diagram of a sense amplifier.**

high gain amplifier, and another stage includes the threshold and reference circuits. An output stage shapes the waveform so that it can be used to drive the system logic gates.

An externally generated strobe pulse is used with most sense amplifier circuits to serve two purposes: to enable the circuit to take advantage of high signal-to-noise ratio, and to inhibit the circuit at all other times, including the write cycle time. The maximum signal-to-noise ratio is reached during the memory read cycle, and so the strobe is timed accordingly.   The strobe gate is included with the output logic in Fig. 10-19.

The differential amplifier is the fundamental circuit of most sense amplifiers.   It is used for both the input and reference amplifiers.   There are two good reasons for using this circuit: it can handle pulses of either polarity generated by the memory; it eliminates common-mode signals.

Core memories require that several lines be associated with each core in order that biasing and read-write functions can be performed.   These windings are excellent noise sources, and the noise level can present a considerable obstacle to accurate sensing.

A differential amplifier with a high common-mode rejection ratio suppresses the winding noise.   The common-mode signals, while not entirely eliminated, are at least reduced to a level where they will not interfere with the desired signals.   Threshold setting circuits also help to minimize noise by excluding those signals that fall below a certain predetermined amplitude.   Differential amplifier operation has already been discussed in detail; only the aspects which are pertinent to the sensing function will be presented.

One example of a sense amplifier, which is a Texas Instrument Series 7520N sense amplifier, is illustrated in Fig. 10-20.   It is seen that both the

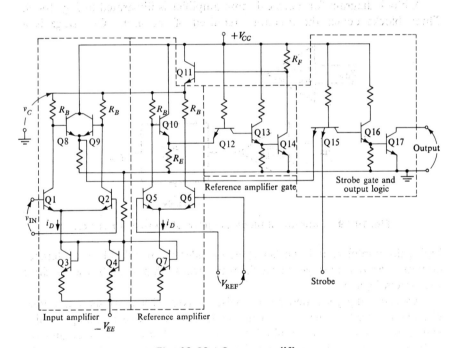

Fig. 10-20   Sense amplifier.

input and the reference amplifiers are differential amplifiers. The threshold level is determined by the external voltage applied to the reference amplifier.

The output logic is T^2L type, and is similar to the gates discussed in Chapter 9. Likewise, the fanout is the same as for the other T^2L circuits.

The reference amplifier of the figure has gating that is identical to that of the output logic. The only difference between the two circuit portions is a feedback loop around the gate of the reference amplifier, which serves to maintain that gate in the active region for all usable reference voltages. Voltage is coupled to the base of the emitter-follower transistor Q10 through the feedback amplifier, comprised of Q11 and R_F.

An emitter-follower involving Q10 and R_E couples the differential amplifier output to the reference amplifier output gate transistors Q12, Q13, and Q14. Notice that the input differential amplifier has an emitter-follower connected to each collector. This arrangement allows for both positive and negative pulses on the sense lines from the cores. This bipolar capability is most desirable.

A relationship between the threshold and reference voltages will now be derived. The potential at the emitter of Q15 is the critical voltage, because it drives the output gate to saturation or cutoff. Equation (5-27) for the case of $\alpha_{DC} = 1$, gives an expression for the voltage across R_B. Then for an input voltage less than the threshold voltage, an equation for $v_E(Q15)$ is

$$v_E(Q15) = v_C - i_D R_B\left(\frac{1}{1 - \epsilon^{\Delta V_{\text{IN}}}}\right) - V_{BE}(\text{Q8 or Q9}), \qquad (10\text{-}20)$$

where v_C is determined by the reference amplifier, and is given as

$$v_C = V_{BE}(Q14) + V_{BE}(Q13) + V_{BC}(Q12) - V_{BE}(Q12) + V_{BE}(Q10)$$

$$+ i_D R_B \frac{1}{(1 + \epsilon^{\Delta V_{\text{REF}}})} \qquad (10\text{-}21)$$

When the threshold voltage at the emitter of Q15 has been reached, then the output gate will turn ON, saturating Q15, Q16, and Q17. The voltage at the emitter of Q15 for the saturated condition is

$$v_E(Q15) = V_{BE}(Q17) + V_{BE}(Q16) + V_{BC}(Q15) - V_{BE}(Q15). \qquad (10\text{-}22)$$

For the case where the input voltage is at the threshold level, $v_{\text{IN}} = V_{TH}$, Eq. (10-21) is substituted into Eq. (10-20), and then Eqs. (10-20) and (10-22) are combined. This is the boundary condition between the active and saturated regions of the output logic. It is now assumed for integrated circuits that the resistors, transistor parameters, temperature coefficients, and so on, can be matched so that all saturation voltages and resistances are equal. The equations then reduce to

$$I_D R_B \frac{1}{(1 + \epsilon^{\Delta V_{REF}})} = I_D R_B \frac{1}{(1 + \epsilon^{\Delta V_{TH}})},\qquad (10\text{-}23)$$

or, finally,

$$V_{TH} = V_{REF}.\qquad (10\text{-}24)$$

This equation shows that the threshold level is fixed by the external reference voltage. The system designer has the option of choosing a reference level that is best suited for his particular design requirements.

Comparators

A comparator circuit senses when an input voltage has a specific magnitude. The output signal of the comparator circuit is in digital form and generally appears as a change in a logic voltage level.

Comparators are used in systems where the level of an analog signal is of interest. A detection system with a photo cell as the transducer is one example. The detector is designed so that there is an analog voltage developed across the transducer that is proportional to the light intensity at the surface of the photo cell. This analog signal forms one input to the comparator. The other input is often a reference direct voltage.

Designs for comparators are available where the functional circuit is a differential-input operational amplifier. These designs are capable of providing the required comparison function. They are acceptable for many applications, even though their response speed is limited.

From this discussion we conclude that the sense amplifier is useful for comparator applications. The amplifier has a differential input, variable threshold, and the output logic is compatible with T^2L circuits.

10-7   DRIVERS AND RECEIVERS

Standard logic gates are usually not designed with sufficient power capability for transmitting digital data and also actuating memories, readouts, displays, and other peripheral equipment associated with digital systems. Rather than manufacturing every gate with a high output power capability, thereby employing overrated gates for routine logic functions, special gates are utilized. These special gates are called *drivers*.

Drivers interface between logic circuitry and system components, such as cables, clock lines, incandescent lamps, relays, memory cores, neon bulbs, readout tubes, and magnetic write heads. Digital drivers, available in IC form, are capable of driving capacitive loads from sources of low impedance, matching transmission lines, providing high current, high voltage, or high power levels.

Many drivers are simply logic gates with single-transistor inverter amplifiers added to the output circuitry.   In other instances, the driver circuit is much the same as that of a standard logic gate.

Bias Driver.   Many circuits, especially in the ECL line of digital IC's, require a fixed voltage level that can be used as a reference voltage.   The voltage must be temperature-compensated over the same range as the associated ECL gates.   The bias driver circuit of Fig. 10-21 provides such a

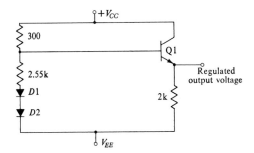

Fig. 10-21   Motorola MC354 bias driver.

regulated voltage.   This circuit is a regulated power supply, and at 25°C, if $V_{CC} = 0$ and $V_{EE} = -5.2$ volts, the regulated output is -1.15 volts.

Clock Driver.   As was mentioned, drivers may be made from standard circuits by slightly altering the component values and the fabrication method. During manufacture of the clock driver, the circuit components are carefully located so that lead lengths are minimized.   By strictly controlling every phase of the fabrication process, a better response-time–power-output product can be achieved.   The gate will then have a higher fanout at the same operating speed.

A driver that is a variation of a standard gate circuit is shown in Fig. 10-22. The reader will recognize that the circuit configuration is the same as the ECL gate in Fig. 9-26.   Only a few component values have been changed.   The clock driver circuit shown is one-half of Motorola's MC1023 dual four-input clock driver.   The circuit is especially designed to yield a high power output without slowing the switching speed.   Such a device is invaluable for simultaneously clocking a string of logic gates, such as a long shift register that must receive simultaneous clock pulses at every stage so synchronism will not be lost and the register will function properly.

A comparison between the driver and the standard logic gate can best be made on the basis of the rise time and the propagation delay as functions of

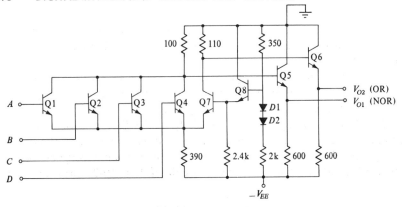

Fig. 10-22   ECL clock driver.

loading.   The curves illustrated in Fig. 10-23 show how much improvement is obtained over the standard circuit.   The propagation delay is about 2.5 nanoseconds faster for the driver circuit.   Likewise, the rise time for this specially designed circuit is shortened by about 3 nanoseconds.

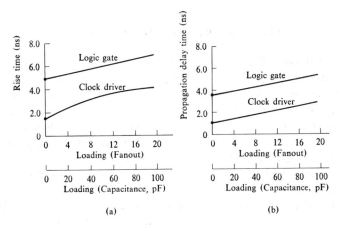

(a)   (b)

Fig. 10-23   (a) Rise times; (b) propagation delay times for logic gate and clock driver.[13]

Long Line Drivers and Receivers

Complex digital systems require large numbers of electronic components.   It is not possible to locate them in close proximity.   Therefore, in many cases, binary data must be transmitted more than just a few inches.   For extended distances of several feet, the environmental noise level is usually so great that special precautions must be taken to insure error-free transmission.

The noise level can be reduced somewhat by selecting a special type of cable and by terminating the cable properly.   For short runs of a few inches, regular coaxial cable is satisfactory; for longer runs, twisted-pair cable is employed as the transmission line because the twisting affects a cancellation of voltages that are induced in the line.   Also, by terminating the line in its characteristic impedance (typically less than 100 ohms), signal reflections are eliminated.

A block diagram depicting a long-line data-transmission system is shown in Fig. 10-24.   Two-wire systems are desirable because differential circuits can

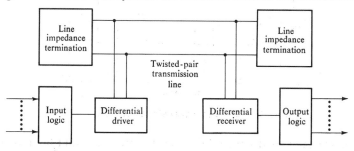

Fig. 10-24   Block diagram of data transmission system.

be used as the *line driver* and the *line receiver* for suppression of common-mode signals.   The receiver differential-input stage has a high common-mode rejection ratio.   Further, the output impedance of the driver and the input impedance of the receiver are designed to match the transmission line characteristic impedance.

10-8 TRANSLATORS

Digital systems that employ saturated and nonsaturated logic gates require a circuit that will interface between the two.   The voltage level of the binary data must be shifted when data flow from one type of logic to the other.   A *translator* provides the necessary voltage shift.

A basic translator circuit for translating from nonsaturated-to-saturated logic is shown in Fig. 10-25.   This circuit is an ECL gate with the addition of a transistor switch Q1 acting as a load for one collector.   The circuit values are chosen so that the base-to-emitter voltage of Q1 is held at proper bias. If a high voltage is applied to the base of Q2 and a low voltage to the base of Q3, then Q2 will draw collector current while Q3 will not.   As the collector of Q3 is high, transistor Q1 will be forward-biased and draw base current through R_{C3}.   This base current saturates Q1, lowering its collector voltage to a level typical of saturated logic gates.   A high level output is obtained by applying inputs so that Q2 is cutoff and Q3 is saturated.

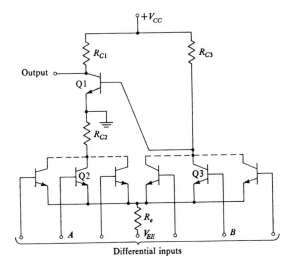

Fig. 10-25   Basic translator circuit.

A practical integrated circuit ECL-to-saturated logic translator is illustrated in Fig. 10-26. A positive logic NOR function is available when the circuit is wired as shown on the schematic diagram. The logic function depends on which side of the differential amplifier the bias driver is connected. If it were connected on the opposite side of the diff amp, the logic function would become OR.

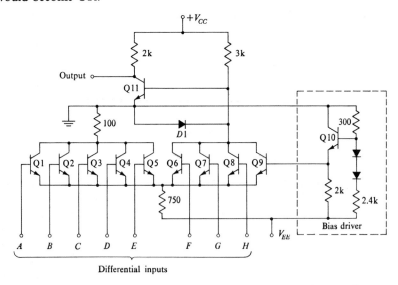

Fig. 10-26   Motorola MC1018 ECL-to-saturated logic translator.

The bias driver controls the operating point of the input transistors. The bias voltage for these transistors must be held constant so that the correct saturated-level output voltage is realized.

The circuit operates in a manner that was explained for the simpler version of Fig. 10-25.   Diode D1 limits the collector excursion of transistors Q6, Q7, Q8, and Q9.   By holding these transistors out of saturation, the circuit switching speed is increased.

With supply voltages of $V_{CC} = 5.0$ V and $V_{EE} = -5.2$ V and an operating temperature of 25°C, the bias driver output is -1.15 C.   A 1 input voltage of -0.86 V causes an output voltage of 0.3 V and for a 0 input voltage of -3.3 V, Q11 is cut off, so that the output is near 5.0 V.   The propagation delay is typically 19 ns.

A saturated-to-nonsaturated logic circuit is illustrated in Fig. 10-27.   The nucleus of this circuit is the differential amplifier, Q1-Q2.   The diff amp

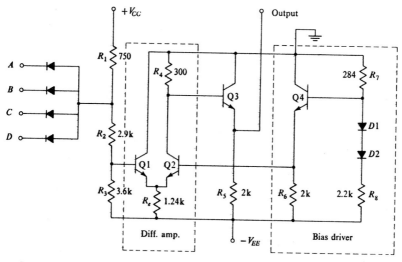

Fig. 10-27   Motorola MC1017 saturated logic-to-ECL translator.

compares the input level to a reference level supplied by a bias driver.   The output of the amplifier drives an emitter-follower circuit that yields the voltage level characteristic of nonsaturated circuitry.   For $V_{CC} = 5.0$ V and $V_{EE} = -5.2$ V, an input of 1.9 V results in an output of -0.775 V, and 1.0 V at the input makes the output -1.65 V.   The propagation delay for the circuit is 15 ns.

The input voltage is gated by the input diodes and then reduced in magnitude by the resistive voltage divider R_2 and R_3.   This fraction of the input voltage is applied to the base of Q1, and since the reference voltage is applied to the base of Q2, the output at the collector of Q2 is determined by the difference between the two voltages.

Output transistor Q3 reacts to the voltage applied to its base and saturates or cuts off accordingly. The output taken at the emitter of Q3 has the phase of the input voltage, and the voltage level has been appropriately shifted.

The input diodes have voltage levels that are typical of saturated logic gates, and they provide the logical AND function. The circuitry following the diode gate serves to shift the level of the logic voltage without altering the logic function of the diodes.

10-9 EXPANDERS

Logic functions cannot always be conveniently performed with the limited number of input lines available with standard logic gates. When more inputs are needed for implementation of a particular logic function, an *expander* can be used. The expander serves to increase the fanin capability.

It is not possible to expand every logic gate of a particular line of integrated circuits. The manufacturer will specify those gates that are expandable. Typically, provisions are made to expand the EXCLUSIVE OR (AND-OR-INVERT) gate. The logic diagram for this gate is shown in Chapter 9, Fig. 9-31.

Additional OR terms, if needed for a logic function, may be obtained with an expander coupled to an EXCLUSIVE OR gate. A typical number of expanders that may be wired into a T^2L EXCLUSIVE OR gate is 4.

A T^2L expander and expandable gate are shown in Fig. 10-28. Combining these circuits the logic an expanded OR capability. The reader will recognize that the circuit for the EXCLUSIVE OR gate is very similar to the modified T^2L circuit of Fig. 9-23. The AND logic functions are performed by the multiple-emitter transistors Q1 and Q4. Transistors Q2 and Q3 are in the NOR gate configuration. The combination of AND-NOR makes the EXCLUSIVE OR gate.

When Q2 or Q3 is saturated because of the base drive from Q1 or Q4, the output transistor Q6 is also driven into saturation, so that the output voltage at the collector of Q6 is at a low value, about 0.3 V. As the base drive for Q2 and Q3 is removed, they go high and the output transistor Q5 turns ON, pulling the output toward V_{CC}. The diode D insures that both output transistors will not be ON simultaneously. This point can be illustrated by examination of a voltage equation. As was mentioned, when either Q2 or Q3 is driven to collector saturation, the base-to-emitter junction of Q6 is also saturated. The voltage V_A at the collector of Q2 is equal to the sum of two saturation voltages: the collector-to-emitter voltage of Q2, and the base-to-emitter voltage of Q6:

$$V_A = V_{CE2} + V_{BE6}. \tag{10-25}$$

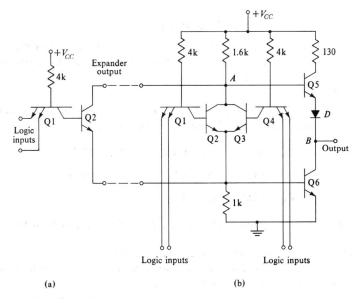

Fig. 10-28   (a) Expander; (b) Texas Instrument 7450 expandable EX-
CLUSIVE OR.

The saturation voltage at the collector of Q6 is V_{CE6}. For this saturated
condition, the voltage V_{AB} from the base of Q5 to the cathode of the diode
equals

$$V_{AB} = V_A - V_{CE6}. \tag{10-26}$$

Now, as a good approximation, it can be assumed for IC transistors that the
saturation voltages will be the same order of magnitude; then Eq. (10-26)
reduces to

$$V_{AB} \cong V_{BE6}. \tag{10-27}$$

If the diode were not in the voltage path from point A to point B, Q5 might
be forward-biased base-to-emitter.   As Eqs. (10-25) and (10-26) show, if the
saturation voltages are not identically matched or if V_{BE6} is greater than V_{BE5},
then Q5 would be saturated, or at least partially ON, when it should be cut off.

The diode in the emitter circuit of Q5 raises the threshold voltage of Q5
by 0.7 V.   This additional voltage insures that Q5 will not inadvertently be
turned ON.   Transistor Q5 will turn ON when $V_{AB} \geq V_{BE5} + V_D$, where V_D
is the saturation voltage of the diode.

The switching response of expanded gates may be impared if load capacitance is excessive.   One method in general practice is to locate the expander
close to the expanded gate so that lead lengths are short, thereby maintaining
the system switching speed.

Problems

10-1   Derive an expression similar to Eq. (10-11) for the case where the ground terminal of R_e in Fig. 10-14 is returned to a source V_{EE}. Comment on the resulting relationship between V_{IU} and V_{EE}.

10-2   Derive Eq. (10-15).

10-3   For the Schmitt trigger circuit of Fig. 10-14, the transistors are matched and all other values except that of resistor R_{C2} are known.

$$V_{cc} = 10 \text{ V} \qquad R_2 = 6 \text{ k}\Omega \qquad V_{BE} = 0.7 \text{ V}$$

$$R_{C1} = 3 \text{ k}\Omega \qquad R_e = 0.5 \text{ k}\Omega \qquad V_V = 0.5 \text{ V}$$

$$R_1 = 3 \text{ k}\Omega \qquad h_{FE} = 20$$

(a) Calculate the magnitude of R_{C2} that is required to just hold Q2 out of saturation.

(b) Solve for the output voltage levels V_{OU} and V_{OL}.

(c) If $v_{IN} = 8 \sin 2\pi 10^3 t$, determine the time with respect to v_{IN} when the output voltage goes to V_{OU} and when it returns to V_{OL}.

10-4   Find the trigger levels for the Schmitt circuit that was designed in Section 10-5 if the transistor base-to-emitter saturation voltages are 0.7 V. Also, for this problem, consider that the cut-in voltages are 0.5 V.

10-5   Redraw the ECL circuit of Fig. 10-17 so that it has the form of Fig. 10-14.

10-6   Assume that Q5 in your solution to Problem 10-5 is fully saturated when Q1, Q2, and Q3 are OFF and that saturation resistances and voltages can be ignored. Determine the Thevenin voltage and resistance, V_T and R_T, as defined in the text.

10-7   A flip-flop that has application in digital systems in the T or *trigger* flip-flop. It is a single input device. If the input is a 1 the output changes state or flips. For a 0 at the input the output remains unchanged. Derive a truth table for the T flip-flop and write the functional equation that describes the circuit action.

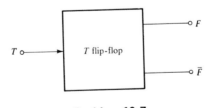

Problem 10-7

10-8   A discrete *astable multivibrator* is shown. This circuit is an oscillator in the sense that it has no stable state. The transistors are generally driven from saturation to cutoff and visa versa, repetitively. The waveforms generated at the collectors are almost square waves, and are used for timing pulses in digital systems.

Equations for the collector and base voltages of Q2, v_B, and v_C, respectively, are given by the following:

$$v_B = +V_{cc} - 2V_{cc}\epsilon^{-t/R_1C_1}$$
$$v_C = +V_{cc}(1 - \epsilon^{-t/R_{C2}C_2})$$

These equations are written for the case where device saturation and cut-in voltages and saturation resistances are ignored.   Plot the waveforms defined by the equations for t in the range from 0 to T.   Assume that $R_1 = R_2 = 10R_{C1} = 10R_{C2}$ and $C_1 = C_2$. T is the time that $v_B = 0$ and is also the width of the collector voltage pulse.

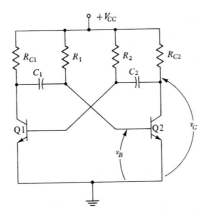

Problem 10-8

10-9   Modify the circuit of Fig. 10-3 as suggested in the diagram and recognize that this circuit is an astable multivibrator.   Derive equations that are similar to those of Problem 10-8 for the base and collector voltages of Q2.

Show that in the limit $R_{B2} \to 0$ and $R_{B1} \to 0$, $v_B' = v_B$ and $v_C' = v_C$, where the primed voltages are defined in the figure and v_B and v_C are given in Problem 10-8.

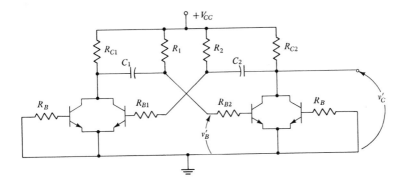

Problem 10-9

10-10   *Monostable multivibrators* are used to delay digital pulses, to generate pulses with controlled widths, and for waveshaping in digital systems. When the circuit is triggered by a small pulse, it generates a pulse whose width is controlled by the circuit designer.

A discrete monostable circuit is shown in the diagram. For the stable state, Q1 is OFF and Q2 is ON. Upon triggering, Q1 turns ON and Q2 is driven OFF by the regenerative action of the circuit. This unstable condition remains until the voltage at the base of Q2 becomes slightly positive. At that time the circuit returns to a stable condition. The base waveform at Q2 is similar to that discussed in Problem 10-8, so that the same equation for base voltage applies.

Assume that the circuit is triggered by the pulse on Q1 base at $t = 0$ and when $V_B(Q2) = 0$ it reverts to the stable state. The period of time that the circuit is in an unstable state determines the output pulse width T. Solve for this time in terms of the circuit component values. Ignore active device saturation values.

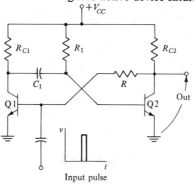

Problem 10-10

10-11   Derive an equation for the pulse width of the monostable circuit of Problem 10-10 in terms of the saturation and cut-in voltages. The pulse width is defined as the time from the occurance of a trigger pulse until transistor Q2 starts conducting.

10-12   The monostable circuit is available as an integrated circuit. The diagram shows a circuit that was designed by Texas Instruments, SN7380. Recognize that this circuit is very similar to the one of Problem 10-10, where Q4 and Q7 are the transistors characteristic of the basic monostable circuit. Transistors Q1, Q2, and Q3 are input pulse amplifiers. Transistor Q6 exercises control over the charging rate of capacitor C_3. Once the circuit is triggered, C_3 has a discharge path through the emitter circuit of Q5 rather than the base circuit of Q4.

Output transistors Q8, Q9, and Q10 make the output circuit compatible with T^2L type logic.

(a) Assume that the circuit is in a stable condition, $v_{IN} = 0$, and determine the state of each transistor, whether ON or OFF.

(b) Use active device saturation voltages typical of Si transistors, $V_{CE} = 0.3$ V, $V_{BE} = 0.7$ V, and determine the voltage across C_3 for the stable state of the circuit.

(c) Apply a trigger pulse at $t=0$ to the input, $v_{IN}=4.0$ V, and determine the state of each transistor, whether ON or OFF, at $t=0+$.

(d) Explain the circuit action for the time after the application of a trigger pulse until the monostable returns to the quiescent condition.

(e) Assume that $V_V=0.5$ V. Using these values and the information from part (b), derive an equation that expresses the OFF time, T of Q7.

(f) Assume saturation voltages for Silicon and calculate the power supply current drain I_{CC} for parts (a) and (c).

(g) A parasitic capacitance from collector to ground of Q2 C_d and resistor R_d combine to cause a delay in the occurrence of circuit switching action. Explain why this delay occurs.

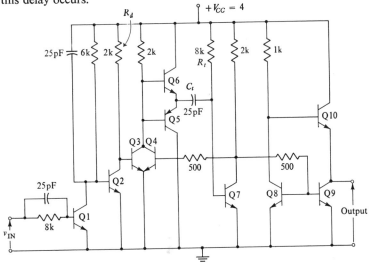

Problem 10-12

10-13 The bias driver circuit of Fig. 10-21 is directly related to the constant-current source of Section 5-2. Calculate the regulated output voltage of the circuit using $V_{CC}=0$, $V_D=V_{BE}=0.7$ V, $V_{EE}=-5.2$ V and the typical resistive values of Fig. 10-21.

10-14 For the bias driver circuit of Fig. 10-21, use the information of Section 5-2 to derive an equation for the regulated output voltage. Note: $V_{CC}\neq0$. Let $\alpha\to1$ for your final answer.

10-15 List the logic outputs of the expandable EXCLUSIVE OR circuit of Fig. 10-28 with the following logic inputs applied. Write your answer to (a) through (d) in the form of a truth table.

(a) All inputs 0
(b) All inputs 1
(c) One input 0; the remainder 1
(d) One input 1; the remainder 0

(e) Use the information gained in parts (a) through (d) to write a logic function equation for the circuit.

(f) For the case of part (a) with $V_{cc} = 5.0$ V, $V_{CE} = 0.3$ V, and $V_D = 0.7$ V, determine the short-circuit current available at the output terminal.

(g) Assume that the diode D is shorted and rework (f).

11
DIGITAL
IC
MODULES

The logic gates and bistable devices discussed in the two previous chapters can now be combined to synthesize complex digital networks or subsystems used to perform certain important digital operations. These subsystems in modular form include shift registers, counters, decoders, adders, plus others, and are available as off-the-shelf items.

To understand the operation of complex subsystems, additional background information in number systems, digital logic, and sequential circuits is required. These fundamentals will be presented as the need arises. The first topic discussed in the chapter is clock circuits. When transferring data or performing computations, we frequently wish to continue the operation at a fixed or clocked rate. Hence, a stable electronic clock is necessary. Section 11-2 discusses shift registers which are later applied to form counters and serial adders. Ripple counters and shift counters are discussed in Sections 11-3 and 11-4, respectively. Decoders are presented in Section 11-5; the examples discussed are the BCD-to-decimal and BCD-to-seven segment decoders. The chapter closes with a discussion of binary adders.

11-1 ELECTRONIC CLOCKS

At this point, the discussion takes on an additional degree of complexity. Instead of considering individual pulses or levels, we must appreciate that we have a sequence of pulses.

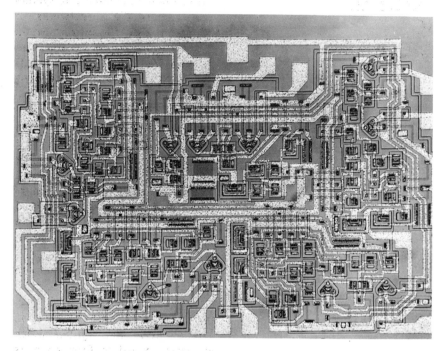

The 60 × 85 mil chip shown is a four-bit shift register. It contains the equivalent of 175 discrete components. This register can perform conversion, storage, delay, and shifting operations in digital computers and digital control systems. Note that some transistors are heart-shaped, some are parallel stripes, and some appear as "L" shapes. (Courtesy Sylvania Electric Products, Inc.)

In digital computers, control systems, and communication systems, the electrical signals which constitute the data usually take on only two voltage or current levels. Such signals are referred to as binary signals or binary data. Examples of binary signals are shown in Fig. 11-1.

The waveform in Fig. 11-1(a) is an example of a clock signal. Such signals are used to establish the proper time references necessary for an organized approach to the communication and computation of data. The period of one cycle of the square wave in Fig. 11-1(a) is commonly referred to as the *bit time*. The frequency of the signal in Fig. 11-1(b) is one-fourth of the bit rate. Its period, which is equal to four bit times, is referred to as the *word time*. There are usually four or more bits per word.

The waveform in Fig. 11-1(c) is an example of a data signal. Note that

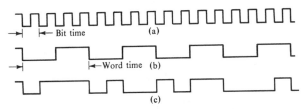

Fig. 11-1   Binary signals: (a) clock signal; (b) output of divide-by-four counter; (c) data signal.

transitions from the high to low and low to high levels occur coincidentally with high to low transitions in the clock signal. Logic circuits can be designed to operate with descending (high to low) clock transitions or with ascending (low to high) clock transitions.   Both are used in IC's.

A 1 MHz crystal controlled clock is illustrated in Fig. 11-2.   Two NOR

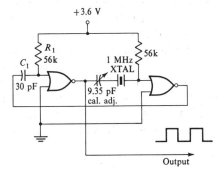

Fig. 11-2 Motorola Semiconductor Products, Inc., 1 MHz crystal-controlled clock.[17]

gates are cross-coupled to form an astable multivibrator.   Excellent frequency stability is obtained by placing the crystal in one of the feedback legs. Because a higher gain transistor is required for an oscillator than for a gate circuit, care must be taken to select devices with sufficiently high β.

11-2 SHIFT REGISTERS

Binary data are acquired by converting a signal available from nature or man-made systems into a discrete form.   Certain data sources are readily monitored by converting the input signal to a shaft position.   The position of the shaft can be detected by using brushes which make contact with the shaft.   The information is obtained by coating the shaft with an insulating or conducting surface, depending upon whether the output is to be binary 0 or binary 1, respectively.   Using this convention, the output of the shaft encoder in Fig. 11-3 for the position illustrated is $A = 0$, $B = 1$, and $C = 0$.

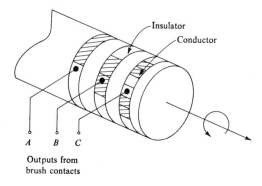

Insulator

Conductor

A   B   C

Outputs from
brush contacts

Fig. 11-3   Shaft encoder.

Suppose that it is necessary to transmit data from a shaft encoder over a telephone line. Using four brush contacts and a binary code such as that given in Table 9-2, sixteen different shaft positions can be identified. As is evident from Fig. 11-3, the data are outputed from the encoder in parallel form. However, before it can be transmitted over a single channel, it must be converted to serial form. Data in parallel form require as many wires or lines to transmit as there are bits; data in serial form require but one line independent of the number of bits. A device which implements this operation is called a *parallel-to-serial converter*. It is illustrated generically in Fig. 11-4.

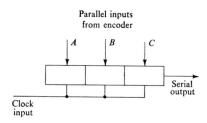

Parallel inputs
from encoder

A   B   C

Serial
output

Clock
input

Fig. 11-4   A generic parallel-to-serial data converter.

The heart of the parallel-to-serial converter is a data storage device called a *register*.

A basic register that can be augmented to provide a variety of operations is called a *shift register*. It is characterized by the property that all data in the register are shifted with each clock pulse. In Fig. 11-5, the flip-flops FF1, FF2, FF3, and FF4, are RS flip-flops connected to form a four-bit circulating shift register. A low level applied to the PRESET input of FF1 and the CLEAR inputs of FF2, FF3 and FF4 sets the F output of FF1 to 1 and the remaining $\bar{F}$ outputs to 1. Let the FF's be edge-triggered devices

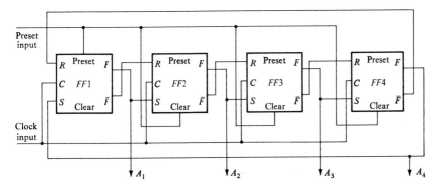

Fig. 11-5 Four bit circulating shift register.

with information transferred to the outputs on the negative edge of the clock pulse. Then, as shown in Fig. 11-6, the 1 initially stored in FF1 is transferred to FF2 with the first clock pulse. The one is advanced with each clock pulse, and with the fourth clock pulse it is shifted back to FF1.

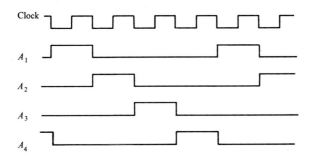

Fig. 11-6 Input-output waveforms of shift register in Fig. 11-5.

The waveforms appearing in Fig. 11-6 illustrate the sequence of events described above. For the example illustrated, a shift occurs coincident with the falling edge of a clock pulse. Not all devices are designed to shift with the falling edge of the clock; some shift with the rising edge.

The circulating shift register is incorporated with NAND gates as shown in Fig. 11-7 to realize a parallel-to-serial converter. In terms of the variables C_1, C_2, C_3, and C_4, at the outputs of the NAND gates, we have

$$\text{Output} = \overline{C_1 C_2 C_3 C_4}$$
$$= \overline{\overline{A_1 B_1}\ \overline{A_2 B_2}\ \overline{A_3 B_3}\ \overline{A_4 B_4}}. \tag{11-1}$$

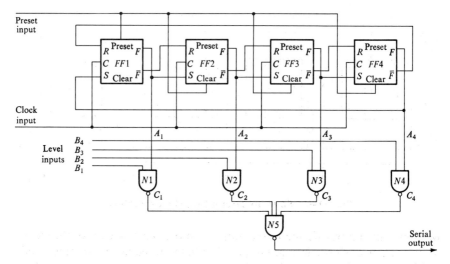

Fig. 11-7   Parallel-to-serial converter.

Using de Morgan's laws, Eq. (11-1) reduces to

$$\text{Output} = A_1 B_1 + A_2 B_2 + A_3 B_3 + A_4 B_4. \qquad (11\text{-}2)$$

The parallel-to-serial converter in Fig. 11-7 serves to convert level inputs in parallel form into serial data with a bit rate equal to the clock frequency. In an operating system, additional bits, called synchronization bits, are required to synchronize the logic circuitry at the receive terminal.

This example has served to introduce shift registers. Two widely used registers are discussed in greater detail. However, before presenting these examples, let us digress to present symbology frequently used by IC manufacturers in their logic diagrams.

This particular system of logic symbols shown in Fig. 11-8 is based on military specification, MIL-STD-806-B.* Five of the logic symbols were previously presented in Fig. 9-9. These are the positive AND, positive OR, positive NAND, positive NOR, and positive INVERT. The positive NONINVERT gate is simply a unit gain amplifier. Note that the small circle on the output signal line of the positive NAND and positive NOR gates indicates a signal inversion. The small circles at the inputs to the negative gates have the same function. As an example, consider the negative NOR gates. Input 0's are first changed to 1's, and these OR'd. Likewise, input 1's are changed to 0's prior to the OR operation. Consequently, for A and B, both 1, the output is 0.

---

* Military Standard, Graphic Symbols for Logic Diagrams, MIL-STD-806-B. Available from U.S. Naval Supply Depot, 5801 Tabor Avenue, Philadelphia, Pennsylvania 19120.

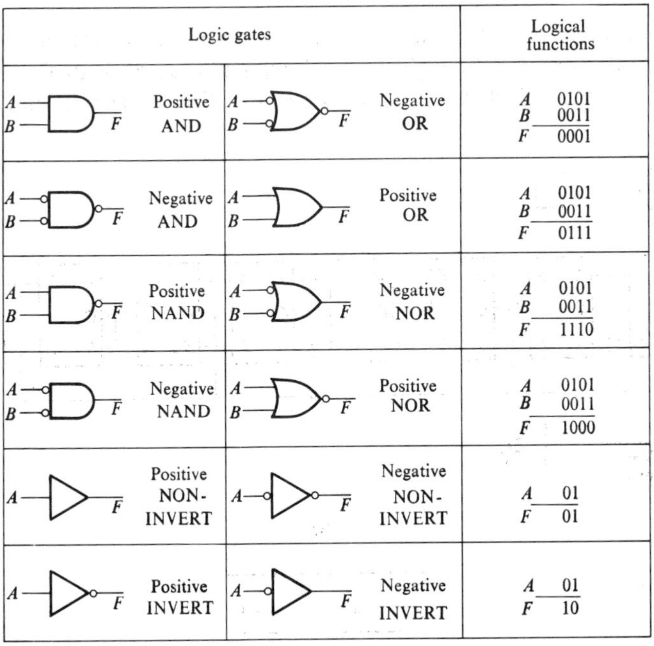

Logic gates				Logical functions	
A B → F	Positive AND	A B → F	Negative OR	A B F	0101 0011 0001
A B → F	Negative AND	A B → F	Positive OR	A B F	0101 0011 0111
A B → F	Positive NAND	A B → F	Negative NOR	A B F	0101 0011 1110
A B → F	Negative NAND	A B → F	Positive NOR	A B F	0101 0011 1000
A → F	Positive NON- INVERT	A → F	Negative NON- INVERT	A F	01 01
A → F	Positive INVERT	A → F	Negative INVERT	A F	01 10

Fig. 11-8   Equivalent logic operations in terms of MIL-STD 806-B symbols.

Generally, output circles and input circles are *matched* to accomplish a logic operation.   However, the NOT operation can be realized without an INVERT circuit by utilizing an assignment *mismatch*, that is, where a circled output connects to a noncircled input, or vice versa.   A logical 1 is sometimes referred to as a "high" or "positive" signal, and a logical 0 is also referred to as a "low" or "negative" signal.

Serial-in, Serial-out Shift Register

The *serial-in, serial-out shift register*, useful as a temporary storage device, requires few external connections.   Therefore, it appears very attractive in integrated form.   An eight-bit shift register made by Texas Instruments, Inc.,[18] is illustrated in Fig. 11-9(a).   This monolithic device uses T^2L circuits.   The shift register is composed of eight R-S master-slave flip-flops, an input gate, and a clock driver.   The input data and input control are applied to a NAND gate through terminals A and B.   At the output of the NAND gate, an inverter is used to provide complementary inputs to the first stage of the shift register.   The complement of the clock signal is inputed, and an inverting clock driver provides drive for the internal common clock line.   The

flip-flops used in this circuit are designed to operate with a rising clock pulse.
The truth table for the eight-bit shift register illustrated in Fig. 11-9(a) is presented in Fig. 11-9(b). Positive logic is utilized. Typical input-output

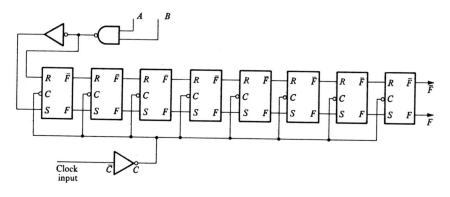

(a)

	t_n		t_{n+8}
	A	B	F
	0	0	0
	0	1	0
	1	0	0
	1	1	1

t_n = bit time before clock pulse
t_{n+8} = bit time after 8 clock pulses

(b)

Fig. 11-9   Eight-bit, serial-in, serial-out shift register—Type SN7491AN by Texas Instruments, Inc.:[18] (a) logic diagram; (b) truth table.

waveforms are presented in Fig. 11-10.   Input B is tied to a high level, 1, so that when a positive pulse, 1, appears at input A, the NAND gate is enabled and a 1 is placed at the input of the first flip-flop of the shift register.   It is entered into the first flip-flop with the subsequent rising edge of an input clock pulse.

The typical release or hold time on this particular device is −12 ns.   This accounts for the fact that the first data pulse at input A is shifted into the register even though the pulse switches off coincident with the rising edge of the first clock pulse.   The second data pulse shown in Fig. 11-10 has a rising

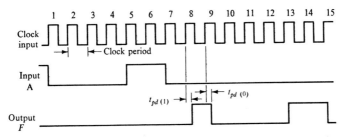

Fig. 11-10   Typical input-output waveforms for eight-bit shift register in Fig. 11-8 with B input high.

edge coincident with the rising edge of the fifth clock pulse.   Because of the setup time required by the input gate, the first flip-flop will not change state until the positive-going edge of the sixth clock pulse arrives.   The typical propagation delays (t_{pd}) for this device are 25 ns.   These are illustrated relative to a 10 MHz clock signal.

The advantages of having a shift register in integrated form are apparent from the schematic diagram presented in Fig. 11-11.   This circuit, which is a realization of Texas Instruments, Inc. type SN7491AN shift register, shown in block diagram form in Fig. 11-9(a), utilizes seventy-six transistors, twenty-nine diodes, and fifty-nine resistors.   It requires but seven external connections.   The increase in reliability and decrease in assembly cost of systems incorporating such devices is apparent.

Logical components shown in Fig. 11-9(a) are easily identified in circuit form in Fig. 11-11.   The final stage is modified somewhat to provide a fanout capability of 10 from the F and $\bar{F}$ outputs.   Data can be stored or transferred at rates up to 18 MHz while maintaining a typical noise immunity of 1 volt. Power dissipation is typically 175 milliwatts.

Parallel-in, Parallel-out Register

As pointed out earlier, it is frequently necessary to convert data from parallel form to serial form or from serial form to parallel form.   These operations can be readily implemented with a shift register by using additional external connections.

Recall from Section 9-4 that each digit between 0 and 9 can be represented in binary form.   Decimal data are frequently encoded by assigning four bits to each digit.   One such code, referred to as the 8421 binary coded decimal (BCD) code, is given in Table 11-1.   Using this encoding method, the decimal number 719 is encoded as

$$719 = 0111 \quad 0001 \quad 1001$$

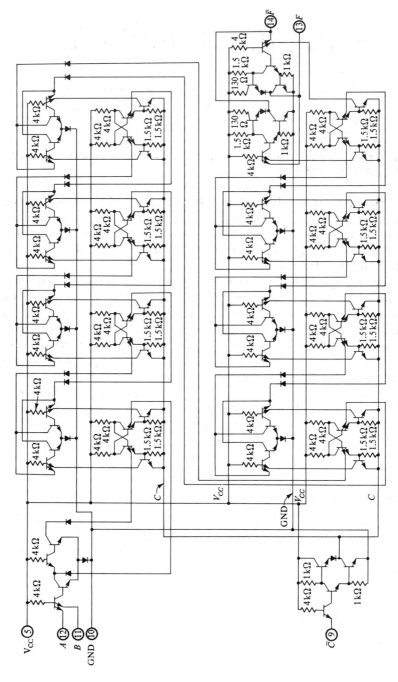

Fig. 11-11 Schematic diagram of Texas Instruments, Inc.,[18] Type SN7491AN eight-bit, serial-in, serial-out shift register.

TABLE 11-1   8421 BINARY CODED DECIMAL (BCD) CODE

Decimal	Binary Coded Decimal
	8421
0	0000
1	0001
2	0010
3	0011
4	0100
5	0101
6	0110
7	0111
8	1000
9	1001

These data arriving at a computing center via telephone line could be entered serially into a shift register in blocks of four bits each.

After each block is entered, it is frequently desirable to shift the four bits of data out in parallel. The four-bit shift register shown in Fig. 11-12 accomplishes this operation. The device is type SN7495N manufactured by

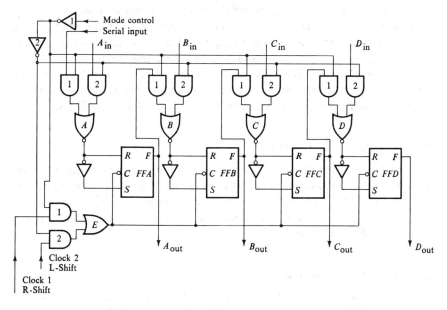

Fig. 11-12   Logic diagram of Texas Instruments, Inc.,[18] Type SN7495N four-bit, right-shift, left-shift register.

Texas Instruments, Inc.,[18] and utilizes T^2L in a fourteen-pin plastic dual-in-line package.   This monolithic shift register is composed of four R-S master-slave flip-flops, four AND-OR-INVERT gates, one AND-OR gate, and six inverter-drivers.   The device is extremely versatile.   By providing the proper logical input level to the mode control, it will perform either a right-shift or a left-shift operation.   A right-shift operation consists of shifting all data stored in the register one position to the right.   This allows for a single bit to be entered in the left-most stage while the bit in the rightmost stage is shifted out of the register.   A left-shift operation is analogous, but in the opposite direction.

The right-shift and left-shift operations are accomplished through proper use of the mode control.   With a low level or 0 at the mode control, AND gates 1 are enabled and AND gates 2 are inhibited.   This allows for serial data to be entered into FFA, and couples the output of each flip-flop to the R-S inputs of the succeeding flip-flop.   The right-shift operation is performed by applying a clock signal to the clock 1 input.   In this mode, the device functions as a serial-to-parallel converter by sensing output terminals A_{out} through D_{out}.

With a 1 applied to the mode control, AND gates 2 are enabled and AND gates 1 are inhibited.   This decouples the successive stages, and allows for parallel entry of data through inputs A_{in} through D_{in} and of a clock signal at the clock 2 input.   Parallel-in, parallel-out operation is accomplished directly, and, with external connections, the shift-left operation is implemented.   To shift data to the left, each flip-flop output is tied to the input of the previous stage (D_{out} to C_{in}, and so on), and serial data are entered at D_{in}.

A number of these registers can be connected in series to form an n-bit right-shift or left-shift register.   Such connections provide for n-bit serial-to-parallel and parallel-to-serial converters or n-bit storage registers.

11-3   RIPPLE COUNTERS

Counting is the operation of adding or subtracting unity from the least significant digit of a number.   Electronic counters are useful in data acquisition, timing, computing, and so on.   In this section, three kinds of ripple counters, the binary ripple counter, the decade counter, and the divide-by-twelve counter, will be discussed.

Binary Ripple Counter

A straightforward counter is the *binary ripple counter*.   The name describes the action of the counter, as shown in Fig. 11-13(a).   J-K flip-flops are placed in series somewhat similar to a shift register.   However, instead of sending the

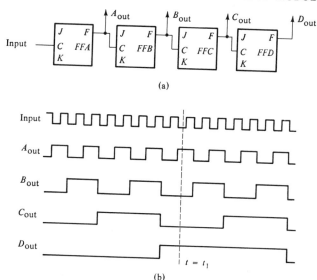

(a)

(b)

Fig. 11-13   Binary ripple counter: (a) logic diagram; (b) input-output waveforms.

input clock pulse to all flip-flops simultaneously, as in a shift register, the input data are fed to the clock pulse input of the first flip-flop in the chain. The output of FFA is connected to the input of FFB and so on.   The J and K terminals are connected to a logical 1 level.

As is illustrated in Fig. 11-13(b), FFA changes state for every input pulse, and the remaining flip-flops change state whenever a negative-going edge is present at their clock pulse inputs.   The total count is present at the output terminals in binary form.   Output A is the coefficient of 2^0, output B is the coefficient of 2^1, output C is the coefficient of 2^2, and output D is the coefficient of 2^3.   Consider for example the point in time $t = t_1$.   This count is present in parallel form at the outputs in the binary number 1001.   Note that after a binary 15 is present, the counter resets to zero.

The counter illustrated in Fig. 11-13(a) is also referred to as a divide-by-16 counter, because the D output has only one negative-going edge for sixteen input pulses.   The term "ripple counter" stems from the fact that owing to propagation delay, the flip-flops change state sequentially.   Hence, when observing the outputs, a rippling effect is present.

As an example of an integrated *binary counter*, consider the Texas Instruments, Inc.,[18] type SN7493N, IC shown in Fig. 11-14(a).   It is a high-speed monolithic binary counter consisting of four master-slave flip-flops.   Interconnections provide for a divide-by-2 counter and a divide-by-8 counter.   By connecting output A to input B, a divide-by-16 or a four-bit binary ripple

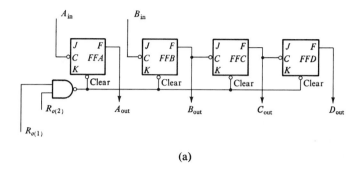

(a)

Count	Output			
	D_{out}	C_{out}	B_{out}	A_{out}
0	0	0	0	0
1	0	0	0	1
2	0	0	1	0
3	0	0	1	1
4	0	1	0	0
5	0	1	0	1
6	0	1	1	0
7	0	1	1	1
8	1	0	0	0
9	1	0	0	1
10	1	0	1	0
11	1	0	1	1
12	1	1	0	0
13	1	1	0	1
14	1	1	1	0
15	1	1	1	1

(b)

Fig. 11-14   Four-bit binary counter—Type SN7493N by Texas Instruments, Inc.:[18] (a) logic diagram; (b) truth table.

counter is obtained.   The operation of the counter in this mode is presented in the truth table in Fig. 11-14(b).

A gated direct reset line is provided which inhibits the count inputs and simultaneously returns the four flip-flop outputs to the logical 0 state.   To effect the reset of all outputs to logical 0, both $R_{o(1)}$ and $R_{o(2)}$ inputs must be at logical 1.

Decade Counter

It is frequently advantageous to count directly in the decimal system instead of dividing by some power of 2. A four-stage binary divide-by-16 counter can be readily adapted to a divide-by-10 counter. The type SN7490N decade counter manufactured by Texas Instruments, Inc.,[18] which is shown in Fig. 11-15(a), is an example. It utilizes three J-K and one R-S flip-flop. They are all of the master-slave type and are internally connected to provide for a divide-by-2 counter and a divide-by-5 counter. Output A must be externally

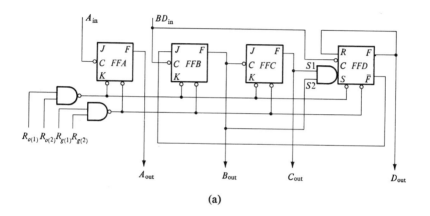

(a)

	Output			
Count	D_{out}	C_{out}	B_{out}	A_{out}
0	0	0	0	0
1	0	0	0	1
2	0	0	1	0
3	0	0	1	1
4	0	1	0	0
5	0	1	0	1
6	0	1	1	0
7	0	1	1	1
8	1	0	0	0
9	1	0	0	1

(b)

Fig. 11-15   BCD decade counter—Type SN7490N by Texas Instruments, Inc.:[18] (a) logic diagram; (b) truth table.

connected to the BD input to obtain a binary coded decimal counter. The BCD code utilized is the familiar 8421 code, and the count sequence is presented in the truth table in Fig. 11-15(b).

Operationally, a decade counter is identical to the binary divide-by-16 counter for the first nine counts. However, with the tenth count, it must reset all flip-flops to the binary 0 state.

The SN7490N decade counter operates in the following manner. Start with all flip-flops in the logical 0 state. The J-K terminals with no inputs are internally wired to a logical 1 level. Therefore, FFA changes state for every input pulse. For BCD operation, the signal is connected to the A_{in} terminal and FFA output is connected to input BD; a logical 1 is at the J input of FFB owing to the state of FFD. The arrival of the second input causes FFA to change from 1 to 0, causing a negative edge at the clock pulse input of FFB. FFD does not change state because of logical 0's at both the R and S inputs. Counting continues in this manner identical to that of the binary counter until the count of seven. For a seven, logical 1's are present at both the $S1$ and $S2$ inputs, enabling the AND gate and setting FFD. Since FFD is still in the 0 state, a logical 0 is at the reset input. Therefore, the next negative edge causes FFD to change from 0 to 1. The ninth input pulse affects only FFA, but the tenth presents a negative edge at the BD input. FFB now has a 0 at the J input and a 1 at the K input, so that the output remains the same. However, FFD has a 1 at the R input and a 0 at the S input, which causes the output to change to the 0 state. All four flip-flops are now back to the 0 state.

Gated direct reset lines are provided to inhibit count inputs and return all outputs to a logical 0 or to a binary coded decimal 9. The reset to a BCD 9 is useful in certain decimal computational methods discussed in the literature. Decimal readout is obtained by using a BCD-to-decimal decoder to interface the BCD counter with a decimal numeral readout device. Decoders are discussed in Sections 11-4 and 11-5.

Divide-by-12 Counter

Both counters previously presented have the count output in the familiar 8421 code. The 8421 code is the most natural code, since it expresses a decimal number in the binary system. This convenience is not always possible and, in fact, may not always be desirable. A counter may have its own unique code. The only requirement is that the code must be uniquely decodable.

An example is Texas Instruments, Inc.,[18] type SN7492N divide-by-12 IC counter shown in Fig. 11-16(a). To effect the divide-by-12 mode, A_{out} must be connected to input BC. From the truth table presented in Fig. 11-16(b), we note that the code words for the decimal numbers 0 through 5 are those

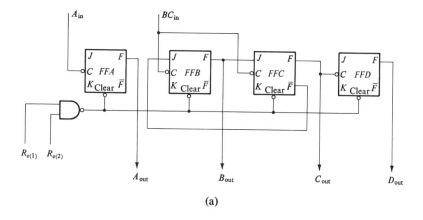

(a)

	Output			
Count	D_{out}	C_{out}	B_{out}	A_{out}
0	0	0	0	0
1	0	0	0	1
2	0	0	1	0
3	0	0	1	1
4	0	1	0	0
5	0	1	0	1
6	1	0	0	0
7	1	0	0	1
8	1	0	1	0
9	1	0	1	1
10	1	1	0	0
11	1	1	0	1

(b)

Fig. 11-16   Divide-by-twelve counter—Type SN7492N by Texas Instruments, Inc.:[18] (a) logic diagram; (b) truth table.

of the 8421 code.   At that point the pattern changes.   Again, the flip-flops are designed to operate on the falling edge of the clock pulse.   By applying the input signal at the BC input, a divide-by-6 counter is realized (see Problem 8).   Divide-by-6 and divide-by-12 counters are useful in time measurement applications.

We have presented three counters, a binary divide-by-16 counter, a BCD counter, and a divide-by-12 counter.   All are ripple-through counters. Other counters which do not have the ripple-through feature can be implemented using shift registers.

11-4   SHIFT REGISTER COUNTERS

The shift registers discussed in Section 11-2 augmented with external logic gates make excellent counters. Two types of counters, the ring counter and Johnson counter, are presented in this section.

Ring Counters

A rather simple counter can be realized from a shift register, such as that shown in Fig. 11-5. More generally, an n-stage shift register is obtained by connecting the output of the nth stage back to the input of the first stage, as is illustrated in Fig. 11-17(a), to form a *ring counter*. We shall see in the analysis that follows that only n-distinguishable code words are generated. Therefore, the n-bit ring counter can count only to n.

Using the PRESET and CLEAR terminals, a pulse on the start line will set FFA to 1 and the remaining n-1 flip-flops to 0. The 1 in FFA is shifted to FFB with the arrival of the first clock pulse. As more clock pulses occur,

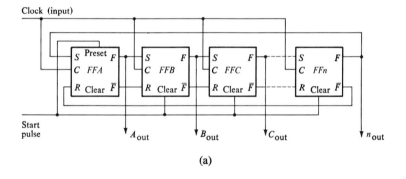

(a)

Clock (Input) Count	Outputs		
	A	B	$C \ldots n$
1	1	0	0 0
2	0	1	0 0
3	0	0	1 0
$\vdots$			
n	0	0	0 1
1	1	0	0 0

(b)

Fig. 11-17   Ring counter: (a) logic diagram; (b) truth table.

it is shifted along the ring from FFB to FFC, and so on, until the nth pulse, which shifts it back to FFA, as shown in the truth table in Fig. 11-17(b).

The circuit has an advantage as a counter, because it does not require decoding in order to acquire separate signal outputs for each decimal number. It is used where decoded signals are required, but it is not desirable to use a logic decoder.   The circuit also has merit when clocked by a controlled signal because it performs as a *commutator*.   For example, each output terminal can be tied to the base of an *npn* transistor.   As the 1 circulates around the ring it sequentially gates each transistor ON.   An application in analog-to-digital conversion is presented in Section 12-2.

Johnson Counter

Another form of the n-stage feedback shift register counts to the number $2n$. This is accomplished by cross-coupling the output of the last stage to the input of the first.   Consider the case where $n = 5$ for the ring counter in Fig. 11-17(a) and where the feedback leads are switched so that F goes back to R, and $\bar{F}$ back to S of FFA.   This circuit, often referred to as the *Johnson* or *switch tail ring counter*, is illustrated in Fig. 11-18(a).

The inverse of the signal stored in FFE is shifted into FFA each time a clock pulse occurs.   Consider that all flip-flops are initially in the 0 state. The first clock pulse transfers a 1 from the $\bar{F}$ output of FFE into FFA and a 0 into the remaining four stages.   For the next four clock pulses, 1's are fed back from FFE to FFA.   Hence, after five clock pulses, the register is filled with 1's.   At this time a 0 is shifted back to the S input of FFA, and the sixth through the tenth clock pulses store 0's in FFA.   The tenth input pulse returns the register to its original state.   This sequence of events is summarized in Fig. 11-18(b).

Only ten of the $2^5 = 32$ possible five bit sequences are obtained in the counting sequence of the Johnson counter.   Should a sequence other than one of the ten given in Fig. 11-18(b) be initially present in the register, that sequence will recur every ten counts.   Two alternate solutions to this situation are available.   The first is to reset all flip-flops to the 0 state before entering data by inserting a 1 on the start pulse line in Fig. 11-18(a).   The second is to provide logic circuitry to force the register into a desired combination of 0's and 1's.   Figure 11-19 shows a circuit which forces the register into one of the sequences listed in Fig. 11-18(b).   A 1 is cross-coupled back to reset FFA only when both FFD and FFE outputs are in the 1 state.   This occurs when input pulses 5, 6, 7, and 8 are stored in the register.   For input pulse 9, feedback signals DE and $\bar{E}$ are both at the 0 state, causing FFA to remain in its present state at input pulse 10.

The Johnson counter is readily implemented in integrated circuit form

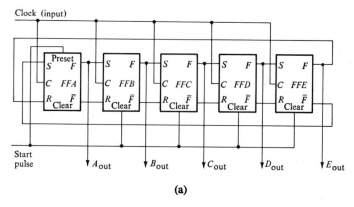

(a)

	Outputs				
Clock (Input) Count	A	B	C	D	E
0	0	0	0	0	0
1	1	0	0	0	0
2	1	1	0	0	0
3	1	1	1	0	0
4	1	1	1	1	0
5	1	1	1	1	1
6	0	1	1	1	1
7	0	0	1	1	1
8	0	0	0	1	1
9	0	0	0	0	1

(b)

Fig. 11-18   Johnson counter: (a) logic diagram; (b) truth table.

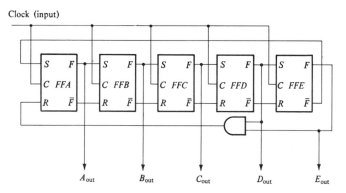

Fig. 11-19   Johnson counter with feedback logic to direct counter into desired sequence.

using a five-bit shift register. Consider, for example, the logic circuit shown in Fig. 11-20. A type SN7496N integrated circuit by Texas Instruments, Inc.,[19] is used to realize the counter. The inverter in the feedback lead is required because of the internal inverter drivers that feed FFA.

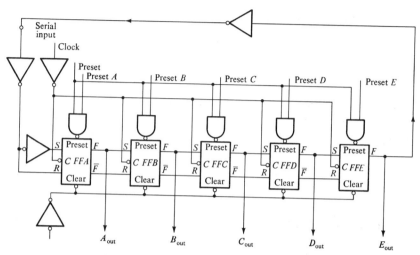

Fig. 11-20   Logic diagram of Johnson counter realized from one Texas Instruments, Inc.,[19] Type SN7496 IC and one inverter.

It is evident from Fig. 11-18(b) that the five-bit sequences representing the ten true distinct output states of the Johnson counter are not binary expressions for the decimal numbers 0 through 9. Apparently, some method for decoding these five-bit sequences into the decimal numbers is necessary. A straightforward method is to use five-input AND gates. The logic expression for the decimal zero output, obtained from the first line in Fig. 11-18(b), is

$$F_0 = \bar{A}\,\bar{B}\,\bar{C}\,\bar{D}\,\bar{E}. \tag{11-4}$$

However, this expression is not in its simplest form, which is verified by noting that another unique expression for the decimal zero is

$$F_0 = \bar{A}\,\bar{E}. \tag{11-5}$$

A technique for simplifying logic expressions is presented in the next subsection. The realization of a decoder for a Johnson counter will be used to illustrate the method. Decoders will be discussed further in Section 11-5.

Simplifying Boolean Functions

At this point, we digress to present a method for simplifying Boolean functions. We will use the laws for manipulating Boolean algebraic expressions which were presented in Section 9-4.

Any given Boolean expression may contain one or more variables or terms which could just as well be omitted.   Such expressions exhibit redundancy. Consider the function

$$F = AB + ABC. \tag{11-6}$$

The distributive and unit laws reduce F:

$$F = AB(1 + C)$$
$$= AB. \tag{11-7}$$

Therefore, the term ABC in Eq. (11-6) is redundant.

An expression may contain no redundant terms or variables but still not be in its simplest form.   For this discussion, a function will be considered to be in its simplest form when a minimal number of input gates is required to implement the logic circuits.   The functions

$$F_1 = \bar{A}\,\bar{B}\,CD + \bar{A}\,BCD + \bar{A}\,BC\,\bar{D} \tag{11-8a}$$

$$F_2 = \bar{A}\,CD + \bar{A}\,BC \tag{11-8b}$$

have the same truth table and are therefore logically equivalent to each other. However, Eq. (11-8a) requires three four-input AND gates and one three-input OR gate to implement, while Eq. (11-8b) requires but two three-input AND gates and one two-input OR gate.

Karnaugh maps.   One method for minimizing Boolean expressions utilizes the *Karnaugh map*.   A brief discussion of this method occurs here.   The discussion is not rigorous, but, rather, is intended to point out features helpful in simplifying Boolean expressions.

A Karnaugh map consists of a rectangle divided into *squares* or *cells*, as shown in Fig. 11-21(a).   The variables of the function are listed on the left side and top.   Note that all values of AB are obtained from the row headings and that all values of CD are obtained from the column headings.   A specific cell is designated by a combination of all four variables.   For example, the area corresponding to $ABCD = 1010$ is in the lower right-hand corner.

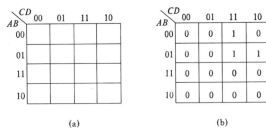

(a)   (b)

Fig. 11-21   Karnaugh maps for 4 variables: (a) general form; (b) map of $F = \bar{A}CD + \bar{A}BC$.

An important feature of the map is that for any cell the adjacent four cells represent adjacent states of the variables.   An adjacent state of a code word is one which differs in the value of only one variable.   These adjacencies exist not only on the interior of the map but also from end to end of each row and each column.   That is, the bottom cell in any column is adjacent to the top cell in that same column, and the right-most cell in any row is adjacent to the left-most cell in that same row.

A Boolean function is *mapped* into the rectangle in the following manner. All possible values of the variables are substituted into the function.   A 1 is entered into each cell which corresponds to a combination of values that produced a 1 in the function, and 0's are entered into all others.   As an example, a map of the function in Eq. (11-8a) is shown in Fig. 11-21(b).

The advantage of the Karnaugh map for simplifying Boolean expressions now becomes evident.   Consider the two adjacent cells located in the right half of the second row.   An expression for these cells is

$$F_3 = \bar{A} \, BCD + \bar{A} \, BC \, \bar{D}. \qquad (11\text{-}9)$$

By factoring, it reduces to

$$F_3 = \bar{A} \, BC(D + \bar{D}). \qquad (11\text{-}10)$$

And, since $D + \bar{D} = 1$, we have

$$F_3 = \bar{A} \, BC. \qquad (11\text{-}11)$$

This illustrates that two adjacent cubes can be represented by only three, instead of four, variables.

Groupings of one or more adjacent cells in a map which all have the same value are called *subcubes*.   Much of the convenience in using the map depends on our ability to recognize these subcubes.   To assist in identifying subcubes, all cubes within the subcube are enclosed in a loop.   For example, consider this function:

$$G = \bar{A} \, \bar{B} \, \bar{C} \, \bar{D} + \bar{A} \, B \, \bar{C} \, \bar{D} + AB \, \bar{C} \, \bar{D} + A \, \bar{B} \, C \, \bar{D}$$
$$+ ABC \, \bar{D} + \bar{A} \, \bar{B} \, \bar{C} D + A \, \bar{B} \, \bar{C} D + A \, \bar{B} \, CD, \qquad (11\text{-}12)$$

for which the map is presented in Fig. 11-22.   One possible grouping is illustrated in Fig. 11-22(a), which results in the simplification

$$G = \bar{A} \, \bar{C} \, \bar{D} + AB\bar{D} + \bar{B} \, \bar{C} D + A\bar{B}C. \qquad (11\text{-}13)$$

Another possible grouping, illustrated by the solid lines in Fig. 11-22(b), gives

$$G = \bar{A} \, \bar{B} \, \bar{C} + B \, \bar{C} \, \bar{D} + A\bar{B}D + AC\bar{D}. \qquad (11\text{-}14)$$

The expressions given by Eq. (11-13) and Eq. (11-14) are both minimal.

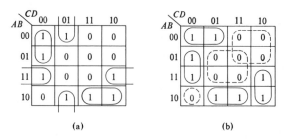

Fig. 11-22   Karnaugh maps of Eq. (11-12): (a) G only; (b) G and $\bar{G}$.

Another advantage of the Karnaugh map is that the complement of a function is also evident in the map. Every combination of input variables which results in a 0 in the map for the function G produces a 1 for the function $\bar{G}$. A simplification of the function $\bar{G}$ is obtained from the subcubes contained in the dashed loops in Fig. 11-22(b). The equation is

$$\bar{G} = \bar{A}\,\bar{B}\,\bar{C}\,\bar{D} + BD + \bar{A}C. \tag{11-15}$$

Because of the ready availability of inverters, complemented functions are frequently desirable.

Subcubes in a four variable map may contain one, two, four, or eight cells. As observed in Eq. (11-15), the number of variables required to identify a subcube is reduced by one for a two-cell subcube, by two for a four-cell subcube, and by three for an eight-cell subcube.

Four-variable Karnaugh maps will be utilized in the discussion of decoders in Section 11-5. However, recall that the Johnson counter had five variables in the output expressions. These expressions may also be plotted on Karnaugh maps.

One method of expressing five variables in a Karnaugh map is shown in Fig. 11-23. This map utilizes eight rows and four columns to represent all 32 combinations of the five variables. Each cell differs from all adjacent cells by one variable. Also, there exist outlying cells which differ by only one variable. Some of these are indicated by the connected dots in Fig. 11-23. To better understand the five-variable Karnaugh map, assume the presence of a mirror between the cells representing $\bar{A}\,B\,\bar{C}$ and $A\,B\,\bar{C}$. Then adjacent cells and those which are mirror images of each other are combinable.

EXAMPLE. From Fig. 11-18(b), which gives the truth data of a Johnson counter, we can write separate logic expressions for each decimal output. These logic expressions are given in the second column of Table 11-2. By labeling the binary function for the decimal 0, F_0, we map $F_0 = \bar{A}\,\bar{B}\,\bar{C}\,\bar{D}\,\bar{E}$ into the five variable map shown in Fig. 11-24 by placing a 1 in the upper left cell. We previously placed 0's in all cells of a map not containing 1's. However, in this case, only ten of the thirty-two cells are required to decode

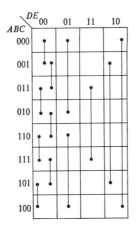

Fig. 11-23 Karnaugh map for five variable functions (the connected dots indicate rows which differ by only one variable).

the decimal digits. We accordingly place 0's in the remaining nine cells identified in the second column of Table 11-2.

It is not critical what the remaining twenty-two cells contain. Such cells or terms are frequently referred to as optional terms or " don't-care " terms. They are identified by the symbol ϕ. Using the don't-cares inside the loop shown in Fig. 11-24 as 1's, we obtain the simplification

$$F = \bar{A}\,\bar{E}.$$

The remaining nine expressions can be simplified in a similar manner to obtain the expressions in the third column of Table 11-2. Consequently, the outputs of the Johnson counter shown in Fig. 11-20 can be decoded into ten

TABLE 11-2 LOGIC EXPRESSIONS FOR JOHNSON COUNTER

Decimal Number	Logic Expression	Simplified Logic Expression
0	$\bar{A}\bar{B}\bar{C}\bar{D}\bar{E}$	$\bar{A}\bar{E}$
1	$A\bar{B}\bar{C}\bar{D}\bar{E}$	$A\bar{B}$
2	$AB\bar{C}\bar{D}\bar{E}$	$B\bar{C}$
3	$ABC\bar{D}\bar{E}$	$C\bar{D}$
4	$ABCD\bar{E}$	$D\bar{E}$
5	$ABCDE$	AE
6	$\bar{A}BCDE$	$\bar{A}B$
7	$\bar{A}\bar{B}CDE$	$\bar{B}C$
8	$\bar{A}\bar{B}\bar{C}DE$	$\bar{C}D$
9	$\bar{A}\bar{B}\bar{C}\bar{D}E$	$\bar{D}E$

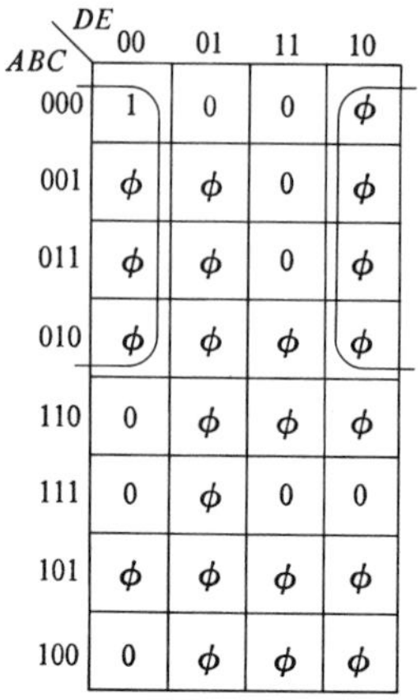

ABC＼DE	00	01	11	10
000	1	0	0	ϕ
001	ϕ	ϕ	0	ϕ
011	ϕ	ϕ	0	ϕ
010	ϕ	ϕ	ϕ	ϕ
110	0	ϕ	ϕ	ϕ
111	0	ϕ	0	0
101	ϕ	ϕ	ϕ	ϕ
100	0	ϕ	ϕ	ϕ

Fig. 11-24   Karnaugh map of $F_0 = \bar{A}\bar{B}\bar{C}\bar{D}\bar{E}$.

distinct terms using ten inverters and then two-input AND gates.   Also, in light of deMorgan's laws, ten inverters and ten two-input NOR gates will suffice.

To complete our introduction of Karnaugh maps, we point out that the map method is developed for AND and OR logic.   Consequently, it is more suitable for two-level AND-OR or OR-AND realizations than for NAND or NOR realizations.   Frequently though, through deMorgan's laws, AND-OR realizations can be converted to NAND-NAND realizations, and OR-AND realizations can be converted to NOR-NOR realizations.

11-5   DECODERS

A *decoder* is a device which associates an output message to a block of binary data.   An application where such decoding is necessary is for driving readout devices.   For example, the number of pulses inputed to a counter during a one-second period is frequently displayed by light bulbs.   A lens in front of the light bulb is scribed with the proper numeral.   If the count is always nine or less, then 10 bulbs and lenses might be arranged vertically. A single BCD counter, such as the one described in Section 11-3, will provide the count, but an interfacing network is required between the counter and the readout.

Many decoders are the one-of-a-kind type designed for special situations. However, some applications, such as the BCD-to-decimal conversion described above, occur with sufficient frequency to warrant production in integrated form.   Two of these, the BCD-to-decimal decoder and the BCD-to-seven segment decoder are discussed in this section.

BCD-to-Decimal Decoder

BCD-to-decimal decoders are also referred to as a one-of-ten or four-line-to–ten-line decoders. Their function is to provide ten separate outputs on ten different lines from four-bit binary data arriving on four input lines.

A truth table for the BCD-to-decimal decoder is presented in Table 11-3.

TABLE 11-3 TRUTH TABLE FOR A BCD
DECODER

BCD Input				Decimal Output									
D	C	B	A	0	1	2	3	4	5	6	7	8	9
0	0	0	0	1	0	0	0	0	0	0	0	0	0
0	0	0	1	0	1	0	0	0	0	0	0	0	0
0	0	1	0	0	0	1	0	0	0	0	0	0	0
0	0	1	1	0	0	0	1	0	0	0	0	0	0
0	1	0	0	0	0	0	0	1	0	0	0	0	0
0	1	0	1	0	0	0	0	0	1	0	0	0	0
0	1	1	0	0	0	0	0	0	0	1	0	0	0
0	1	1	1	0	0	0	0	0	0	0	1	0	0
1	0	0	0	0	0	0	0	0	0	0	0	1	0
1	0	0	1	0	0	0	0	0	0	0	0	0	1

Let F_0 be the binary function for the decimal 0. A Karnaugh map of F_0, obtained from the truth table, is shown in Figure 11-25(a). Note that only ten of the sixteen states determined by the four variables need to be defined. The remaining six are optional, or "don't care," states. It is apparent from the map that the optional states cannot be used to simplify the expression for F_0. The same is true for F_1, the binary function for decimal 1. However,

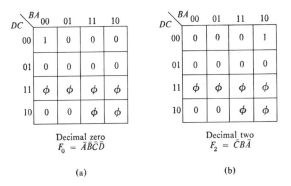

Decimal zero
$F_0 = \bar{A}\bar{B}\bar{C}\bar{D}$

(a)

Decimal two
$F_2 = \bar{C}B\bar{A}$

(b)

Fig. 11-25 Karnaugh maps for: (a) the decimal zero; (b) decimal two of a BCD-to-decimal decoder.

note from Fig. 11-25(b) that, by using the optional state in the lower right corner as a 1, the function F_2 simplifies to

$$F_2 = \bar{A}\,B\,\bar{C}. \qquad (11\text{-}17)$$

Karnaugh maps can be plotted for the eight remaining binary functions specified in Table 11-3. The simplified functions obtained using the maps are listed in Table 11-4. These equations can be implemented as shown in

**TABLE 11-4 SIMPLIFIED
LOGIC EXPRESSIONS
FOR BCD DECODER**

Decimal Output	Simplified Functions
0	$F_0 = \bar{A}\bar{B}\bar{C}\bar{D}$
1	$F_1 = A\bar{B}\bar{C}\bar{D}$
2	$F_2 = \bar{A}B\bar{C}$
3	$F_3 = AB\bar{C}$
4	$F_4 = \bar{A}\bar{B}C$
5	$F_5 = A\bar{B}C$
6	$F_6 = \bar{A}BC$
7	$F_7 = ABC$
8	$F_8 = \bar{A}D$
9	$F_9 = AD$

Fig. 11-26. Four inverters and ten AND gates are employed. The AND gates required are two four-input gates, six three-input gates, and two two-input gates.

Certain circuit innovations such as those used in Texas Instruments, Inc.,[18] type SN7441AN BCD-to-decimal decoder/driver further simplify the design. A schematic of this integrated circuit is shown in Fig. 11-27. Two inverters in series are used to provide sufficient drive capability for both the inputs and their complements. From Table 11-4, we see that A or $\bar{A}$ is present in each of the ten simplified functions. This fact is exploited to reduce the number of AND gates from ten to four by incorporating the output transistors into the decode operation. Consequently, the output transistors play a dual roll: that of decoding the variable A, and that of providing the output drive capability.

Figure 11-28 illustrates an application of the integrated decoder/driver. The output transistors must be capable of withstanding voltages and sinking currents required to operate a gas-filled indicator tube. These IC's can also be used to drive miniature lamps and small relays. The values of the B^+ supply voltage and the load resistor R_L depend on the readout device. For

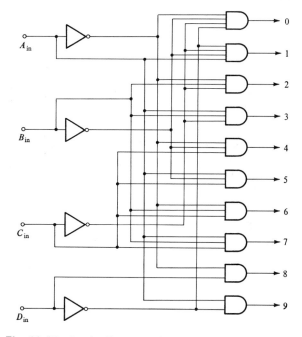

Fig. 11-26   Logic diagram of BCD-to-decimal decoder.

the case of gas-filled tubes, a B^+ supply voltage greater than V_{CC} is required. Output diodes are used to provide protection against over voltages accidentally applied to the output terminals.

BCD-to-Seven Segment Decoders

The *BCD-to-seven decoder* provides a seven line output for driving seven segment numeric displays like the one shown in Fig. 11-29(a). The decoder accepts a four-bit code input. Using seven segments, each of which can be in either of two states (OFF or ON) allows for $2^7 = 128$ different display patterns. Ten of these are used to illustrate the ten digits. Since four input lines allow for but $2^4 = 16$ input data code words, we have six more of the remaining 118 available patterns to specify. These might be chosen as shown in Fig. 11-29(b). Other designations for certain numerals as well as the patterns corresponding to the numbers 10 through 15 may be preferred. However, once the numeral designations are determined, the corresponding truth table can be recorded.

Generally, a driver stage is required to light the segment. The *npn* transistor amplifier shown in Fig. 11-30 is an example. Using positive logic, a high level or 1 at the output of the decoder will turn the transistor ON and light the lamp.

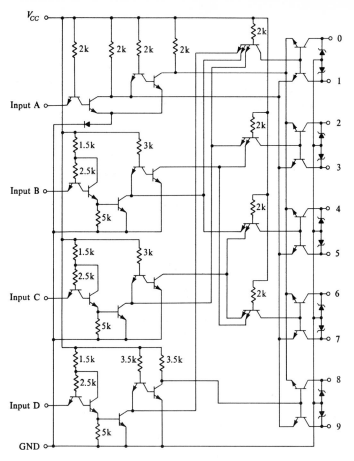

Fig. 11-27 Schematic diagram of Texas Instruments, Inc.,[18] Type SN7441AN BCD-to-decimal decoder and driver.

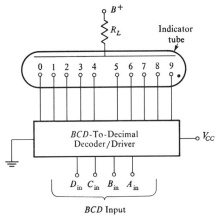

Fig. 11-28 BCD-to-decimal decoder/driver applied to gas-filled indicator tube.

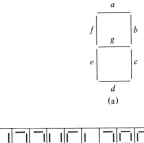

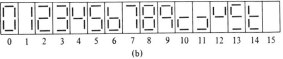

Fig. 11-29   Seven-segment display: (a) segment designation; (b) numeral designations.

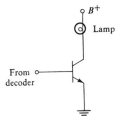

Fig. 11-30   Lamp driver.

The truth table corresponding to the display patterns shown in Fig. 11-29(b) is presented in Table 11-5.   Logical functions for the seven output lines a through g can be written from the truth table.   This is done by specifying the states required of all input variables to make the output variable 1. As an example, consider the a element in the seven segment display.   Note from Table 11-5 that it is on for the decimal 0.   This condition is specified by including the term $\bar{A}\,\bar{B}\,\bar{C}\,\bar{D}$ in the logical expression for the ON state of the element a, F_a.   Continuing down the column in Table 11-5, we note that the a element is ON for the decimal numerals 2, 3, 5, 7, 8, 9, and 13. Consequently the binary function F_a is

$$F_a = \bar{A}\,\bar{B}\,\bar{C}\,\bar{D} + \bar{A}B\bar{C}\,\bar{D} + AB\bar{C}\,\bar{D} + A\bar{B}C\bar{D}$$
$$+ ABC\bar{D} + \bar{A}\,\bar{B}\,\bar{C}D + A\bar{B}\,\bar{C}D + A\bar{B}CD \tag{11-18}$$

The Karnaugh map of F_a is plotted in Fig. 11-31.   By choosing the grouping shown by the loops, Eq. (11-18) reduces to

$$F_a = \bar{A}\,\bar{B}\,\bar{C} + B\bar{C}\,\bar{D} + A\bar{B}D + AC\bar{D}. \tag{11-19}$$

TABLE 11-5 TRUTH TABLE FOR BCD-TO-SEVEN SEGMENT DECODER

Decimal or Function	D	C	B	A	a	b	c	d	e	f	g
0	0	0	0	0	1	1	1	1	1	1	0
1	0	0	0	1	0	1	1	0	0	0	0
2	0	0	1	0	1	1	0	1	1	0	1
3	0	0	1	1	1	1	1	1	0	0	1
4	0	1	0	0	0	1	1	0	0	1	1
5	0	1	0	1	1	0	1	1	0	1	1
6	0	1	1	0	0	0	1	1	1	1	1
7	0	1	1	1	1	1	1	0	0	0	0
8	1	0	0	0	1	1	1	1	1	1	1
9	1	0	0	1	1	1	1	0	0	1	1
10	1	0	1	0	0	0	0	1	1	0	1
11	1	0	1	1	0	0	1	1	0	0	1
12	1	1	0	0	0	1	0	0	0	1	1
13	1	1	0	1	1	0	0	1	0	1	1
14	1	1	1	0	0	0	0	1	1	1	1
15	1	1	1	1	0	0	0	0	0	0	0

Eq. (11-19) can be realized using four three-input AND gates and a four-input OR gate. However, another design approach using AND-OR-NOT gates proves to be more efficient.

The logical function $\bar{F}_a$ can also be written from the map in Fig. 11-31. Using the groupings indicated by the dashed loops, we have

$$\bar{F}_a = \bar{A}C + BD + A\bar{B}\,\bar{C}\,\bar{D}. \tag{11-20}$$

Hence only three AND gates are required for the AND-OR-NOT realization. Both realizations are shown in Fig. 11-32. The procedure used to determine Eq. (11-20) can be employed to obtain expressions for $\bar{F}_b$ through $\bar{F}_g$. These results are:

$$
\begin{aligned}
\bar{F}_b &= BD + A\bar{B}C + \bar{A}BC \\
\bar{F}_c &= CD + \bar{A}B\bar{C} \\
\bar{F}_d &= A\,\bar{B}\,\bar{C} + \bar{A}\,\bar{B}C + ABC \\
\bar{F}_e &= A + \bar{B}C \\
\bar{F}_f &= AB + B\bar{C} + A\bar{C}\,\bar{D} \\
\bar{F}_g &= ABC + \bar{B}\,\bar{C}\,\bar{D}.
\end{aligned}
\tag{11-21}
$$

Texas Instruments, Inc.,[18] type SN7449 integrated circuit is a BCD-to-seven segment decoder which behaves in accordance to Table 11-5. A

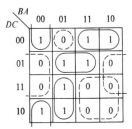

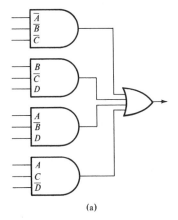

Fig. 11-31   Karnaugh map of Eq. (11-18).

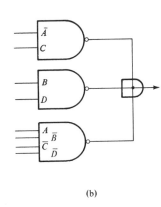

(a)   (b)

Fig. 11-32   Logic realizations for F_a: (a) using AND and OR gates; (b) using AND-OR-NOT gates.

schematic diagram of the device is presented in Fig. 11-33. It includes a blanking input that can be used to blank out all seven segments regardless of the state of any other input condition. The open collector terminal can be connected externally through a 2 kΩ resistor to V_{CC} to provide positive logic outputs at the terminals. An open collector allows for current-sourcing applications to drive logic circuits or discrete, active components. Power dissipation for the device is typically 165 milliwatts.

11-6   BINARY ADDERS

A basic operation in computing systems is *binary addition*. Consequently, much effort has been directed toward the implementation of this operation. As a result two widely used circuits, the *half adder* (HA) and the *full adder* (FA) have emerged. In this section the HA and FA logic blocks will be developed and IC realizations of them will be discussed. As with shift registers it is possible to carry out arithmatic operations in either parallel or serial modes. The parallel mode is the faster of the two at the cost of more complex circuitry. Serial adders are implemented using shift registers in conjunction with adder circuits. Both methods are discussed in this section. Prior to discussing circuitry we will review the basic elements of addition.

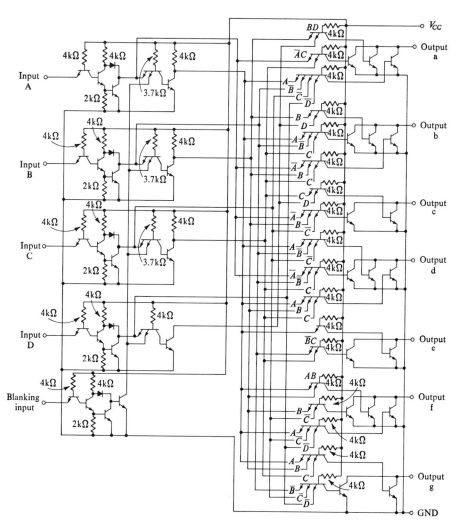

Fig. 11-33   Schematic diagram of Texas Instruments, Inc.,[18] Type SN7449 BCD-to-seven segment decoder/driver.

Binary Addition

Binary addition proceeds in a manner similar to decimal addition. The procedure starts in the right-hand column where the *augend digit* (the number to which another number is to be added) and the *addend digit* (the number which is added to the augend) are added. This operation produces two digits: a *sum digit* and a *carry digit*. In decimal addition, a carry digit is generated whenever the position sum exceeds 9, which is 1 less than the base of the decimal number system. Analogously, in binary addition, a carry digit is generated whenever the position sum exceeds 1, which is 1 less than the base of the binary number system.

We continue the addition process by next adding the augend and addend digits in the second column from the right. This operation is different than that of the right-most column because of the possibility of the generation of a carry digit in the right-most column. We call this digit that is transferred into a column from the column on the right the *carry-in* digit. A carry digit generated in a column and transferred out to the column on the left is called the *carry-out* digit.

As an example, consider the addition of $A = 19$ to $B = 27$, shown in Table 11-6. By adding the powers of two that have the coefficient 1, we observe that the result is the same as for decimal addition.

TABLE 11-6   BINARY AND DECIMAL ADDITION

Weight	Binary						Decimal	
	2^5	2^4	2^3	2^2	2^1	2^0	10^1	10^0
Carry-in	1	0	0	1	1	—	1	—
A	0	1	1	0	1	1	1	9
B	0	1	0	0	1	1	2	7
Sum	1	0	1	1	1	0	4	6

Note that there is no carry-in into the right-most or least significant position. This position is labeled the zero position because the A and B entries are coefficients of 2^0. For the example in Table 11-6, the carry-out of the 0 position is a 1, which is, in turn, the carry-in at the 1 position. All positions except the right-most have a carry-in digit. The logic circuitry used to implement the sum and carry-out operations at the ith position ($i > 0$) has three input variables. The HA is the logic block used to implement the sum and carry-out operation on the least significant digit. An FA is used for each of the remaining positions.

Half Adder

The *half adder* (HA) is developed for adding the least significant bits of two binary expressions. As shown in Fig. 11-34(a), the HA has two inputs and two outputs. Its truth table is presented in Fig. 11-34(b). From the truth

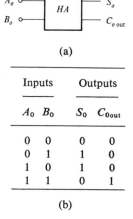

(a)

Inputs		Outputs	
A_0	B_0	S_0	C_{0out}
0	0	0	0
0	1	1	0
1	0	1	0
1	1	0	1

(b)

Fig. 11-34   Half adder: (a) logic diagram; (b) truth table.

table, we determine the logical expressions for S_0 and C_{0out} to be

$$S_0 = A_0 \bar{B}_0 + \bar{A}_0 B_0 = A_0 \oplus B_0 \qquad (11\text{-}22a)$$

$$C_{0_{out}} = A_0 B_0. \qquad (11\text{-}22b)$$

Recall from Section 9-12 that the $\oplus$ symbol denotes the exclusive $-$OR operation.

DeMorgan's Laws allow us to write Eqs. (11-22) in other forms. One form amenable to integration will be developed. From Eq. (11-22a) we obtain

$$S_0 = (\overline{\overline{S}_0}) = \overline{(A_0 B_0 + \bar{A}_0 \bar{B}_0)} = (\bar{A}_0 + \bar{B}_0)(A_0 + B_0), \qquad (11\text{-}23a)$$

and from Eq. (11-22b) we obtain

$$C_{0_{out}} = (\overline{\overline{C}_{0_{out}}}) = \overline{(\bar{A}_0 + \bar{B}_0)}. \qquad (11\text{-}23b)$$

An RTL IC implementation of Eq. (11-23a) and (11-23b) is Motorola's[12] MC904 device.   From the logic diagram shown in Fig. 11-35(a), we observe that it requires that both the variables and their complements be inputed. Two NOR gates and an AND gate with inverted inputs are used.   The schematic diagram is shown in Fig. 11-35(b).   The AND gate consists of the two transistor stages in the left center of the schematic.

Full Adder

The *full adder* (FA), which is capable of adding three bits, is illustrated in block form in Fig. 11-36(a).   The truth table written with the aid of the example worked out in Table 11-6 is given in Fig. 11-36(b).   From the truth

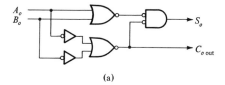

(a)

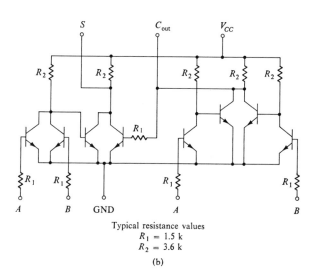

Typical resistance values
R_1 = 1.5 k
R_2 = 3.6 k

(b)

Fig. 11-35 Motorola Semiconductor Products, Inc.,[18] Type MC904 half adder: (a) logic diagram; (b) schematic diagram.

table, we write the logical expressions for the sum and carry-out digits as

$$S = \bar{A}\,\bar{B}C_{in} + \bar{A}B\,\bar{C}_{in} + A\bar{B}\,\bar{C}_{in} + ABC_{in} \qquad (11\text{-}24)$$

and

$$C_{out} = \bar{A}BC_{in} + A\bar{B}C_{in} + AB\bar{C}_{in} + ABC_{in}. \qquad (11\text{-}25)$$

The resulting three-variable Karnaugh maps are shown in Fig. 11-37. From the map of S in Fig. 11-37(a), we see that the function cannot be simplified from the form in Eq. (11-24). However, it is possible to manipulate Eqs. (11-24) and (11-25) so that a realization using HA's is evident. The expressions are:

$$
\begin{aligned}
S &= \bar{A}\,\bar{B}C_{in} + \bar{A}B\,\bar{C}_{in} + A\bar{B}\,\bar{C}_{in} + ABC_{in} \\
&= (\bar{A}B + A\bar{B})\bar{C}_{in} + (AB + \bar{A}\,\bar{B})C_{in} \\
&= (\bar{A}B + A\bar{B})\bar{C}_{in} + \overline{(\bar{A}B + A\bar{B})}C_{in} \qquad (11\text{-}26) \\
&= (A \oplus B)\bar{C}_{in} + \overline{(A \oplus B)}C_{in} \\
&= (A \oplus B) \oplus C_{in}.
\end{aligned}
$$

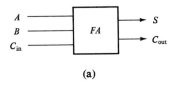

(a)

	Inputs			Outputs	
	A	B	C_{in}	S	C_{out}
	0	0	0	0	0
	0	0	1	1	0
	0	1	0	1	0
	0	1	1	0	1
	1	0	0	1	0
	1	0	1	0	1
	1	1	0	0	1
	1	1	1	1	1

(b)

Fig. 11-36   Full adder: (a) block diagram; (b) truth table.

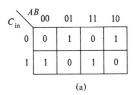

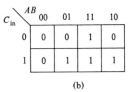

(a)   (b)

Fig. 11-37   Karnaugh maps for: (a) sum bit; (b) carry-out bit of the full adder.

Various forms of the expression for C_{out} can be written from its map in Fig. 11-37(b):

$$C_{out} = AB + BC_{in} + AC_{in}$$
$$= AB + \bar{A}BC_{in} + A\bar{B}C_{in} \qquad (11\text{-}27)$$
$$= AB + (\bar{A}B + A\bar{B})C_{in}.$$

Equations (11-26) and (11-27) are implemented using HA, inverter, and NOR gates, as shown in Fig. 11-38.

Another implementation of the FA can be developed by writing the expressions $\bar{S}$ and $\bar{C}_{out}$ from the maps in Fig. 11-37. They are

$$\bar{S} = \bar{A}\,\bar{B}\,\bar{C}_{in} + \bar{A}BC_{in} + AB\bar{C}_{in} + A\bar{B}C_{in} \qquad (11\text{-}28)$$

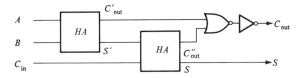

Fig. 11-38   Full adder realized from half adders.

and

$$\bar{C}_{out} = \bar{A}\bar{B} + \bar{A}\bar{C}_{in} + \bar{B}\bar{C}_{in}. \tag{11-29}$$

By complementing both sides of Eq. (11-28) and (11-29) and applying de Morgan's laws, we obtain

$$S = (A + \bar{B} + \bar{C}_{in})(\bar{A} + B + \bar{C}_{in})(\bar{A} + \bar{B} + C_{in})(A + B + C_{in}) \tag{11-30}$$

$$C_{out} = (A + B)(A + C_{in})(B + C_{in}) \tag{11-31}$$

This is sometimes referred to as converting from the minimum sum-of-products form to the minimum product-of-sums form. Equations (11-30) and (11-31) are implemented directly using two-level NOR logic. An IC example shown in Fig. 11-39 is type MC996 by Motorola Semiconductor

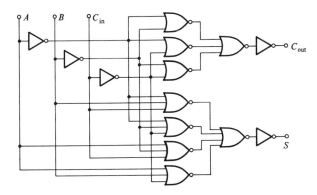

Fig. 11-39   IC full adder realized with NOR gates (one-half of Motorola Semiconductor Products, Inc.,[13] MC996 dual full adder).

Products, Inc.[13]   Though this realization may appear wasteful in terms of gates, it is extremely fast. Its propagation delay is typically only half that for the realization in Fig. 11-38. This completes our discussion of HA and FA IC logic blocks. The next two topics present methods by which these blocks can be combined to carry out multidigit binary addition.

Parallel Adder

The *parallel adder* is a device consisting of one HA plus n-1 additional FA's interconnected to provide the capability for adding two n-bit binary numbers. A four-bit parallel adder is illustrated in Fig. 11-40. Two four-bit binary

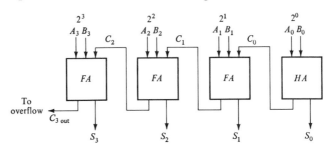

Fig. 11-40 Four-bit parallel "Ripple Carry" adder (inverters for $\bar{A}$, $\bar{B}$, $\bar{C}$, and $\bar{D}$ are assumed present in each adder).

numbers are applied to the input simultaneously. However, the correct output sum bits are not generated simultaneously because the higher order carry-in bits are not available simultaneously.

The operation of the adder proceeds in the following manner. Preliminary sums are generated immediately for all positions; then, if a 1 is generated as a carry bit, it augments the sum at the next higher order stage and possibly generates another carry bit. This sequential generation of carry bits is the reason why the parallel adder is also known as the ripple adder. In the worst-case condition, a carry bit generated at the zero position effects the generation of a carry bit at each successive stage. Consequently, the maximum time required the addition of n bits is n times the carry propagation time per stage.

IC binary full adders are available. Shown in Fig. 11-41 is Texas Instru-

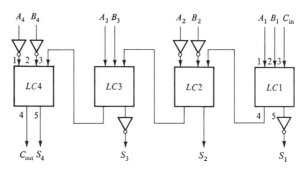

Fig. 11-41 Texas Instruments, Inc.,[18] Type SN7483 four-bit binary full adder.

ments, Inc.,[18] type SN7483N four-bit binary full adder implemented with T²L circuitry. The logic diagram for the logic circuits (LC) used in the SN7483N device is presented in Fig. 11-42. For analysis purposes, the input and output variables are defined as shown in the figure.

Output G in Fig. 11-42 is readily computed as

$$G = \overline{XZ + YZ + XY} \qquad (11\text{-}32)$$

$$= \overline{XZ} \cdot \overline{YZ} \cdot \overline{XY}. \qquad (11\text{-}33)$$

As is evident from Fig. 11-41, the value of the binary function is propagated to the left. However, from Fig. 11-42 we see that it is also used to compute the function F:

$$F = \overline{ZG + XG + YG + XYZ}. \qquad (11\text{-}34)$$

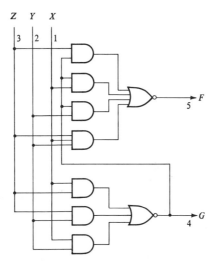

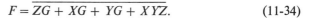

Fig. 11-42   Logic diagram of the logic circuits (LC's) shown in Fig. 11-41.

Using the Karnaugh maps of XG, YG, and ZG shown in Fig. 11-43, it follows that

$$F = \overline{XY\overline{Z} + \overline{X}Y\overline{Z} + \overline{X}YZ + XYZ}. \qquad (11\text{-}35)$$

By comparing Eqs. (11-35) and (11-32) with Eqs. (11-26) and (11-27), respectively, we note that

$$F = \bar{S}$$
$$G = \bar{C}_{\text{out}}, \qquad (11\text{-}36)$$

where S and C_{out} are the outputs of a FA.

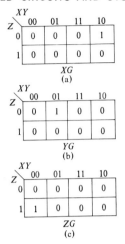

Fig. 11-43   Karnaugh maps of: (a) XG; (b) YG; (c) ZG.

Consequently, the outputs of LC1 and LC3 in Fig. 11-41 are $\bar{S}_i$ and $\bar{C}_{i_{out}}$ ($i = 1, 3$), while, in light of Eqs. (11-28) and (11-29), the outputs of LC2 and LC4 are S_i and $C_{i_{out}}$ ($i = 2, 4$).   A number of the four-bit parallel adders can be connected in series to obtain a four-n-bit parallel adder.

Serial Adder

Serial binary addition is a sequential operation performed one bit at a time. The binary digits A_0 and B_0, the least significant bits (LSB) of the numbers A and B, are added first.   A sum bit S_0 and carryout bit $C_{0_{out}}$ are generated. The S_0 bit is stored in an output storage register, and $C_{0_{out}}$ is fed back for computing S_1 and $C_{1_{out}}$.   This was illustrated previously in the example presented in Table 11-6.

A serial adder in block diagram form is shown in Fig. 11-44.   The two

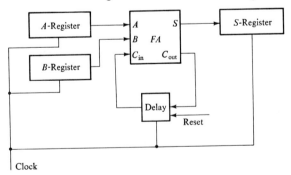

Fig. 11-44   Serial n-bit adder.

numbers to be added are placed in registers A and B.   Register A stores the augend number, and register B stores the addend number.   These are shifted out to the right one digit at a time, as controlled by the clock signal.   After a clock pulse, the S_i and $C_{i_{out}}$ bits are determined in the FA and made available at the output.   With the next clock pulse, S_i is shifted into the sum register, $C_{i_{out}}$ is fed through the delay element into the FA as $C_{(i+1)_{in}}$, and A_{i+1} and B_{i+1} are shifted into the FA.

All shift registers operate in the shift-right mode, as explained in Section 11-2, and the FA behaves as explained previously in this section.   A J-K flip-flop and an inverter can be used to implement the delay element which handles the carry-transfer operation.   This device also serves as an overflow monitor.   Register S in Fig. 11-44 is not always needed.   The sum bits could be transferred back into the augend register (register A).

Serial adders are favored over parallel adders for applications in which time is not critical and the numbers are large.   Parallel adders are faster than serial adders but more expensive for adding large numbers.

Problems

11-1   Level inputs to the circuit in Fig. 11-7 are $B1 = 1$, $B2 = 0$, $B3 = 0$, $B4 = 1$.   For a clock frequency of 1 kHz, sketch the serial output signal for the first six milliseconds immediately after a preset pulse is applied.

11-2   If the propagation delay and set-up times are 24 and 15 ns, respectively, for the eight-bit shift register in Fig. 11-9(a), what is the maximum clock frequency?

11-3   The shift register in Fig. 11-9 can be used as a delay element.   Determine the clock frequency required to realize eight milliseconds of delay using that device.

11-4   Design an eight-bit left-shift register utilizing type SN7495N IC's shown in Fig. 11-12.

11-5   Design a twelve-bit serial-to-parallel converter using type SN7495N IC's shown in Fig. 11-12.

11-6   Design a divide-by-6 ripple counter using a type SN7490N IC shown in Fig. 11-15.   No additional components are required.

11-7   Design a four-decimal digit counter using type SN7490N IC's.

11-8   Show that the type SN7492N IC can be used as a divide-by-3 ripple counter and a divide-by-6 ripple counter by applying the input signal at the BD input and using the appropriate outputs.

11-9   Use the algebraic relations of Section 9-2 to reduce Eq. (11-8a) to Eq. (11-8b).

11-10   Draw the Karnaugh maps and determine the simplified expressions for the decimal numbers 1 to 9 for the Johnson counter.

11-11   Use deMorgan's laws to show the equivalence of the logical function for the two realizations in Fig. 11-32.

11-12   Determine the Karnaugh maps for outputs b-g of the BCD-to-7 segment decoder.   Use the truth data in Table 11-5.

11-13   Design a decoder to convert the output of the divide-by-12 counter shown in Fig. 11-16 to decimal readout.

11-14   Verify that the expressions for S and C_{out} for the logic diagram in Fig. 11-35 are the same as those given by Eqs. (11-24) and (11-25).

11-15   A commutator is a device which sequentially samples the outputs of various signal sources.   Use a circulating shift register plus the appropriate gates to design a commutator.

11-16   Use an inverter and J-K flip-flop to implement the delay element in Fig. 11-44.

11-17   Design a thirty-two–bit serial adder using commercially available IC's. Specify the clock requirements.

12
IC
APPLICATIONS

The previous chapters discussed various analog and digital IC's available in individual chip form. This chapter shows by means of selected examples how IC's may be combined to perform complex operations. The operations considered are analog-to-digital conversion, multiplexing, sampling, majority voting, numerical control and digital communications. Another area of interest, active memories, is considered in order to illustrate the use of IC's for digital memories and to demonstrate the application and uses of LSI and discretionary wiring.

In order to take advantage of digital computation systems in analyzing and controlling analog operations, and also to utilize digital communications systems for the transmission of analog data, we must provide an analog-to-digital interface between the analog and digital parts of each system. Techniques for this conversion are discussed in this chapter, along with the allied operations of sampling and multiplexing used to accomplish accurate analog-to-digital conversion of more than one simultaneous signal with one converter.

Error reduction has become more economical with the use of integrated circuits to perform majority voting functions in logic systems where redundancy is employed to counteract errors due to noise disturbances or malfunctioning units. A section of this chapter discusses the logic and techniques used to perform the voting operation and also uses the voting blocks in different logic situations.

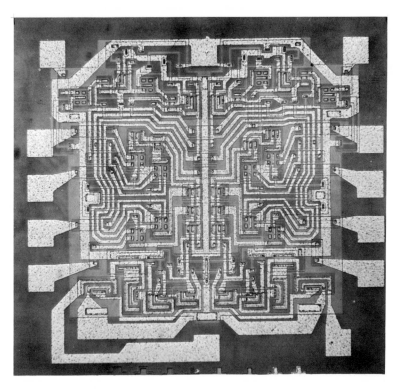

This type SM210 device is designed to function as a four-bit multiplexer. Control lines select one of four inputs for presentation at the output. The device can directly feed type D flip-flops, shift registers, adders, and so on. (Courtesy Sylvania Electronic Products, Inc.)

Finally, we consider the integration of a complete digital communication system.

12-1 DIGITAL-TO-ANALOG CONVERSION

The conversion of digital values to proportional analog values is a necessary task in order that results of digital computations can be used and easily understood in the analog world. Two schemes for digital-to-analog (D/A) conversion are commonly used: (1) a summing operational amplifier with binary-weighted input resistances; (2) a constant resistance ladder network with branches switched in and out depending on the digital value to be converted. A D/A converter utilizing the constant resistance ladder network to convert straight binary codes will be described here. Because this particular ladder network uses only two different values of resistors, it is easily implemented in integrated circuit form. Since only two values of resistance are used in the ladder network, constant resistance ratios, and therefore, accurate output voltage values, can easily be maintained over large operational ranges of temperature.

Analog Output Voltage

The constant resistance ladder network converter is composed of the ladder network shown in Fig. 12-1(a) together with a transistor switching circuit for each stage; a single stage is shown in Fig. 12-1(b). Each stage of the ladder network is associated with a binary bit position, and if that position is in the ON state, a regulated positive or negative voltage reference, V_{REF}, is applied to the corresponding stage via the associated transistor switch. Fig. 12-1(b)

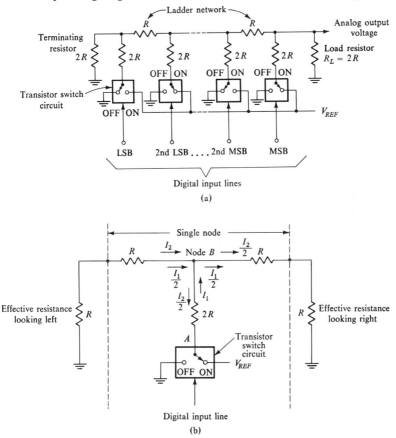

Fig. 12-1   Circuit diagrams of resistance ladder network: (a) constant resistance ladder network; (b) single stage analysis of ladder network.

illustrates the effect of each stage on the analog output voltage across the load resistor in Fig. 12-1(a). Point A is connected to ground if the associated bit is in the OFF state; if the bit is in the ON state, point A is connected to V_{REF}.

Utilizing superposition, we can first analyze the effect when one stage is connected to a positive reference voltage, V_{REF}, and all others are connected to ground, and expand our analysis to include all other combinations. From Fig. 12-1(b) we note that the resistance from node B to ground, proceeding either to the right or left from any stage, is $2R$. Therefore, if V_{REF} is applied to point A in Fig. 12-1(b) the current I_1 will be $V_{REF}/3R$. However, since the resistance from point A to ground is near zero, whether point A is at ground potential or V_{REF}, the input resistance at point A of any particular stage is always $3R$ no matter how the other stages are connected. At node B, the current I_1 will divide into two equal currents of $I_1/2$ amps. Now, consider the current $I_1/2$ flowing to the right, into node B of the next most significant bit (MSB). This current is represented by I_2 in Fig. 12-1(b). Again, owing to equal resistance values from node B to ground through point A and also to the right of point A, I_2 divides into two equal currents of $I_2/2$ amps. The current continuing to the right is halved at each stage until it flows through the load resistor R and produces a voltage corresponding to its particular binary weight. For example, if V_{REF} is applied to the stage of the MSB in Fig. 12-1(a), the current through the load resistor is $(V_{REF}/3R)(1/2)$ or $V_{REF}/6R$. The current contribution of the second MSB is $V_{REF}/12R$ since the current is halved twice before reaching the load resistor. Therefore, the total output voltage will be the sum of each stage current contribution multiplied by the load resistor $2R$, or

$$v_{OUTPUT} = 2R\left[\sum_{n=0}^{N-1} \frac{a_n}{6R(2^n)} V_{REF}\right],$$

where

$$a_n = \begin{cases} 1 \text{ if } n\text{th bit is ON} \\ 0 \text{ if } n\text{th bit is OFF} \end{cases}; \tag{12-1}$$

also n corresponds to the stage of the MSB, and n increases as the bit weight decreases. N is the number of significant digital input bits to be converted. For the four-bit digital number 1010, with weights of 8-4-2-1, N is 4 and,

$$v_{OUTPUT} = \frac{2R}{6R}\{V_{REF}(1 + 0 + 1/4 + 0)\}$$

$$= \left(\frac{V_{REF}}{3}\right)\left(\frac{5}{4}\right) = 0.4167 V_{REF}. \tag{12-2}$$

The actual value of the load resistance need not necessarily equal $2R$, since its only effect is on the relative output voltage level.[20]

Typical Circuit Operation

The schematic of a five-bit D/A converter fabricated by TRW Semiconductors, Inc., as an integrated circuit is shown in Fig. 12-2. Operation can be

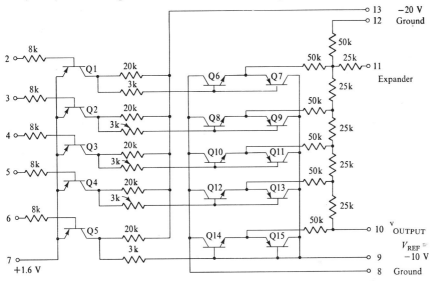

Fig. 12-2   Schematic diagram of TRW Semiconductors, Inc., DAC 100.

investigated by considering a typical ON state of 2.4 volts applied to the MSB position, pin 6. This positive voltage fed through the 8 kΩ resistor turns Q5 OFF. This brings the voltage at the collector of Q5 to near -20 volts. Thus Q15 is turned ON and connects the reference voltage, V_{REF}, equal to -10 volts, through its collector-to-emitter path to the ladder network. The output voltage, v_{OUTPUT} will be $(-10)(1/3) = -3.33$ volts, which corresponds to the digital equivalent of a 1 in the MSB position and 0 for all other bits. A 50 kΩ load resistor is assumed.

Transistor Q14, which connects the ladder network to ground or pin 8, is turned ON only when a voltage less than 0.4 volt is applied to pin 6; Q5 is then ON, and the collector voltage of Q5 is near zero. This corresponds to a 0 in the MSB position corresponding to pin 6.

We have therefore shown that the digital representation of a quantity, consisting of a given number of bits (in this case, five), can be converted to an analog voltage value which uniquely corresponds to the digital representation. This is accomplished by selectively switching branches of a resistor network either to a reference voltage or ground.

Pin 11, labeled as an expander, can be connected to pin 10 of another D/A converter for greater than five-bit operation. V_{REF} can be changed, within limits, to any convenient value. Usually V_{REF} is made as high as possible to make switching transistor offset voltages and thermal noise voltages small when compared to full-scale output voltages in order to reduce errors.

The values of R in the ladder network are chosen as a compromise between static errors and speed of conversion. As R increases, the error introduced by the ON state resistance of the transistor switches will decrease. On the other hand, a large resistance will increase both the conversion time and the difficulty in loading the network properly. A good compromise is R equal to 20 to 25 kΩ.

Temperature Effects

The analog output voltage will be effected by changes in temperature. The maximum analog voltage change, Δv, will correspond to the most significant bit and for a temperature change of ΔT and a load resistance equal to $2R$;

$$\Delta v = \frac{C_T \, V_{REF} \, \Delta T}{9}, \qquad (12\text{-}3)$$

where

C_T = temperature coefficient of ladder network in parts per million per degree C

ΔT = temperature change in degrees C

V_{REF} = reference voltage.

The temperature coefficient C_T is on the order of twentyfive parts per million per degree centigrade in IC packages.[21]

12-2   ANALOG-TO-DIGITAL CONVERSION

Analog-to-digital (A/D) converters are encoders that convert analog current or voltage signals to digital codes compatible with digital systems such as computers, telemetry links, simulators, decimal readout devices, or control networks. The converter is therefore a key part of many industrial, commercial, and military systems because it is the interface between analog systems and digital systems. The use of integrated circuits has reduced the size, increased the capabilities, and also decreased the cost of A/D converters. These developments have subsequently expanded the practical uses of digital equipment in many areas, such as process control, aircraft control, telemetry, and so on. The following discussion will cover the terminology and the methods used in A/D conversion and will analyze some typical A/D conversion circuits. Analog-to-digital conversion methods can be classified into two groups: open-loop methods, and closed-loop methods. In the open-loop method, the input signal magnitude information is converted to another form, such as a frequency variation. This frequency is counted for a fixed time interval, and a digital output is produced. An example of open-loop conversion is shown in block diagram form in Fig. 12-3. The method uses the input signal to

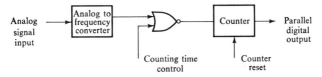

Fig. 12-3 Analog-to-frequency open-loop conversion method.

determine the frequency of an oscillator signal which is applied to a NOR gate. If positive logic is assumed, when the counting time control signal is zero, the NOR gate produces output pulses during the negative portions of the frequency waveform and the signal counter counts these pulses. The total count is digitally processed to drive a readout display or to produce a digital output for other systems. This type of conversion is relatively slow, and is used in instruments, such as digital voltmeters, where speed of conversion may not be critical. The discussion to follow describes a popular type of closed-loop A/D converter which has better accuracy and speed characteristics.

Successive-Approximation A/D Converter

The *successive-approximation technique,* sometimes called the successive-comparison or the "put-and-take" technique, performs a conversion by comparing the analog input voltage with successively better approximations which are generated by a programmed register and decoder or D/A converter Referring to Fig. 12-4, the most significant bit of the register is initially set to 1 by the programmer. This generates a voltage at the D/A converter output which is one-half the full-scale input voltage. This voltage is compared with the input. If the input is still larger than this approximation, the second MSB in the register is set to 1. Then the input is compared with three-fourths of the full-scale voltage and so on. If the input is less than the number registered and converted, the last significant bit which was set to 1 is reset to 0 while the next bit is set to 1. In this manner the analog input is compared to successively better approximations until the least significant bit has been determined.

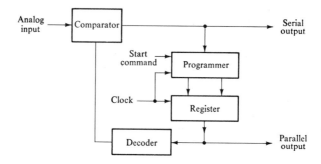

Fig. 12-4   Successive approximation A/D conversion technique.

A twelve-bit successive-approximation converter diagram is shown in Fig. 12-5(a). The comparison function is performed by the summing operational amplifier, labeled DC, and the zener diode. The analog input signal v_X is added to $-v_F$, the output of the D/A converter, and the resulting difference is amplified to produce an output voltage state of ON or OFF, which is determined by the sign of the difference, $(v_X - v_F)$. The programming circuitry consists of the timing generator and the associated NAND gates. The register is made up of twelve R-S flip-flops. The decoder is a twelve-bit D/A converter utilizing a negative reference voltage, $-V_{REF}$. Parallel outputs from the R-S flip-flops are gated to external circuits by the gating circuitry when conversion is complete.

The gain of the operational amplifier must be sufficient to produce an output large enough to equal the ON level sensed by NAND gate C, for an error voltage magnitude, $|v_X - v_F|$, equivalent to the voltage corresponding to the least significant bit. For a twelve-bit converter and a maximum full scale voltage, v_{MAX}, of 10 volts, this least significant bit error voltage, v_{LSB}, is

$$v_{LSB} = \frac{v_{MAX}}{2^n} = \frac{10}{2^{12}} \cong 2.5(10^{-3}) \qquad \text{volts,} \qquad (12\text{-}4)$$

where n is the number of bits available in the digital code. If the necessary ON state voltage is 2.5 volts, the gain must be at least 1000. In this case, the amplifier output is limited by the 3 volt zener diode to values between -0.6 volt and $+3.0$ volts. With a gain of $-K$, an output of -0.6 volt indicates v_X is larger than v_F, and $+3.0$ volts indicates the opposite. With these levels, the comparator can drive bipolar logic circuits. If MOS logic is used, a 0 to -12 volt range is needed. This can be obtained by using a 12 volt zener diode in place of the 3 volt zener and connecting it in the opposite direction. Also, the gain of the operational amplifier must be increased to accommodate the larger signal needed to detect the least significant bit errors.

The fourteen-stage timing generator must produce fourteen mutually

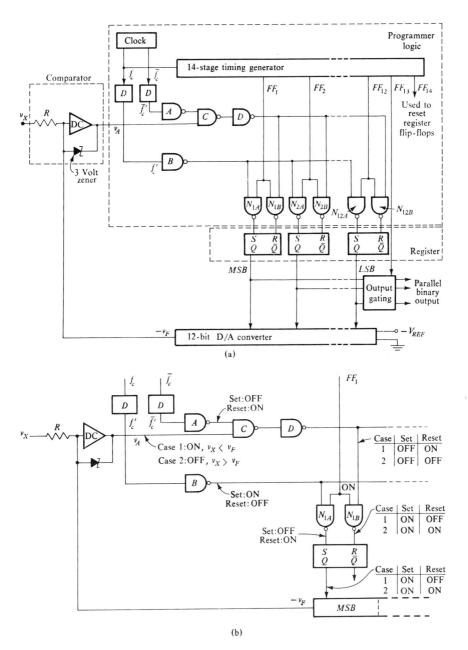

**Fig. 12-5 A successive approximation analog-to-digital converter:
(a) complete logic diagram; (b) logic diagram for MSB determination.**

exclusive signals, one during each timing interval, twelve to set the R-S flip-flops, one to gate out the digital code, and one to reset all register flip-flops for the next conversion.   Several methods are available to do this, and Fig. 12-6(a) shows one solution: a ring counter consisting of fourteen flip-flops with additional logic to set the first stage to ON when all others are OFF and thereby start the next conversion sequence.

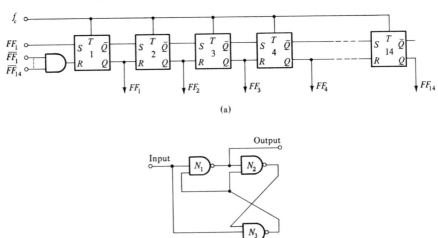

(a)

(b)

Fig. 12-6   Timing generator and block D: (a) shift-register used as a fourteen-stage timing generator; (b) logic for block D, Fig. 12-5.

In order to make successively better digital approximations to the input signal, the R-S flip-flops of the A/D converter must be set in a predetermined sequence of steadily increasing or decreasing digital code values, depending on whether the signal is above or below the initial or subsequently generated codes.   If each of the timing interval signals generated by the ring counter of Fig. 12-6(a) is one clock period wide, then the Q output of each flip-flop circuit is SET at the start of the corresponding timing period beginning with the MSB flip-flop.   In this case, when the SET input of the R-S flip-flop is OFF, and the RESET is ON, the Q terminal will be ON.   Conversely, when the SET is ON and the RESET is OFF, Q will be OFF.

For conversion of one analog value, the sequence is as follows.

(1) Initially, the clock pulse f_c triggers the output FF_1 of the timing generator ON for t_1 seconds, as shown in Fig. 12-7.   All other timing outputs remain OFF.   During the time interval t_1, the MSB flip-flop output Q must be SET.   Then, after v_X and v_F are compared, Q must be RESET to either the ON or OFF state if v_X is greater or less than v_F, respectively.   Fig. 12-5(b)

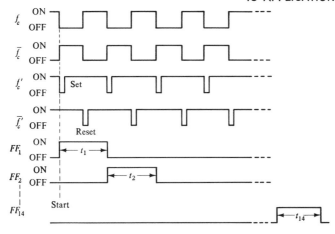

Fig. 12-7   Typical waveforms for A/D converter.

shows only the logic necessary for determining the MSB; other bit determinations are similar.

(2) To SET the MSB flip-flop, the clock pulse f_c is modified by the circuitry of Fig. 12-6(b), labeled D in Fig. 12-5(a), to produce a short OFF period in its output f_c'. This OFF period, labeled SET in Fig. 12-7, occurs at the falling edge of each clock pulse.

(3) The output of the inverting NAND gate B is then ON; with FF_1 also ON, N_{1A} is OFF. The inversion function is accomplished by connecting one input of a NAND gate to an ON state voltage and varying the remaining input. Gates A, B, and D are inverting NAND gates.

(4) Since N_{1A} is OFF, the SET or S terminal of the MSB flip-flop is OFF; the RESET or R terminal is ON. This can be determined by reconstructing the table entries of Fig. 12-5(b). With the S input OFF and the R input ON, the Q output state is ON.

(5) The twelve-bit parallel D/A converter converts the digital code to an analog voltage, $-v_F = -V_{REF}/2$.

(6) Then the comparison amplifier determines the state of v_A. As shown in Figure 12-5(b), there are two cases to consider when determining whether the state of v_A is ON or OFF. Since the amplifier labeled DC has a negative gain, $-K$, the output state will be ON if $v_X < v_F$ and OFF is $v_X > v_F$. The zener diode feedback restricts the excursion of v_A to a small negative voltage, such as -0.6 volt, corresponding to the OFF state, and an ON state voltage, such as 3 volts.

(7) To initiate the RESET operation, the negation of the clock pulse $\bar{f}_c$ is modified by the D block logic to produce the waveform $\bar{f}_c'$ in Fig. 12-7. The OFF pulse, labeled RESET in the figure is inverted at gate A to the ON state and applied together with the state of v_A to gate C.

(8) Considering case 1, where $v_X < v_F$, the state of v_A, will be ON. During the RESET pulse period, the output of NAND gate C will be OFF. This is inverted through D to give an ON state at one input of N_{1B}. This, together with the existing ON state of FF_1, results in an OFF at the R terminal of the MSB flip-flop. An ON state is present at the S terminal during this short period; therefore the output state of Q will be OFF. If v_X would have been greater than v_F, the output Q would have remained ON. Therefore the voltage v_X is determined to be less than $V_{REF}/2$ and the next bit, corresponding to a voltage level equal to $V_{REF}/4$, will be compared with v_X. If it is less than v_X, that Q output will remain ON. This process will continue until all twelve bits have been determined.

(9) During timing interval t_{13}, the output gating circuits channel the digital code corresponding to the analog signal v_X to the output circuitry.

(10) The clock pulse FF_{14}, during timing interval t_{14}, resets all the flip-flops for the next conversion sequence.

The components which determine the accuracy of the A/D conversion are the comparison amplifier and D/A converter. The comparison amplifier is limited in its slew rate, or the speed at which it can change its output voltage from one level to another. The speed of conversion, or conversion time, is limited by the slew rate of the comparison amplifier and the D/A conversion rates. These limitations are usually the dominant factors, because the digital circuits will operate at much faster rates than the comparison amplifier and the D/A converter. For example, most of the required digital circuits will operate in less than 0.1 microsecond, and available D/A converters require from 0.3 to 1 microsecond for conversion. A comparison amplifier output change from -0.6 volt to 3 volts in less than 0.3 microsecond requires a slew rate larger than 10 volts/microsecond. This rate is not easy to achieve with monolithic amplifiers, but is available at a cost premium.

If the conversion rates are low and switching transients and amplifier settling times are negligible, the accuracy of the A/D converter is dependent on the accuracy of the D/A converter and the precision of the input sensing circuit of the comparator. However, at high conversion rates, switching transients, amplifier slew rates, and D/A converter delays all create errors in addition to low frequency errors. Another important error is the uncertainty of when the input signal was at the voltage corresponding to the digital code produced. This uncertainty is called the *aperture time*, and is usually equal to the total conversion time required by the specific A/D converter used. The next sections will consider a sample-and-hold circuit and a multiplexer, or electronic switching circuit, which help to reduce these errors and provide for one A/D converter to convert more than one analog input signal. Problem 12-7 discusses two different systems which use sample-and-hold and multiplexer circuits for the above purpose.[22]

12-3   SAMPLE-AND-HOLD

In many A/D conversion systems, the analog signal may be varying at a rate high enough to produce considerable error in the time associated with the final digital code for a particular signal sample point.   This problem occurs because a finite length of time is required to accomplish the A/D conversion. This uncertainty in sample time is called the aperture time error, and can be reduced by using a sample-and-hold device.   *A sample-and-hold circuit samples the analog signal at a given time and holds that voltage at its output terminal over a sufficient period of time for the A/D converter to make a conversion.*   The block diagram of a typical sample-and-hold circuit is shown in Fig. 12-8.   It consists of a switch to connect the analog signal to a

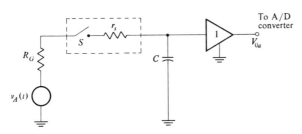

Fig. 12-8   A sample-and-hold circuit with $v_A(t)$ and R_G simulating an analog source, switch S and r_s representing a switch, a storage capacitor, C, and a unity gain operational amplifier.

capacitor C, which stores the voltage for subsequent use, and an IC amplifier to maintain the capacitor voltage and provide a buffer between the capacitor and the A/D converter.

Specifications which must be considered when analyzing a sample-and-hold circuit are the following:

(1) *Acquisition time* is the time required to charge the capacitor, used as the voltage memory device, to the input analog signal value within a tolerable error limit.   The error limit can be specified in percent of full-scale input voltage for example.

(2) *Output decay rate* indicates how fast the voltage on the capacitor is decaying in the hold mode.   This decay is due to capacitor leakage currents, operational amplifier loading, and leakage currents in the switch when junction devices are used.   A high-input impedance operational amplifier is desirable in this application, since it serves to decrease the decay of voltage across the capacitor.

(3) *Input impedance* in a sample-and-hold circuit fluctuates as the sample function changes from a sample mode to the hold mode.   For Intronics FS101 sample-and-hold unit, the sample mode input impedance is equivalent

to 500 pF in series with 100 ohms, but the hold mode impedance is 10^{10} ohms. This impedance variation will have some effect on the sampled circuit operation, and should be considered.[23]

12-4 MULTIPLEXING

When more than one analog signal must be converted, a decision must be made as to whether an A/D converter should be used for each signal or whether one A/D converter should be time-shared or *multiplexed* between several signals.   An *analog multiplexer*, then, is a circuit that serially switches a number of different analog signals onto a single line or channel.   Fig. 12-9 shows the block diagram of one such device which is compatible with DTL/T²L sources of switching waveforms at the drive inputs.

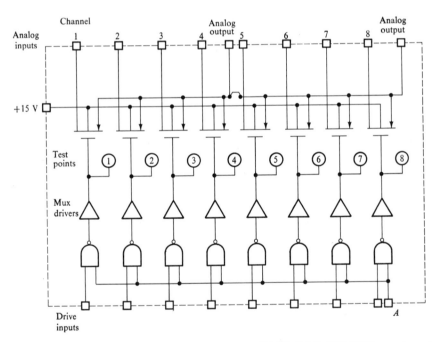

Fig. 12-9 An eight-channel multiplexer (MOD-PAK MP-MX8 produced by California Systems Components, Inc., Chatsworth, California).

The device of Fig. 12-9 accepts switching waveforms at the drive inputs. These switching pulses should be close to zero volts.   When they are applied simultaneously with a zero or grounded pulse at point *A*, they turn ON the respective NAND gate at which the zero is applied.   The ON output from the gate is processed through driver amplifiers to turn ON the MOSFET switches and thereby allow the analog signal to pass from the respective input channel to the analog output terminals in the figure.   In this particular case, the typical ON resistance between the analog input channel and an output terminal is 400 ohms, while the OFF resistance is 100 megohms.   This multiplexer will therefore allow the use of one A/D converter to digitize eight different signals and put their digital representations on one digital channel.

Terminals labeled 5 and OUT may be disconnected to allow for a four-channel differential multiplexer whereby channels 1 and 5, 2 and 6, 3 and 7, and so on can be simultaneously switched to apply four double-ended inputs to one differential amplifier.

Another type of multiplexer, called a *digital multiplexer*, transfers serial digital codes from a number of different input channels or serial registers to a single output channel.   In this case, direct connections are not completed from individual inputs to the output terminal, but logical elements perform the transfer operation.   Fig. 12-10 shows a Fairchild 9309 integrated circuit dual

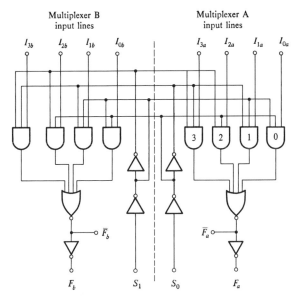

Fig. 12-10 Logic diagram of Fairchild 9309 dual four-input digital multiplexer.

four-input digital multiplexer circuit with input selection logic terminals, S_1 and S_0.   The multiplexer is the logical implementation of a two-pole four-position switch.   The position of the switch is controlled by the logic level codes supplied to the two selection terminals.   The logic equations relating each set of four inputs to their respective outputs are

$$F_a = I_{0a} \cdot \bar{S}_1 \cdot \bar{S}_0 + I_{1a} \cdot \bar{S}_1 \cdot S_0 + I_{2a} \cdot S_1 \cdot S_0 + I_{3a} \cdot S_1 \cdot S_0$$
$$F_b = I_{0a} \cdot S_1 \cdot S_0 + I_{1b} \cdot \bar{S}_1 \cdot S_0 + I_{2b} \cdot S_1 \cdot S_0 + I_{3b} \cdot S_1 \cdot S_0 .$$
(12-5)

Table 12-1 is the truth table for the above equations.

TABLE 12–1   TRUTH TABLE FOR DUAL FOUR-INPUT DIGITAL MULTIPLEXER

Select Inputs		Inputs				Outputs	
S_0	S_1	I_{0a}	I_{1a}	I_{2a}	I_{3a}	F_a	$\bar{F}_a$
0	0	0	X	X	X	0	1
0	0	1	X	X	X	1	0
1	0	X	0	X	X	0	1
1	0	X	1	X	X	1	0
0	1	X	X	0	X	0	1
0	1	X	X	1	X	1	0
1	1	X	X	X	0	0	1
1	1	X	X	X	1	1	0
S_0	S_1	I_{0b}	I_{1b}	I_{2b}	I_{3b}	F_b	$\bar{F}_b$
0	0	0	X	X	X	0	1
0	0	1	X	X	X	1	0
1	0	X	0	X	X	0	1
1	0	X	1	X	X	1	0
0	1	X	X	0	X	0	1
0	1	X	X	1	X	1	0
1	1	X	X	X	0	0	1
1	1	X	X	X	1	1	0

0 = low voltage level
1 = high voltage level
X = either 1 or 0

To understand the operation of the circuit, consider the select inputs of S_0 and S_1 to be both 0. Then, considering multiplexer A, we see that all AND gates except the 1_{0a} channel gate will have at least one 0 input. Therefore the only 1 that can be transmitted to the NOR gate will be generated when I_{0a} goes to 1, a 0 from I_{0a} will produce a 0 output, and the NOR gate output will be 1. An inverter is added to return the signal to its 0 level if desired. The negation of I_{0a} is available at terminal $\bar{F}_a$.

12-5   NUMERICAL CONTROL

Many manufacturing operations today have been automated, utilizing *numerical* or *digital control*.   Numerical control is a method of controlling the motions of machine components by numbers.   The operations primarily consist of positioning objects or parts with respect to other objects.   Numerically controlled machines are used in such operations as drilling, milling, assembly, packaging, inspection, or any other operation in which an object is moved in a predetermined, repetitive path.

The motion to be controlled can be in any of the xyz coordinate directions, circular, parabolic, in spirals, arcs, squares, and so on, depending on the complexity of the digital control equipment and the specific job.   The more complex motions usually require digital computer systems to control the movements of the moving object, whereas simpler movements, such as linear motion along the x, y, or z axes, can be done by small logic modules designed specifically for each problem.   Linear motion prescribes a direct path from the initial to the final point as shown in the two dimensional system of Fig. 12-11(a), while nonlinear motion can follow any random path from the initial to final position.   Typical machines using nonlinear motion usually change the x and y axes positions in equal increments until the object has reached the final x and y coordinate position, as shown in Fig. 12-11(b).

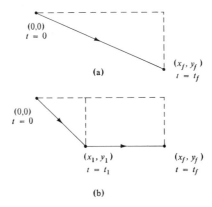

Fig. 12-11   Types of motion used to move an object from an initial position to a final position: (a) linear motion; (b) nonlinear motion.

Drive Systems

Two types of drive systems are commonly used.   One is an open-loop system which uses an output register or counter signal to drive a stepping motor for a fixed number of counts.   Another system is a feedback system where a DC motor is continuously driven, while the object position is monitored by detectors in each coordinate direction and compared with the final desired position.   An open-loop drive system which utilizes counter stepping logic is shown in Fig. 12-12.   Digital numbers corresponding to the number of x and

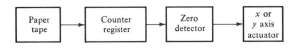

Fig. 12-12   Open-loop drive system for linear motion.

y steps to be made are read into down counters from a tape reader.   Then the counters count down to zero, and the zero detectors increment the x and y drives on each count until the counter outputs are zero.   The x and y drives move the xy table or the object accordingly.

Linear motion from any initial position can also be accomplished with a system such as shown in Fig. 12-13.   Initially, the Δx and Δy registers are set

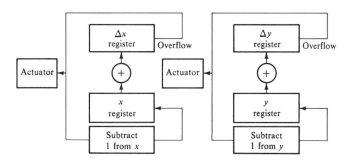

Fig. 12-13   Modified open-loop drive system for linear motion.

to zero, and the x and y registers are set to the digital equivalent of the final required positions.   The value x is added to Δx until an overflow occurs in the Δx register.   This overflow signal is used to drive an x coordinate actuator, which moves the object one preset increment in the positive x direction. The same signal initiates the subtraction of a 1 from the x register.   This sequence of events continues until x is zero.   The same sequence occurs in the y drive control.   This system has the feature that the frequency of drive automatically decreases as the object reaches the final x and y values.   This is due to the number of times x must be added to Δx in order to produce an overflow in Δx.   For example if the final x coordinate value is to be three units to the right, x in the binary system would initially be 011.   The following numerical operations would be performed

000	Initial Δx
011	Add initial $x = x_1 = 3$
$\overline{011}$	$x_1 + \Delta x$
011	Add x_1
$\overline{110}$	$x_1 + \Delta x + x_1 = 2x_1 + \Delta x$
011	Add x_1
$\overline{001}$ →	Overflow
010 Add	$x_2 = (x_1 - 1) = 2$
$\overline{011}$	$x_2 + \Delta x$, where Δx equals 001
010	Add x_2
$\overline{101}$	$2x_2 + \Delta x$
010	Add x_2
$\overline{111}$	$3x_2 + \Delta x$
010	Add x_2
$\overline{001}$ →	Overflow
001 Add	$x_3 = (x_2 - 1) = 1$
$\overline{010}$	$x_3 + \Delta x$
001	Add x_3
$\overline{011}$	$2x_3 + \Delta x$
001	Add x_3
$\overline{100}$	$3x_3 + \Delta x$
001	Add x_3
$\overline{101}$	$4x_3 + \Delta x$
001	Add x_3
$\overline{110}$	$5x_3 + \Delta x$
001	Add x_3
$\overline{111}$	$6x_3 + \Delta x$
001	Add x_3
$\overline{000}$ →	Overflow

$$x_4 = x_3 - 1 = 0.$$

Since the arithmetic operations are clocked at a fixed rate and additional operations are required to produce overflow as the object nears its destination, the frequency of overflow occurs at a steadily decreasing rate until the final position is attained.[24]

12-6 MAJORITY VOTING

Noise or some other disturbance or component failure can cause an incorrect output signal in a digital system. A *majority voting scheme* compares the simultaneous outputs of several identical logic systems, which may individually be in error, and then determines the correct answer based on the output which occurred for more than a majority of the systems.   This is accomplished by employing redundant circuitry.   In the example discussed here, we will consider two-out-of-three voting.   Redundancy is achieved by realizing three inputs which represent three signals that are processed equivalently.   The system will tolerate an error or incorrect functioning of one of the three processing systems, because it will accept the indication of the majority.

The three signals are given the symbols A, A', and A''.   Table 12-2

TABLE 12-2   TRUTH   TABLE   FOR   MAJORITY   VOTING LOGIC FUNCTION

A	A'	A''	AA'	AA''	$A'A''$	$AA' + AA'' + A'A''$
0	0	0	0	0	0	0
0	0	1	0	0	0	0
0	1	0	0	0	0	0
0	1	1	0	0	1	1
1	0	0	0	0	0	0
1	0	1	0	1	0	1
1	1	0	1	0	0	1
1	1	1	1	1	1	1

considers the various possibilities.   It is evident that if two or more signals are zero, $AA' + AA'' + A'A''$ is zero, and if two or more are 1, the function is 1.

Therefore, for any two signals in agreement, $AA' + AA'' + A'A''$ indicates that fact.   Implementation is shown in Fig. 12-14.   This two-out-of-three

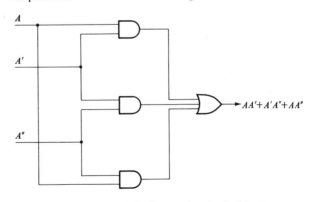

Fig. 12-14   Majority-voting logic block.

voting logic block can be connected to three parallel, identical logic function modules as shown in Fig. 12-15(a).

Since the voting logic blocks can themselves introduce errors, additional redundancy can be achieved by interconnecting three blocks for more redundancy.   Fig. 12-15(b) indicates how three voting blocks can be interconnected

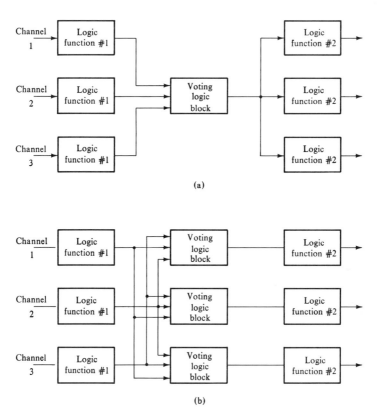

Fig. 12-15   Logic organization for single and triplicate voting: (a) single-voting; (b) triplicate-voting.

in order to eliminate errors introduced in the voting logic blocks.   In this case, if two channels of Logic Function #1 are correct and one is in error, all voting blocks should output the result of the correct two channels.   If one voting block fails, and if the other two blocks indicate the proper output, the results of the second tier of logic function blocks will be correct for two blocks and incorrect for one.   This difference will be corrected if the next set of voting blocks indicate properly.[25]

12-7   ACTIVE MEMORIES

An *active memory* is a digital memory that employs a bistable electronic circuit to store each bit.   Other names used to describe an active memory are semiconductor memory, integrated-circuit memory, transistor memory, and flip-flop memory.   Digital memories are usually organized to store W words, each word containing B bits.   The total storage capacity is, then, $W \times B$ bits.   The semiconductor active memory is of importance now, since large-scale integration of such devices reduces connector wiring and system packaging costs, increases system speed and reliability and reduces the parts count. The disadvantages of LSI memories are now cost-per-gate or memory cell, volatility of stored information in case of power failures, and standby power dissipation.   Improved fabrication methods and circuit design innovations will eliminate many of these objections in the future.

Active Memory Cell

One type of active memory cell, the *multiple-emitter bipolar cell* is shown in Fig. 12-16.   The cell can be in one of two states.   If Q1 is conducting, it is assumed for this discussion that this state corresponds to a binary 1.   If Q2

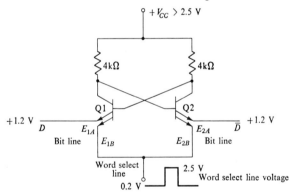

Fig. 12-16   Bipolar transistor memory cell.

is conducting, a binary 0 has been stored in the cell.   This cell can be READ, to determine the state of the cell, or a WRITE operation can be performed whereby the cell is switched to a desired logic state.   To understand the READ operation, consider Q1 to be conducting through emitter E_{1B}, therefore the collector voltage at Q1 is 0.3 volt above the word-select line voltage.   The base of Q1 and collector of Q2 are at 0.9 volt with respect to ground; this is the base-emitter voltage drop plus the 0.2 volt on the word-select line.   If bit line D and bit line $\bar{D}$ are held at a voltage of 1.2 volts and the word-select line voltage is raised to 2.5 volts, as shown by the pulse in Fig. 12-16, emitter E_{1B} of Q1 becomes reverse-biased. But, emitter E_{1A} is forward-biased, and current flows in bit line D. This current is sensed and considered an indication of the 1 state. Since the collector of Q1 is then at 1.5 volts with respect to ground and emitter E_{2A} of Q2 is at 1.2 volts, the $\bar{D}$ bit line-emitter junction does not conduct current, and a 0 state is indicated. This is the logical negation of the D bit line.

To WRITE information into the cell, the word-line voltage is increased to 2.5 volts, and the bit line $\bar{D}$ is simultaneously held at 2.5 volts or 0.2 volt, depending on whether a 1 or 0 is stored. If the bit line is at 2.5 volts, Q2 is turned OFF and Q1 ON; therefore a 1 is stored. If the bit line is held at 0.2 volt, Q2 will conduct and remain conducting when the word line returns to its normal voltage of 0.2 volt.

This circuit fulfills memory cell requirements, such as high switching speed, low parasitic capacitance, and low leakage currents, but does not provide low power dissipation. Typical power dissipation values are from 2 to 10 milliwatts per cell. If many of these cells are put on a single chip, the cost of packaging and cooling will be high because of the amount of power which must be dissipated.

To reduce this problem, MOSFET's have been used as active devices, because they will operate at lower power dissipation values. Typical power dissipation values for FET cells are in the nanowatt range. The main disadvantage of using FET devices is the reduction in speed of READ and WRITE operations, because higher threshold voltages are necessary and the intrinsic gain of FET's is lower than that of bipolar transistors.[26]

Memory Organization

The actual organization of a memory and its external control circuitry, such as drivers, sense amplifiers, and decoding circuitry, depends on the type of storage element used and the memory function to be performed. Three common types of memory will be briefly discussed, and an example of a read-only memory will be used to illustrate one possible memory organization.

The *associative memory* or content-addressable memory (CAM) is one from which data are fetched in terms of the data content rather than the data location. Because this type of memory requires logic functions (to compare stored data with input information that identifies it) and storage (to "remember" the data), the size and number of circuits needed were prohibitive until the advent of LSI. With LSI, the memories can be made sufficiently large, and they will operate at speeds necessary for efficient use as auxiliary input data processors or scratch-pad memories. These are used for small storage requirements where fast access times are advantageous.

A *random access memory* (RAM) utilizes logically coded address lines to select the location in storage of each piece of information and separate bit lines to transfer the information from the memory as well as to read new information into the memory. This type of memory could be used for temporary storage in computer input/output processors, buffer storage, or as a register in the central processing unit of a computer.

Another type of memory, the *read-only memory* (ROM) is constructed with the data "stored" by physically modifying the memory cells. These data can be read only from the memory, and not changed. In MOS arrays, storage is established by depositing a thin oxide layer between the gate contact and any externally connected wire. This cell can then signify the existence of a 0 in that particular bit location. An example of a read-only memory will be discussed to show how the information is initially stored in the array and also how it is read out of the array. These memories are useful in such applications as control of high-speed systems, table look-up, code conversion, and display drivers.

Read-only Memory

A read-only memory scheme developed by the Fairchild Semiconductor Division of Fairchild Camera and Instrument Corp. is shown in Fig. 12-17. This is a sixteen-word memory, with N bits per word, depending on the particular application. A four-bit address code can be applied to the address lines, which are numbered 1, 2, 4, and 8, corresponding to their binary weights. This code will serve to select any of sixteen words. The memory cells corresponding to a single bit location that is common to all sixteen words are organized in a four-by-four array so that two adress bits select a row and the other two select a column. Since the MOSFET's are all p-channel, the supply voltages are negative, and a negative gate voltage, corresponding to a logical 1, allows current to flow through the MOSFET. During fabrication of the actual memory matrix, the gates are either connected or disconnected to the address lines by depositing an oxide film between the gate and address line if no connection is desired. A connected gate will allow current to flow through its corresponding MOSFET if a negative voltage is applied; this corresponds to storage of a binary 1. If the gate is purposely not connected by depositing a thin oxide insulating layer between the gate lead and the metalized connector, a binary 0 is stored.

Consider the procedure to simultaneously read the N bits contained in word 5. The four-by-four matrix of bit position 1 is shown inside the dotted line; the other $N - 1$ matrices for each location are represented by the box outlined by a solid line. Address 5 in binary form is 0101; this corresponds to address lines 8421. Address lines 4 and 1 are at a negative voltage, and lines

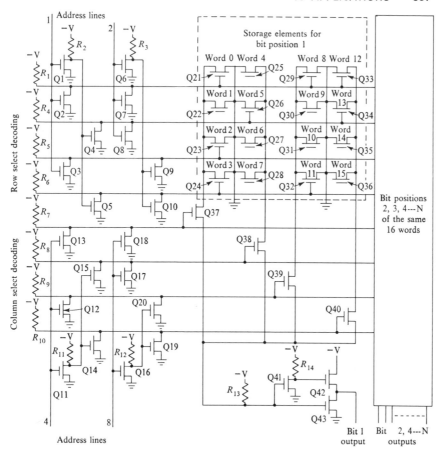

Fig. 12-17   A sixteen-word, *N*-bit, MOSFET read-only memory.

2 and 8 are positive, corresponding to negative logic.   Q6, Q7, and Q8 are OFF, because of the positive voltage on line 2.   Since no current flows through Q6, the gates of Q9 and Q10 are at a negative voltage.   Therefore Q9 and Q10 are conducting.   Line 1 is at a negative voltage, allowing Q1 to conduct. This, in turn, shuts OFF Q4 and Q5.   Since Q4 and Q8 are OFF, the row line connected to their drains is at the negative source voltage, $-V$.   This turns ON Q26, which corresponds to a 1 as the first bit of word 5.   Q38 is ON since Q17 and Q15 are OFF.   Q37, Q38, Q39, and Q40 all have a common load resistor, R_{13}.   Q41, Q42, and Q43 buffer the voltage drop across an external load.   Since Q26 and Q38 are ON, the presence of a binary 1 is sensed as current passes *from* the external circuit through Q42.   If word 6 had been selected, no current would have flowed through Q27, and a current through Q43 *into* the external circuit would represent a binary 0.[27]

Discretionary Wiring

Active memories use integrated circuits with many identical circuits deposited on one chip.   Owing to process problems related to impurities and crystal defects, some of the circuits on one chip may not be functional.   Rather than discard an entire chip because only a few circuits on the chip are bad, a computer-controlled system can be designed to detect defective circuits automatically and determine the proper connection of only functional circuits.   This technique is called *discretionary wiring.*

The use of discretionary wiring in the manufacture of integrated circuits starts at the first stage of metallization, or after the elements, such as resistors, transistors, and capacitors, have been deposited and interconnected to form functional building blocks such as flip-flops and gates.   The next step is to deposit metal interconnections between the functional circuits.   A block diagram of the process is shown in Fig. 12-18.

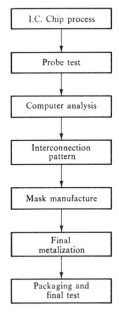

Fig. 12-18   Discretionary wiring process.   The probe test data and the data describing the required logic are fed to a computer, which generates a final wiring diagram and supervises the manufacture of the final metallization mask.

The first-level testing is accomplished by using multipoint automatic probe equipment coupled to an integrated circuit tester.   The integrated circuit tester sequentially makes a number of tests on each circuit and records the results on either punched cards or magnetic tape. Each memory cell is tested for its ability to store a 1 or 0, the 1 current, the word line characteristics, and so on.

The test results are used as input to a digital computer, which generates the discretionary interconnection pattern for the second-level metallization. A photo-mask is developed by utilizing a computer-generated cathode ray tube display and by transferring the picture to photographic film. This image is used to fabricate the actual mask and perform the metallization. This metallization interconnects the good cells and omits the defective cells. The use of this technique greatly reduces the rejected chips during production. But the number of usable circuits per chip will be reduced in comparison to a process whereby all chips with one or more defective circuits are rejected. Therefore the economy of the discretionary wiring method depends upon ingenuity in designing flexible wiring patterns and low-cost mask generating systems.

12-8   A DIGITAL COMMUNICATION SYSTEM

As an example of a system which utilizes many of the components and modules described in the previous chapters, we present a digital communication system.[28]   Specifically, the system is designed to transmit the angular position of a shaft to a remote location. This system could telemeter the velocity of an automobile to a trailing auto or roadside station. The transmitting unit is illustrated by the block diagram shown in Fig. 12-19.

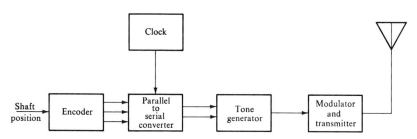

Fig. 12-19   Block diagram of transmitter unit.

Transmitting Unit

Encoder. The encoder is a shaft encoder such as the one shown in Fig. 11-3. The input is the shaft position, an analog variable with an infinity of possible values, and the output is a binary code word consisting of the number of bits required to obtain the desired resolution. As an example, consider the case where a 160° arc, such as the speedometer scale, must be resolved into eight subdivisions, each with a 20° arc.   Each of these eight subdivisions can be uniquely identified by a code word consisting of three bits.   Codes known as Gray codes[29] are well suited to shaft encoders.

Clock.   The clock serves to organize the activity and establish the rate at which the bits are transmitted.   Desired reliability, noise on the communication channel, cost and complexity of equipment, and so on, determine the bit rate and hence the clock frequency.   A clock circuit such as the one shown in Fig. 11-2 could be used.

Parallel-to-serial converter.   The transmission system employed requires that the data be transmitted in a serial stream.   Consequently, it is necessary to employ a parallel-to-serial converter.   This unit might be implemented using a right-shift, left-shift register, such as the one illustrated in Fig. 11-12.

At this point in the system, it is expedient to add extra bits to the bit stream to assist in decoding the data at the receiver.   These are commonly referred to as synchronizing bits, and are used to obtain both bit sync and word sync at the receiver.   Synchronizing bits can be inserted serially into the data stream in an identifiable pattern, or a second wire or line may be used in conjunction with the data line to provide synchronization.   In this example, the latter method is used.   A data word, including the word synchronization bit, is illustrated in Fig. 12-20(a).   The first three bits on the data line correspond to the assumed word 010, and the word synchronization symbol consists of a 1 on each line in the fourth time interval.

Tone generator.   Audio tones corresponding to 0 and 1 voltage levels are used to modulate a carrier frequency, and a third tone is used for synchronization.   Let f_1 be the frequency of the tone corresponding to a 0, f_2 be the

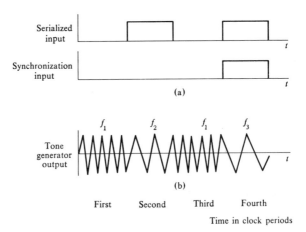

Fig. 12-20   Sample waveform for tone generator.

frequency of the tone corresponding to a 1, and f_3 be the frequency of the tone corresponding to a synchronization symbol, where $f_1 > f_2 > f_3$. The tone generator output waveform is shown in Fig. 12-20(b).

Modulator and transmitter. The data in the form of audio tones can be used to modulate a carrier for subsequent transmission over a radio link or they can be transmitted directly over a telephone channel. Amplitude-, frequency-, or phase-modulation may be employed.

Receiving Unit

A block diagram of the receiver unit is shown in Fig. 12-21(a). The function of the unit is to receive, decode, and display the data.

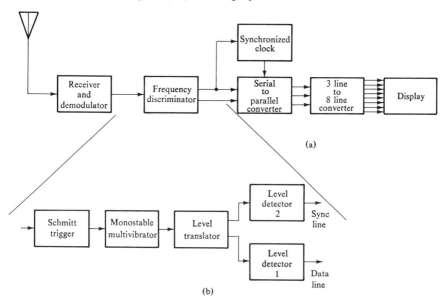

(a)

(b)

Fig. 12-21   (a) Block diagram of receiving unit; (b) expanded diagram of frequency discriminator.

Receiver and demodulator. Functionally, the receiver and demodulator must inverse the operations of the transmitter and modulator. An audio signal which is a near replica of the signal supplied to the modulator is required at the output of the demodulator.

Frequency discriminator. This unit accepts the audio signal provided by the demodulator on a single line and converts it into voltage pulses representing the data and word synchronization symbols, as shown in Fig. 12-20(a).

The discriminator operates in the following manner [see Fig. 12-21(b)]. The audio signal, which at any given time consists of one of three tones, is converted to a series of rectangular pulses by the Schmitt trigger circuit. Next, the resulting wave is fed to a monostable or one-shot multivibrator. This circuit provides a positive pulse of predetermined width, with its leading edge coincident with the falling edge of the input signal. This waveform is shown in Fig. 12-22(b). Note that the average value of this signal is proportional to the frequency of the input signal.

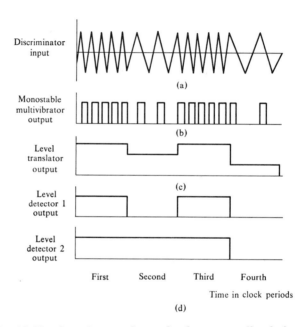

Fig. 12-22   Sample waveforms for frequency discriminator.

An R-C integrator circuit follows the monostable multivibrator and serves as the input stage of the level translator. Translators were discussed in Section 10-8, but the unit used here, though functionally similar, is somewhat different in its detail, as is seen from Fig. 12-23. An emitter-follower minimizes loading on the integrator. The output of the emitter-follower is applied to a series of forward-biased diodes to drop the voltage level to the value required by the level detectors. Two additional emitter-followers are used to pick off the proper voltage levels and drive the level detectors. Note that, except for the integrator, all of the components in the translator circuit are transistors, diodes, and resistors, which makes the device amenable to monolithic integration.

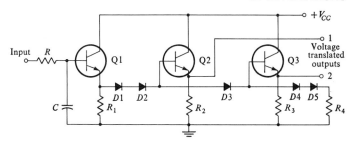

Fig. 12-23   Level translator circuit.

Schmitt triggers are used as level detectors. If the high frequency tone is received, both level detectors will trigger. However, if the middle frequency tone is received, only level detector 2 will be triggered, and if the low frequency tone is received, neither will be triggered. The output signal corresponding to the input word 010 is shown in Fig. 12-22(d). Note that this is the complement of the input signal.

Clock. The receiver clock, which is similar to the transmitter clock, is synchronized with the transmitter clock by the signal from level detector 2. This method requires that the receiver clock remain in tolerable sync for the length of a word.

Serial-to-parallel converter. This unit is implemented by using a circulating shift register with four stages where the data word consists of 3 bits; an example is shown in Fig. 12-24. The shift register serves as a commutator for

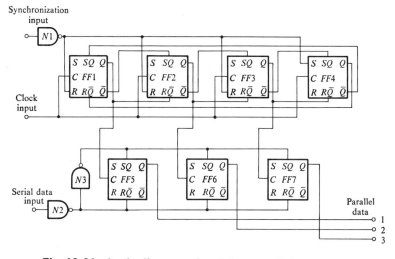

Fig. 12-24   Logic diagram of serial-to-parallel converter.

interrogating the flip-flops FF5, FF6, and FF7, which provide the proper level outputs on the three parallel lines.

3-line-to-8-line decoder and display. For the case where the display consists of a light corresponding to each interval of resolved arc, eight lights are required. An IC such as the Type SN5444N Excess 3 Gray-to-Decimal is available to provide the proper outputs for driving the display.

Problems

12-1 It was mentioned in section 12-1 that the load resistance R_L of the ladder network of Fig. 12-1(a) need not be equal to $2R$ for proper operation.

(a) Consider that only the most significant bit position is ON or connected to V_{REF}, and draw the equivalent circuit which can be used to determine the voltage across R_L.

(b) Show that the voltage across R_L, V_{MSB} will then be

$$V_{MSB} = \frac{V_{REF}}{2} \frac{R_L}{R + R_L}.$$

12-2 Using the D/A converter in Fig. 12-2, determine the change in output voltage, Δv, for a temperature change of 100C°, if $R_L = 50$ kΩ, and $C_T = 25$ parts per million per degree centigrade.

12-3 Connect two D/A converters like the one in Fig. 12-2 together to make a ten-bit converter. Determine what should be connected to pins 10, 11, and 12. Which converter is associated with the MSB?; LSB?

12-4 If the digitizing error for an A/D converter cannot exceed 10 millivolts, how many bits must the converter output contain if the input voltage range is from zero to ten volts?

12-5 What is the minimum gain K needed for the comparator amplifier in Fig. 12-5 if the LSB voltage is 1 millivolt and the ON state voltage levels for the gate logic of an A/D converter must be equal to or greater than 12 volts?

12-6 Explain the operation of the logic block D shown in Figure 12-6(b).

12-7 Two different methods of connecting sample-and-hold circuits, multiplexers, and an A/D converter to convert more than one analog signal with one A/D converter are shown in the figure.

In (a), the sample-and-hold units can be controlled to sample all signals simultaneously and hold the values sampled while the multiplexer switches each value to the A/D converter.

In (b), the sample-and-hold must be synchronized with the multiplexer so that it samples each signal during the period the multiplexer is switched to that signal. If the A/D converter is capable of 100,000 conversions per second, what is the resulting sampling rate for each signal, S_i, in each case? What advantages does method (a) have? What advantages does method (b) have? (Consider the case where it is desirable to sample a number of signals simultaneously.)

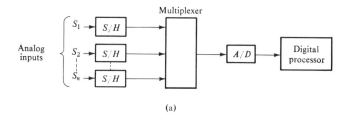

(a)

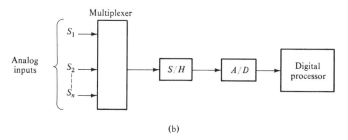

(b)

Problem 12-7

12-8 For a certain sample-and-hold unit, the following specifications are given:
Acquisition time: 2 μsec for 10 V change (0.1 percent accuracy)
Output decay rate: 0.1 volt/sec.
Input range: 0 to 10 volts.

(a) What is the fastest rate at which the sample-and-hold unit will operate and present an output within 0.1 percent of the input?

(b) What is the longest hold time allowable to maintain an accuracy of 0.1 percent of the full range voltage?

(c) From (b), what will be the slowest sample rate allowable considering the held value is used just prior to performing the next sample operation?

12-9 Use the digital multiplexer diagram of Fig. 12-10 to determine which input is transferred to the output F_a, when the S_0 and S_1 terminals are at the 0 and 1 state, respectively. Check the truth table, Table 12-1, then assume a 1 state at the chosen input terminal and determine if the state of F_a is 1 for the above conditions. Do the same for a 0 state at the input.

12-10 In the numerical control system of Fig. 12-13, how many digital operations must take place before the actuator moves to its final position if that position is at $(x = 2, y = 2)$? $(x = 2, y = 1)$? How many mechanical movements of each actuator are involved in each of the above situations?

12-11 Develop the logic equation for a three-out-of-four voting scheme. Construct the logic diagram for the above equation; assume the inputs are A_1, A_2, A_3, and A_4.

12-12   Using the memory of Fig. 12-17, (a) what binary number, when applied to address lines 1, 2, 4, and 8, will allow the contents of word number 9 to be read out?

(b) Show that the output indication for bit 1 will indicate a 0 state if the address of part (a) is used.

REFERENCES

References Cited in the Text

1. "Definitions of Terms for Integrated Electronics," *IEEE Journal of Solid-State Circuits*, Vol. SC-2, No. 1, March, 1967.
2. "IEEE Standard Letter Symbols for Semiconductor Devices," *IEEE Trans. Electron Devices*, ED-11, No. 8, August, 1964.
3. Fitchen, F. C., *Transistor Circuit Analysis and Design*, 2nd Ed., D. Van Nostrand Co., Inc., Princeton, N.J., 1966.
4. Lin, H. C., *Integrated Electronics*, Holden-Day, Inc., San Francisco, 1967.
5. Staff, Integrated Circuit Engineering Corp., *Integrated Circuit Engineering, Basic Technology*, 4th Edition, Boston Technical Publishers, Inc., Cambridge, Mass., 1966.
6. Lynn, D. K., Meyer, C. S., and Hamilton, D. J., *Analysis and Design of Integrated Circuits*, McGraw-Hill Book Co., N.Y., 1967.
7. Camenzind, H. R., *Circuit Design for Integrated Electronics*, Addison-Wesley Publishing Co., Reading, Mass., 1968.
8. Staff, *RCA Linear Integrated Circuits*, Publication IC-41, Radio Corporation of America, Harrison, N. J., 1967.
9. "Transistor Logarithmic Conversion Using an Integrated Operational Amplifier," *Application Note AN-261*, Motorola Semiconductor Products Inc., Phoenix, Ariz., 1968.
10. Staff, *Handbook of Operational Amplifier Applications*, Burr-Brown Research Corporation, Tucson, Ariz., 1963.
11. Berndt, D. F., and Dutta Roy, S. C., "Inductor Simulation Using a Single Unity Gain Amplifier," *IEEE Journal of Solid State Circuits*, Vol. SC-4, No. 3, June, 1969.
12. Giles, J. N., *Fairchild Semiconductor Linear Integrated Circuits Applications Handbook*, Fairchild Semiconductor, Mountain View, Calif. 1967.
13. Staff, *The Integrated Circuit Data Book*, Motorola Semiconductor Products, Inc., Phoenix, Ariz., 1968.
14. Staff, *Fairchild Semiconductor Data Catalog*, Fairchild Semiconductor, Mountain View, Calif., 1968.
15. Millman, J., and Taub, H., *Pulse, Digital and Switching Waveforms*, McGraw-Hill Book Co., N.Y., 1965.
16. Crawford, R. H., *MOSFET in Circuit Design*, McGraw-Hill Book Co., New York, 1967.
17. Botos, B., "A Frequency Counter Using Motorola TRL Integrated Circuits," *AN-451*, Motorola Semiconductor Products, Inc., January, 1969.
18. Staff, *Integrated Circuits Catalog Supplement from Texas Instruments*, Bulletin CC-102, Texas Instruments, Inc., October, 1968.

19. Staff, *TTL Integrated Circuits-Counters and Registers*, Bulletin CA-102, Texas Instruments, Inc., 1968.
20. Penner, L., "Tracking Analog-to-Digital Converter Utilizing MTOS Devices," *General Instrument Company, Microelectronics Division Applications Note,* March, 1967.
21. "TRW Digital to Analog Converter Microcircuits," Bulletin 159-1267, TRW Semiconductors, Inc., Lawndale, Calif.
22. Schmid, H., "An Electronic Design Practical Guide to A/D Conversion," *Electronic Design*, Vol. 16, No. 25, December 5, 1968.
23. Hoeschele, D. F., *Analog-to-Digital/Digital-to-Analog Conversion Techniques,* John Wiley & Sons, New York, 1968.
24. *Integrated Circuit Logic Manual*, Cambion Thermionic Corp., 1968.
25. "Majority Voting Possible with Simple Circuits," *Electronic Engineer*, Vol. 28, No. 4, April, 1969.
26. Canning, M., Dunn, R. S., and Jeansonne, G., "Active Memory Calls for Discretion," *Electronics*, Vol. 40, No. 4, February 20, 1967.
27. Boysel, L., "Memory on a Chip, A Step toward Large-Scale Integration," *Electronics*, Vol. 40, No. 3, February 6, 1967.
28. Unruh, R. R., "A Pulse Code Modulation Vehicular Communications System," M.S. thesis, South Dakota State University, Brookings, S.D. 1969.
29. Kintner, P. M., *Electronic Digital Techniques*, McGraw-Hill Book Co., New York, 1968.

Other References on IC's

Doyle, J. M., *Thin-Film and Semiconductor Integrated Circuitry*, McGraw-Hill Book Co., New York, 1966.
Embinder, J., ed., *Designing with Linear Integrated Circuits*, John Wiley and Sons, Inc., New York, 1969.
Embinder, J., ed., *Linear Integrated Circuits: Theory and Applications*, John Wiley and Sons, Inc., New York, 1968.
Hoffman, G. B., "MOS Static Shift Registers and TTl/DTL Systems," Bulletin CA-114, Texas Instruments, Inc., November, 1968.
Renschler, E., and Hepworth, E., "Binary Addition Using MRTL Integrated Circuits," An-286, Motorola Semiconductor Products, Inc., March, 1967.
Staff, "Integrated Circuits—BCD to Seven Segment Decoder," td91-674, Westinghouse Electric Corporation/Molecular Electronics Division, April, 1968.
Staff, *Integrated Circuits Catalog, 1967–68*, Texas Instruments, Inc., Dallas, Texas, 1967.
Staff, *TTL Integrated Circuits, Series 74N Complex Functions* SC-10512-1067, Texas Instruments, Inc.
Stern, L., *Fundamentals of Integrated Circuits*, Hayden Book Co., Inc., New York, 1968.
Stover, W. A., ed., *TI Series 54/74 Integrated Circuits*, Texas Instruments, Inc., Dallas, Texas, 1966.

"Test Methods and Procedures for Microelectronics," MIL-STD 883, U.S. Department of Defense, Washington D.C., 1 May 1968.

Warner, R. M., Jr., *Integrated Circuits, Design Principles and Fabrication*, McGraw-Hill Book Co., New York, 1965.

Widlar, R. J., "Design Techniques for Monolithic Operational Amplifiers," *IEEE Journal of Solid-State Circuits*, Vol. SC-4, No. 4, August, 1969.

Other References on Analog Circuits and Systems

Manasse, F. K., Ekiss, J. A., and Gray, C. R., *Modern Transistor Electronics Analysis and Design*, Prentice-Hall, Inc., Englewood Cliffs, N. J., 1967.

Mitra, S. K., "Synthesizing Active Filters," *Spectrum*, Vol. 6, No. 1, January, 1969.

Schilling, D. L., and Belove, C., *Electronic Circuits: Discrete and Integrated*, McGraw-Hill Book Co., Inc., New York, 1968.

Other References on Digital Circuits and Systems

Braun, E. L., *Digital Computer Design*, Academic Press, New York and London, 1963.

Delhom, L., *Design and Application of Transistor Switching Circuits*, McGraw-Hill Book Company, New York, 1968.

Richards, R. K., *Electronic Digital Components and Circuits*, D. Van Nostrand Company, Inc., Princeton, N. J., 1967.

Staff, *The Digital Logic Handbook*, Digital Equipment Corporation, Maynard, Mass., 1966.

Wickes, W. E., *Logic Design with Integrated Circuits*, John Wiley and Sons, Inc., New York, 1968.

APPENDIX I

Matrix Analysis

Rules of Matrix Algebra

1. *Matrix Equality.* Two matrices [a] and [b] are equal when each element in [a] is equal to the corresponding element in [b]. If

$$[a] = [b],$$

then

$$a_{11} = b_{11}, a_{12} = b_{12}, \text{ etc.}$$

2. *Matrix Addition and Subtraction.* The sum (or difference) of two matrices is a matrix in which each term is the sum (or difference) of the corresponding terms in the two matrices. Thus if

$$[a] + [b] = [c],$$

then

$$c_{11} = a_{11} + b_{11}, c_{12} = a_{12} + b_{12}, \text{ and so on.}$$

3. *Matrix Multiplication.* The product of [a] and [b] is meaningful only when the number of columns of [a] is equal to the number of rows of [b]. Then

$$[c] = [a] \times [b]$$

and

$$c_{ij} = \sum_{k=1}^{k=n} a_{ik} b_{kj}.$$

Two examples follow:

$$\begin{bmatrix} x_{11} & x_{12} \\ x_{21} & x_{22} \end{bmatrix} \begin{bmatrix} y_{11} & y_{12} \\ y_{21} & y_{22} \end{bmatrix} = \begin{bmatrix} (x_{11}y_{11} + x_{12}y_{21}) & (x_{11}y_{12} + x_{12}y_{22}) \\ (x_{21}y_{11} + x_{22}y_{21}) & (x_{21}y_{12} + x_{22}y_{22}) \end{bmatrix}$$

and

$$\begin{bmatrix} y_{11} & y_{12} \\ y_{21} & y_{22} \end{bmatrix} \begin{bmatrix} v_1 \\ v_2 \end{bmatrix} = \begin{bmatrix} (y_{11}v_1 + y_{12}v_2) \\ (y_{21}v_1 + y_{22}v_2) \end{bmatrix}.$$

The commutative law does not generally hold

$$[a][b] \neq [b][a].$$

But distributive and associative laws are valid in matrix multiplication.

$$([a] + [b] \times [c] = [a] \times [c] + [b] \times [c],$$
$$[a] \times [b] \times [c] = ([a] \times [b]) \times [c] = [a] \times ([b] \times [c]).$$

411

When multiplying a matrix by a constant K, each element of the matrix is multiplied by K. Thus

$$K[a] = [a'],$$

with $a_{11}' = Ka_{11}$, $a_{22}' = Ka_{22}$, and so on.

4. *Matrix Inversion.* It is not possible to divide by a matrix. However, to accomplish this result, matrix inversion is used. To find $[i]$ from the matrix equation

$$[v] = [z][i],$$

we multiply through by $[z]^{-1}$

$$[z]^{-1}[v] = [z]^{-1}[z][i] = [i].$$

$[i]$, therefore, is the product of $[z]^{-1}$ and $[v]$, because $[z]^{-1}[z] = [1] = \begin{bmatrix} 1 & 0 \\ 0 & 1 \end{bmatrix}$.

The inverse of a matrix is symbolized by $[a]^{-1}$. The elements of the inverse matrix are obtained from

$$[a_{ij}]^{-1} = [\Delta_{ij}]^t/D^a.$$

D^a is the determinant of $[a]$ and Δ_{ij} is the *cofactor* of the ith *column*, jth *row*, $[\Delta_{ij}]^t$ represents the cofactor matrix, transposed. Transposition is the interchanging of rows with columns. Thus

$$[a]^{-1} = \begin{bmatrix} a_{11} & a_{12} \\ a_{21} & a_{22} \end{bmatrix}^{-1} = \frac{1}{D^a} \begin{bmatrix} a_{22} & -a_{12} \\ -a_{21} & a_{11} \end{bmatrix}.$$

APPENDIX II

Matrix Parameters

A. Matrix Parameter Conversions

To Find	Given z	y	h	g	a	b
$[z]$	—	$\begin{bmatrix} \dfrac{y_{22}}{\Delta^y} & \dfrac{-y_{12}}{\Delta^y} \\[6pt] \dfrac{-y_{21}}{\Delta^y} & \dfrac{y_{11}}{\Delta^y} \end{bmatrix}$	$\begin{bmatrix} \dfrac{\Delta^h}{h_{22}} & \dfrac{h_{12}}{h_{22}} \\[6pt] \dfrac{-h_{21}}{h_{22}} & \dfrac{1}{h_{22}} \end{bmatrix}$	$\begin{bmatrix} \dfrac{1}{g_{11}} & \dfrac{-g_{12}}{g_{11}} \\[6pt] \dfrac{g_{21}}{g_{11}} & \dfrac{\Delta^g}{g_{11}} \end{bmatrix}$	$\begin{bmatrix} \dfrac{a_{11}}{a_{21}} & \dfrac{\Delta^a}{a_{21}} \\[6pt] \dfrac{1}{a_{21}} & \dfrac{a_{22}}{a_{21}} \end{bmatrix}$	$\begin{bmatrix} \dfrac{b_{22}}{b_{21}} & \dfrac{1}{b_{21}} \\[6pt] \dfrac{\Delta^b}{b_{21}} & \dfrac{b_{11}}{b_{21}} \end{bmatrix}$
$[y]$	$\begin{bmatrix} \dfrac{z_{22}}{\Delta^z} & \dfrac{-z_{12}}{\Delta^z} \\[6pt] \dfrac{-z_{21}}{\Delta^z} & \dfrac{z_{11}}{\Delta^z} \end{bmatrix}$	—	$\begin{bmatrix} \dfrac{1}{h_{11}} & \dfrac{-h_{12}}{h_{11}} \\[6pt] \dfrac{h_{21}}{h_{11}} & \dfrac{\Delta^h}{h_{11}} \end{bmatrix}$	$\begin{bmatrix} \dfrac{\Delta^g}{g_{22}} & \dfrac{g_{12}}{g_{22}} \\[6pt] \dfrac{-g_{21}}{g_{22}} & \dfrac{1}{g_{22}} \end{bmatrix}$	$\begin{bmatrix} \dfrac{a_{22}}{a_{12}} & \dfrac{-\Delta^a}{a_{12}} \\[6pt] \dfrac{-1}{a_{12}} & \dfrac{a_{11}}{a_{12}} \end{bmatrix}$	$\begin{bmatrix} \dfrac{b_{11}}{b_{12}} & \dfrac{-1}{b_{12}} \\[6pt] \dfrac{-\Delta^b}{b_{12}} & \dfrac{b_{22}}{b_{12}} \end{bmatrix}$
$[h]$	$\begin{bmatrix} \dfrac{\Delta^z}{z_{22}} & \dfrac{z_{12}}{z_{22}} \\[6pt] \dfrac{-z_{21}}{z_{22}} & \dfrac{1}{z_{22}} \end{bmatrix}$	$\begin{bmatrix} \dfrac{1}{y_{11}} & \dfrac{-y_{12}}{y_{11}} \\[6pt] \dfrac{y_{21}}{y_{11}} & \dfrac{\Delta^y}{y_{11}} \end{bmatrix}$	—	$\begin{bmatrix} \dfrac{g_{22}}{\Delta^g} & \dfrac{-g_{12}}{\Delta^g} \\[6pt] \dfrac{-g_{21}}{\Delta^g} & \dfrac{g_{11}}{\Delta^g} \end{bmatrix}$	$\begin{bmatrix} \dfrac{a_{12}}{a_{22}} & \dfrac{\Delta^a}{a_{22}} \\[6pt] \dfrac{-1}{a_{22}} & \dfrac{a_{21}}{a_{22}} \end{bmatrix}$	$\begin{bmatrix} \dfrac{b_{12}}{b_{11}} & \dfrac{1}{b_{11}} \\[6pt] \dfrac{-\Delta^b}{b_{11}} & \dfrac{b_{21}}{b_{11}} \end{bmatrix}$
$[g]$	$\begin{bmatrix} \dfrac{1}{z_{11}} & \dfrac{-z_{12}}{z_{11}} \\[6pt] \dfrac{z_{21}}{z_{11}} & \dfrac{\Delta^z}{z_{11}} \end{bmatrix}$	$\begin{bmatrix} \dfrac{\Delta^y}{y_{22}} & \dfrac{y_{12}}{y_{22}} \\[6pt] \dfrac{-y_{21}}{y_{22}} & \dfrac{1}{y_{22}} \end{bmatrix}$	$\begin{bmatrix} \dfrac{h_{22}}{\Delta^h} & \dfrac{-h_{12}}{\Delta^h} \\[6pt] \dfrac{-h_{21}}{\Delta^h} & \dfrac{h_{11}}{\Delta^h} \end{bmatrix}$	—	$\begin{bmatrix} \dfrac{a_{21}}{a_{11}} & \dfrac{-\Delta^a}{a_{11}} \\[6pt] \dfrac{1}{a_{11}} & \dfrac{a_{12}}{a_{11}} \end{bmatrix}$	$\begin{bmatrix} \dfrac{b_{21}}{b_{22}} & \dfrac{-1}{b_{22}} \\[6pt] \dfrac{\Delta^b}{b_{22}} & \dfrac{b_{12}}{b_{22}} \end{bmatrix}$
$[a]$	$\begin{bmatrix} \dfrac{z_{11}}{z_{21}} & \dfrac{\Delta^z}{z_{21}} \\[6pt] \dfrac{1}{z_{21}} & \dfrac{z_{22}}{z_{21}} \end{bmatrix}$	$\begin{bmatrix} \dfrac{-y_{22}}{y_{21}} & \dfrac{-1}{y_{21}} \\[6pt] \dfrac{-\Delta^y}{y_{21}} & \dfrac{-y_{11}}{y_{21}} \end{bmatrix}$	$\begin{bmatrix} \dfrac{-\Delta^h}{h_{21}} & \dfrac{-h_{11}}{h_{21}} \\[6pt] \dfrac{-h_{22}}{h_{21}} & \dfrac{-1}{h_{21}} \end{bmatrix}$	$\begin{bmatrix} \dfrac{1}{g_{21}} & \dfrac{g_{22}}{g_{21}} \\[6pt] \dfrac{g_{11}}{g_{21}} & \dfrac{\Delta^g}{g_{21}} \end{bmatrix}$	—	$\begin{bmatrix} \dfrac{b_{22}}{\Delta^b} & \dfrac{b_{12}}{\Delta^b} \\[6pt] \dfrac{b_{21}}{\Delta^b} & \dfrac{b_{11}}{\Delta^b} \end{bmatrix}$
$[b]$	$\begin{bmatrix} \dfrac{z_{22}}{z_{12}} & \dfrac{\Delta^z}{z_{12}} \\[6pt] \dfrac{1}{z_{12}} & \dfrac{z_{11}}{z_{12}} \end{bmatrix}$	$\begin{bmatrix} \dfrac{-y_{11}}{y_{12}} & \dfrac{-1}{y_{12}} \\[6pt] \dfrac{-\Delta^y}{y_{12}} & \dfrac{-y_{22}}{y_{12}} \end{bmatrix}$	$\begin{bmatrix} \dfrac{1}{h_{12}} & \dfrac{h_{11}}{h_{12}} \\[6pt] \dfrac{h_{22}}{h_{12}} & \dfrac{\Delta^h}{h_{12}} \end{bmatrix}$	$\begin{bmatrix} \dfrac{-\Delta^g}{g_{12}} & \dfrac{-g_{22}}{g_{12}} \\[6pt] \dfrac{-g_{11}}{g_{12}} & \dfrac{-1}{g_{12}} \end{bmatrix}$	$\begin{bmatrix} \dfrac{a_{22}}{\Delta^a} & \dfrac{a_{12}}{\Delta^a} \\[6pt] \dfrac{a_{21}}{\Delta^a} & \dfrac{a_{11}}{\Delta^a} \end{bmatrix}$	—

B. Properties of the Terminated Four-Terminal Network

To find	Given					
	z	y	h	g	a	b
Z_i	$\dfrac{\Delta^z + z_{11}Z_L}{z_{22} + Z_L}$	$\dfrac{1 + y_{22}Z_L}{y_{11} + \Delta^y Z_L}$	$\dfrac{h_{11} + \Delta^h Z_L}{1 + h_{22}Z_L}$	$\dfrac{g_{22} + Z_L}{\Delta^g + g_{11}Z_L}$	$\dfrac{a_{12} + a_{21}Z_L}{a_{22} + a_{21}Z_L}$	$\dfrac{b_{12} + b_{22}Z_L}{b_{11} + b_{21}Z_L}$
Z_o	$\dfrac{\Delta^z + z_{22}Z_G}{z_{11} + Z_G}$	$\dfrac{1 + y_{11}Z_G}{y_{22} + \Delta^y Z_G}$	$\dfrac{h_{11} + Z_G}{\Delta^h + h_{22}Z_G}$	$\dfrac{g_{22} + \Delta^g Z_G}{1 + g_{11}Z_G}$	$\dfrac{a_{12} + a_{22}Z_G}{a_{11} + a_{21}Z_G}$	$\dfrac{b_{12} + b_{11}Z_G}{b_{22} + b_{21}Z_G}$
A_v	$\dfrac{z_{21}Z_L}{\Delta^z + z_{11}Z_L}$	$\dfrac{-y_{21}Z_L}{1 + y_{22}Z_L}$	$\dfrac{-h_{21}Z_L}{h_{11} + \Delta^h Z_L}$	$\dfrac{g_{21}Z_L}{g_{22} + Z_L}$	$\dfrac{Z_L}{a_{12} + a_{11}Z_L}$	$\dfrac{\Delta^b Z_L}{b_{12} + b_{22}Z_L}$
A_i	$\dfrac{-z_{21}}{z_{22} + Z_L}$	$\dfrac{y_{21}}{y_{11} + \Delta^y Z_L}$	$\dfrac{h_{21}}{1 + h_{22}Z_L}$	$\dfrac{-g_{21}}{\Delta^g + g_{11}Z_L}$	$\dfrac{-1}{a_{22} + a_{21}Z_L}$	$\dfrac{-\Delta^b}{b_{11} + b_{12}Z_L}$

Δ is the parameter determinant. For example, if x parameters were to be defined, $\Delta^x = x_{11}x_{22} - x_{12}x_{21}$.

Answers to Selected Problems

1-2 $0 < f < 1/\pi \sqrt{2LC}$

1-4   (b) $v = 5 \times 10^5 \cos 10^6 t$, $R = 10^5 \ \Omega$

1-7 $h_{11} = (0.2 - j0.6)10^3$

1-10 $100 \ | \ 183°$

1-14   (b) 6.4 MHz

2-5 $2.54 \times 10^4 \ \mu$

2-6 $5.2 \ \mu$

2-14   100 Ω, 10 mA

3-3   1500 Ω, 198 kΩ

3-11 $h_{12} = 10^{-2}$

3-15   35.3 MHz

3-27   1.45 μF, 0.64 μF

4-1   0.099, $\pm 0.1\%$

4-9 $R_e = 100$: 14.1 kΩ, -67

4-14 $h_o - h_r h_f/(h_i + R_G)$

5-1   0.5 mA, 0.8 V; 0.5 mA, 3.5 V

5-9   (a) -35.7

5-15 $R_i = 172$ kΩ; $R_o = 30\Omega$

6-3 $V_o = R[-V_1/R_1 - V_2/R_2 - V_3/R_3 + V_4/R_4 + V_5/R_5 + V_6/R_6]$

6-16 $2V_{\text{in}} \beta R_C/(R_R + Z_I + 2r_{b'e})$

7-5   (d) 3000

7-6   (f) 0.7 mA, 0.53 mA

8-5   2.1, 6.1, 5.9

8-10   640 Ω

8-13   0.36 mA, 144

9-2 $2.2T, 2.3T, 3T$

9-8   (a) $A(\bar{B} + \bar{C} + \bar{D})$, (c) $AC + \bar{A}B$

10-4   5.3 V, 5.2 V

10-13 -1.0 V

10-15   (f) 30.8 mA, (g) 36.1 mA

11-2   25.6 MHz

11-3   1 kHz

11-10 $F_1 = A\bar{B}$

11-12 $F_e = \bar{A}\bar{C} + \bar{A}B$

12-2 -2.78 mV

12-4   10

12-5 >12000

12-12   (a) 1001

INDEX